Risk Analysis in Finance and Insurance

Second Edition

T0330919

CHAPMAN & HALL/CRC
Financial Mathematics Series

Aims and scope:
The field of financial mathematics forms an ever-expanding slice of the financial sector. This series aims to capture new developments and summarize what is known over the whole spectrum of this field. It will include a broad range of textbooks, reference works and handbooks that are meant to appeal to both academics and practitioners. The inclusion of numerical code and concrete real-world examples is highly encouraged.

Series Editors

M.A.H. Dempster
Centre for Financial Research
Department of Pure
Mathematics and Statistics
University of Cambridge

Dilip B. Madan
Robert H. Smith School
of Business
University of Maryland

Rama Cont
Center for Financial
Engineering
Columbia University
New York

Published Titles

Proposals for the series should be submitted to one of the series editors above or directly to:
CRC Press, Taylor & Francis Group
4th, Floor, Albert House
1-4 Singer Street
London EC2A 4BQ
UK

Chapman & Hall/CRC FINANCIAL MATHEMATICS SERIES

Risk Analysis in Finance and Insurance

Second Edition

Alexander Melnikov

CRC Press
Taylor & Francis Group
Boca Raton London New York

CRC Press is an imprint of the
Taylor & Francis Group, an **informa** business
A CHAPMAN & HALL BOOK

CRC Press
Taylor & Francis Group
6000 Broken Sound Parkway NW, Suite 300
Boca Raton, FL 33487-2742

First issued in paperback 2019

© 2011 by Taylor & Francis Group, LLC
CRC Press is an imprint of Taylor & Francis Group, an Informa business

No claim to original U.S. Government works

ISBN-13: 978-1-4200-7052-1 (hbk)
ISBN-13: 978-0-367-38286-5 (pbk)

Visit the Taylor & Francis Web site at
http://www.taylorandfrancis.com

and the CRC Press Web site at
http://www.crcpress.com

Preface to 2nd edition

This book deals with the notion of "risk" and is devoted to analysis of risks in finance and insurance. We will study risks associated with financial and insurance contracts, by which we understand *risks* to be uncertainties that may result in financial loss and affect the ability to make payments associated with the corresponding contracts. Our approach to this analysis is based on the development of a methodology for estimating the present value of the future payments given current financial, insurance, and other information. Using this approach, one can appropriately define notions of *price* for a financial contract, of *premium* for insurance policy, and of *reserve* for an insurance company. Historically, financial risks were subject to elementary mathematics of finance and were treated separately from insurance risks, which were analyzed in actuarial science. The development of quantitative methods based on stochastic analysis is a key achievement of modern financial mathematics. These methods can be naturally extended and applied in the area of actuarial science, thus leading to unified methods of risk analysis and management.

The aim of this book is to give an accessible comprehensive introduction to the main ideas, methods, and techniques that transform risk management into a quantitative science. Because of the interdisciplinary nature of this book, many important notions and facts from mathematics, finance, and actuarial science are discussed in an appropriately simplified manner. Our goal is to present interconnections among these disciplines and to encourage our reader toward further study of the subject. We indicate some initial directions in the Bibliographic Remarks.

This edition is reorganized in a way that allows a natural flow of topics covered in the first edition to be combined together with new additions such as: financial markets with stochastic volatility, risk measures, risk-adjusted performance measures, equity-linked insurance, and so forth. The substantial extension of the section regarding the foundations of Probability and Stochastic Analysis makes this book self-contained. Furthermore, an increased number of worked examples and a collection of some 140 problems, which is accompanied by the Instructor's Solutions Manual, make this edition more attractive both from a research and a pedagogical perspective. This book can be readily used as a textbook for a Mathematical Finance course, both at introductory undergraduate and advanced graduate levels. It has been used for teaching Mathematical Finance at both levels at the University of Alberta, and many student comments and recommendations are taken into account in this edition.

The author thanks his graduate students Anna Evstafyeva, Hao Li, and Henry Heung for their help in introducing some new worked examples and problems in this edition. The author is also very grateful to Dr. Alexei Filinkov for translating, editing, and preparing the manuscript.

Alexander Melnikov

University of Alberta, Edmonton, Canada

October 2010

Introduction

Financial and insurance markets always operate under various types of uncertainties that can affect the financial position of companies and individuals. In financial and insurance theories, these uncertainties are usually referred to as risks. Given certain states of the market, and the economy in general, one can talk about risk exposure. It is expected that individuals, companies, and public establishments that aim to accumulate wealth should examine their risk exposure. The process of risk management consists of a sequence of corresponding actions over a period of time that are designed to mitigate the level of risk exposure. Some of the main principles and ingredients of risk management are qualitative identification of risk, estimation of possible losses, choosing the appropriate strategies for avoiding losses and for shifting the risk to other parts of the financial system, including analysis of the involved costs, and using feedback for developing adequate controls.

The first six chapters of this book are devoted to the financial market risks. We aim to give an elementary and yet comprehensive introduction to the main ideas, methods, and stochastic models of financial mathematics. The probabilistic approach appears to be one of the most efficient ways of modeling uncertainties in financial markets. Risks (or uncertainties of financial market operations) are described in terms of statistically stable stochastic experiments, and therefore estimation of risks is reduced to the construction of financial forecasts adapted to these experiments. Using conditional expectations, one can quantitatively describe these forecasts given the observable market prices and events. Thus, it can be possible to construct dynamic hedging strategies and those for optimal investment. The foundations and key concepts of the modern methodology of quantitative financial analysis are the main focus of Chapters 1–6.

Insurance against possible financial losses is one of the key ingredients of risk management. However, the insurance business is an integral part of the financial system. Chapters 7–8 focus on the problems of managing insurance risks. Multiple intrinsic connections between insurance risks and financial risks are also considered.

Our treatment of insurance risk management demonstrates that methods of risk evaluation and management in insurance and finance are interrelated and can be treated using a single integrated approach. Estimations of future payments and their corresponding risks are key operational tasks of financial and insurance companies. Management of these risks requires an accurate evaluation of the present values of future payments, and therefore the adequate modeling of (financial and insurance) risk processes. Stochastic analysis is one of the most powerful tools for such modeling, and it is the fundamental basis of our presentation.

Finally, we note that probabilistic methods were used in finance and insurance since the early 1950s. They were developed extensively over the past decades, especially after the seminal papers by F. Black and M. Scholes and R. C. Merton, published in 1973.

Contents

Chapter 1

Introductory Concepts of Financial Risk Management and Related Mathematical Tools

1.1 Introductory concepts of the securities market

The notion of an *asset* (anything of value) is one of the fundamental notions in mathematical finance. Assets can be *risky* and *non-risky*. Here *risk* is understood to be an uncertainty that can cause loss (e.g., of wealth). The most typical representatives of such assets are the *basic securities*: *stocks* S and *bonds* (bank accounts) B. These securities constitute the basis of a *financial market* that can be understood as a space of assets equipped with a structure for their trading.

Stocks are share securities that accumulate capital required for a company's successful operation. The stockholder holds the right to participate in the control of the company and to receive dividends.

Bonds are debt securities issued by a government or a company for accumulating capital, restructuring debts, and so forth. In contrast to stocks, bonds are issued for a specified period of time. The essential characteristics of a bond include their exercise and maturity time, *face value* (principal or nominal), *coupons* (payments up to maturity), and *yield* (return up to maturity). The zero-coupon bond is similar to a bank account, and its yield corresponds to a bank interest rate.

An *interest rate* $r \geq 0$ is typically quoted by banks as an annual percentage. Suppose that a client opens an account with a deposit of B_0, then at the end of a 1-year period the client's non-risky profit is $\Delta B_1 = B_1 - B_0 = rB_0$. After n years, the balance of this account will be $B_n = B_{n-1} + rB_0$, given that only the initial deposit B_0 is reinvested every year. In this case, r is referred to as a *simple interest*.

Alternatively, the earned interest also can be reinvested (compounded), and then, at the end of n years, the balance will be $B_n = B_{n-1}(1 + r) = B_0(1 + r)^n$. Note that here the ratio $\Delta B_n / B_{n-1}$ reflects the profitability or return of the investment as it is equal to r, the *compound interest*.

1

Now suppose that interest is compounded m times per year, then

$$B_n = B_{n-1}\left(1 + \frac{r^{(m)}}{m}\right)^m = B_0\left(1 + \frac{r^{(m)}}{m}\right)^{mn}.$$

Such rate $r^{(m)}$ is quoted as a *nominal* (annual) interest rate, and the equivalent *effective* (annual) interest rate is equal to $r = \left(1 + \frac{r^{(m)}}{m}\right)^m - 1$.

Let $t \geq 0$, and consider the ratio

$$\frac{B_{t+\frac{1}{m}} - B_t}{B_t} = \frac{r^{(m)}}{m},$$

where $r^{(m)}$ is a nominal annual interest rate compounded m times per year. Then another rate

$$r = \lim_{m \to \infty} \frac{B_{t+\frac{1}{m}} - B_t}{\frac{1}{m} B_t} = \lim_{m \to \infty} r^{(m)} = \frac{1}{B_t} \frac{dB_t}{dt}$$

is called the nominal annual interest rate *compounded continuously*. Clearly, $B_t = B_0 e^{rt}$.

Thus, the concept of interest is one of the essential components in the description of money value time evolution. Now consider a series of periodic payments (deposits) $f_0, f_1, \ldots, f_n$ (*annuity*). It follows from the formula for compound interest that the present value of k-th payment is equal to $f_k(1+r)^{-k}$, and therefore the present value of the annuity is $\sum_{k=0}^n f_k(1+r)^{-k}$.

Worked Example 1.1 *Let an initial deposit into a bank account be $\$10,000$. Given that $r^{(m)} = 0.1$, find the account balance at the end of 2 years for $m = 1, 3$, and 6. Also find the balance at the end of each of years 1 and 2 if the interest is compounded continuously at the rate $r = 0.1$.*

Solution Using the notion of compound interest, we have

$$B_2^{(1)} = 10,000\left(1 + 0.1\right)^2 = 12,100$$

for interest compounded once per year;

$$B_2^{(3)} = 10,000\left(1 + \frac{0.1}{3}\right)^{2\times3} \approx 12,174$$

for interest compounded three times per year;

$$B_2^{(6)} = 10,000\left(1 + \frac{0.1}{6}\right)^{2\times6} \approx 12,194$$

for interest compounded six times per year.

For interest compounded continuously we obtain

$$B_1^{(\infty)} = 10,000\, e^{0.1} \approx 11,052\,, \qquad B_2^{(\infty)} = 10,000\, e^{2 \times 0.1} \approx 12,214\,. \ \square$$

Stocks are significantly more volatile than bonds, and therefore they are characterized as *risky assets*. Similarly to bonds, one can define their *profitability* or *return* $\rho_n = \Delta S_n / S_{n-1}$, $n = 1, 2, \ldots$, where S_n is the price of a stock at time n. Then we have the following discrete equation for stock prices: $S_n = S_{n-1}(1 + \rho_n)$, $S_0 > 0$.

The mathematical model of a financial market formed by a bank account B (with an interest rate r) and a stock S (with profitabilities ρ_n) is referred to as a (B, S)-market.

The volatility of prices S_n is caused by a great variety of sources, some of which may not be easily observed. In this case, the notion of *randomness* appears to be appropriate, so that S_n, and therefore ρ_n, can be considered as *random variables*. Since at every time step n the price of a stock goes either up or down, then it is natural to assume that profitabilities ρ_n form a sequence of independent random variables $(\rho_n)_{n=1,2,\ldots}$ that take values b and a $(b > a)$ with probabilities p and q, respectively $(p + q = 1)$. Next, we can write ρ_n as a sum of its mean $\mu = bp + aq$ and a random variable $w_n = \rho_n - \mu$, which has the expectation equal to zero. Thus, profitability ρ_n can be described in terms of an independent random deviation w_n from the mean profitability μ.

When the time steps become smaller, the oscillations of profitability become more chaotic. Formally, the limit continuous model can be written as

$$\frac{\dot{S}_t}{S_t} \equiv \frac{dS_t}{dt} \frac{1}{S_t} = \mu + \sigma \dot{W}_t\,,$$

where μ is the mean profitability, σ is the volatility of the market, and $\dot{W}_t$ is the Gaussian white noise.

The formulas for compound and continuous interest rates together with the corresponding equation for stock prices define the binomial (Cox-Ross-Rubinstein) and the diffusion (Black-Scholes) models of the market, respectively.

A participant in a financial market usually invests free capital in various available assets that then form an *investment portfolio*. The effective management of the capital is realized through a process of building and managing the portfolio. The redistribution of a portfolio with the goal of limiting or minimizing the risk in various financial transaction is usually referred to as *hedging*. The corresponding portfolio is then called a *hedging portfolio*. An investment strategy (portfolio) that may give a profit even with zero initial investment is called an *arbitrage* strategy. The presence of arbitrage reflects the instability of a financial market.

The development of a financial market offers the participants *derivative securities*, that is, securities that are formed on the basis of the basic securities – stocks and bonds. The derivative securities (forwards, futures, options, etc.) require smaller initial investment and play the role of insurance against possible losses. They also increase the liquidity of the market.

For example, suppose company A plans to purchase shares of company B at the end of the year. To protect itself from a possible increase in shares prices, company A reaches an agreement with company B to buy the shares at the end of the year for a fixed (forward) price F. Such an agreement between the two companies is called a *forward contract* (or simply, *forward*).

Now suppose that company A plans to sell some shares to company B at the end of the year. To protect itself from a possible fall in price of those shares, company A buys a *put option* (seller's option), which confers the right to sell the shares at the end of the year at the fixed *strike price* K. Note that, in contrast to the forwards case, a holder of an option must pay a *premium* to its issuer.

Futures contract is an agreement similar to the forward contract, but the trading takes place on a *stock exchange*, a special organization that manages the trading of various goods, financial instruments, and services.

Finally, we reiterate here that mathematical models of financial markets, methodologies for pricing various financial instruments and for constructing optimal (minimizing risk) investment strategies, are all subject to modern mathematical finance.

1.2   Probabilistic foundations of financial modeling and pricing of contingent claims

Suppose that a non-risky asset B and a risky asset S are completely described at any time $n = 0, 1, 2, \ldots$ by their prices. Therefore, it is natural to assume that the price dynamics of these securities are the essential component of a financial market. These dynamics are represented by the following equations:

$$
\begin{aligned}
\Delta B_n &= r B_{n-1}, & B_0 &= 1, \\
\Delta S_n &= \rho_n S_{n-1}, & S_0 &> 0,
\end{aligned}
$$

where $\Delta B_n = B_n - B_{n-1}$, $\Delta S_n = S_n - S_{n-1}$, $n = 1, 2, \ldots$; $r \geq 0$ is a constant rate of interest and ρ_n represent the level of risk. Quantities ρ_n will be specified later in this section.

Another important component of a financial market is the set of admissible actions or strategies that are allowed in dealing with assets B and S. A sequence $\pi = (\pi_n)_{n=1,2,\ldots} \equiv (\beta_n, \gamma_n)_{n=1,2,\ldots}$ is called an *investment strategy* (portfolio) if for any $n = 1, 2, \ldots$, the quantities β_n and γ_n are determined by prices $S_1, \ldots, S_{n-1}$. In other words, $\beta_n = \beta_n(S_1, \ldots, S_{n-1})$ and $\gamma_n = \gamma_n(S_1, \ldots, S_{n-1})$ are functions of $S_1, \ldots, S_{n-1}$, and they are interpreted as the amounts of assets B and S, respectively, at time n. The *value* of a portfolio π is

$$X_n^{\pi} = \beta_n B_n + \gamma_n S_n ,$$

where $\beta_n B_n$ represents the part of the capital deposited in a bank account and $\gamma_n S_n$ represents the investment in shares. If the value of a portfolio can change only because of changes in assets prices $\Delta X_n^{\pi} = X_n^{\pi} - X_{n-1}^{\pi} = \beta_n \Delta B_n + \gamma_n \Delta S_n$, then π is said to be a *self-financing* portfolio. The class of all such portfolios is denoted SF.

A common feature of all derivative securities in a (B, S)-market is their potential liability (payoff) f_N at a future time N. For example, for forwards, we have $f_N = S_N - F$ and for call options $f_N = (S_N - K)^+ \equiv \max\{S_N - K, 0\}$. Such liabilities inherent in derivative securities are called *contingent claims*. One of the most important problems in the theory of contingent claims is their *pricing* at any time before the expiry date N. This problem is related to the problem of *hedging contingent claims*. A self-financing portfolio is called a *hedge* for a contingent claim f_N if $X_N^{\pi} \geq f_N$ for any behavior of the market. If a hedging portfolio is not unique, then it is important to find a hedge π^* with the minimum value $X_n^{\pi^*} \leq X_n^{\pi}$ for any other hedge π. Hedge π^* is called the *minimal hedge*. The minimal hedge gives an obvious solution to the problem of pricing a contingent claim: the fair price of the claim is equal to the value of the minimal hedging portfolio. Furthermore, the minimal hedge manages the risk inherent in a contingent claim.

Next, we introduce some basic notions from probability theory and stochastic analysis that are helpful in studying risky assets. We start with the fundamental notion of an "experiment" when the set of possible outcomes of the experiment is known, but it is not known *a priori* which of those outcomes will take place (this constitutes the *randomness* of the experiment).

Example 1.1 (Trading on a stock exchange)

A set of possible exchange rates between the dollar and the euro is always known before the beginning of trading but not the exact value. $\square$

Let Ω be the set of all elementary outcomes ω and let $\mathcal{F}$ be the set of all *events* (non-elementary outcomes), which contains the *impossible* event $\emptyset$ and the *certain* event Ω.

Next, suppose that after repeating an experiment n times, an event $A \in \mathcal{F}$ occurred n_A times. Let us consider random experiments that have the following property of *statistical stability*: for any event A, there is a number $P(A) \in [0, 1]$ such that $n_A/n \to P(A)$ as $n \to \infty$. This number $P(A)$ is called the *probability* of event A. Probability $P : \mathcal{F} \to [0, 1]$ is a function with the following properties:

1. $P(\Omega) = 1$ and $P(\emptyset) = 0$;

2. $P\left(\cup_k A_k\right) = \sum_k P(A_k)$ for $A_i \cap A_j = \emptyset$.

The triple $(\Omega, \mathcal{F}, P)$ is called a *probability space*. For the rest of this section, we assume that the set Ω is countable. In this case, $(\Omega, \mathcal{F}, P)$ is referred to as a *discrete probability space*.

Every event $A \in \mathcal{F}$ can be associated with its *indicator*:

$$I_A(\omega) = \begin{cases} 1, & \text{if } \omega \in A \\ 0, & \text{if } \omega \in \Omega \setminus A \end{cases}.$$

Any function $X : \Omega \to \mathbb{R}$ is called a *random variable*. An indicator is the simplest example of a random variable. Any random variable X on a discrete probability space is *discrete* since the range of function $X(\cdot)$ is countable: $(x_k)_{k=1,2,\dots}$. In this case, we have the following representation:

$$X(\omega) = \sum_{k=1}^{\infty} x_k I_{A_k}(\omega),$$

where $A_k \in \mathcal{F}$ and $\cup_k A_k = \Omega$. A discrete random variable X is called *simple* if the corresponding sum is finite. The function

$$F_X(x) := P(\{\omega : X \le x\}), \quad x \in \mathbb{R} \tag{1.1}$$

is called the *distribution function* (or *cumulative distribution function*) of X. For a discrete X, we have

$$F_X(x) = \sum_{k:x_k \le x} P(\{\omega : X = x_k\}) \equiv \sum_{k:x_k \le x} p_k. \tag{1.2}$$

The sequence $(p_k)_{k=1,2,\dots}$ is called the *probability distribution* of a discrete random variable X, and we have $\sum_k p_k = 1$.

Note the following properties of the distribution function:

(D1) $F_X(x)$ are non-decreasing functions of x;

(D2) $F_X(x) \searrow 0$ as $x \to -\infty$ and $F_X(x) \nearrow 1$ as $x \to +\infty$.

The *expectation* (*expected value* or *mean value*) of X is

$$E(X) = \sum_{k \ge 1} x_k p_k.$$

Given a random variable X, for most functions $g : \mathbb{R} \to \mathbb{R}$ it is possible to define a random variable $Y = g(X)$ with expectation

$$E(Y) = \sum_{k \ge 1} g(x_k) p_k.$$

In particular, for any $k = 1, 2, \dots$, the quantity $E(X^k)$ is called the k-th *moment* of X, and the quantity

$$Var(X) = E\left[(X - E(X))^2\right]$$

is called the *variance* of X.

Note the following straightforward properties:

1. *Linearity of the expectation:* for any random variables $X_1, \ldots, X_n$ and any constants $c_1, \ldots, c_n$, we have

$$E\left(\sum_{i=1}^{n} c_i X_i \right) = \sum_{i=1}^{n} c_i E(X_i);$$

2. For any random variable X and any constant c,

$$Var(cX) = c^2 Var(X).$$

Example 1.2 (Examples of discrete probability distributions)

1. Bernoulli:

$$p_0 = P(\{\omega : \ X = a\}) = p, \quad p_1 = P(\{\omega : \ X = b\}) = 1 - p,$$

where $p \in [0, 1]$ and $a, b \in \mathbb{R}$.

2. Binomial:

$$p_m = P(\{\omega : \ X = m\}) = \left(\begin{array}{c} n \\ k \end{array} \right) p^m (1 - p)^{n-m},$$

where $p \in [0, 1]$, $n \geq 1$ and $m = 0, 1, \ldots, n$.

3. Poisson (with parameter $\lambda > 0$):

$$p_m = P(\{\omega : \ X = m\}) = e^{-\lambda} \frac{\lambda^m}{m!}$$

for $m = 0, 1, \ldots$.

Consider a positive random variable $\widetilde{Z}$ on a probability space $(\Omega, \mathcal{F}, P)$. Suppose that $E(\widetilde{Z}) = 1$, then, for any event $A \in \mathcal{F}$, define its new probability:

$$\widetilde{P}(A) = E(\widetilde{Z} I_A). \tag{1.3}$$

The expectation of a random variable X with respect to this new probability is $\widetilde{E}(X) = E(\widetilde{Z}X)$, and this rule is referred to as *change of the probability measure under the expectation sign*. Random variable $\widetilde{Z}$ is called the *density* of the probability $\widetilde{P}$ with respect to P. The proof of this formula is based on the linearity of the expectation:

$$
\begin{aligned}
\widetilde{E}(X) &= \sum_k x_k \widetilde{P}(\{\omega : \ X = x_k\}) = \sum_k x_k E\big(\widetilde{Z} I_{\{\omega: \ X = x_k\}} \big) \\
&= \sum_k E\big(\widetilde{Z} \, x_k \, I_{\{\omega: \ X = x_k\}} \big) = E\Big(\widetilde{Z} \sum_k x_k I_{\{\omega: \ X = x_k\}} \Big) \\
&= E(\widetilde{Z}X).
\end{aligned}
$$

For discrete random variables X and Y with values $(x_i)_{i=1,2,...}$ and $(y_i)_{i=1,2,...}$, respectively, consider the probabilities

$$P(\{\omega : X = x_i, Y = y_i\}) \equiv p_{ij}, \quad p_{ij} \geq 0, \quad \sum_{i,j} p_{ij} = 1.$$

These probabilities form the *joint distribution* of X and Y. Denote $p_{i\bullet} = \sum_j p_{ij}$ and $p_{\bullet j} = \sum_i p_{ij}$, then random variables X and Y are called *independent* if $p_{ij} = p_{i\bullet} \cdot p_{\bullet j}$, which implies that $E(XY) = E(X)E(Y)$.

The quantity

$$E(X|Y = y_i) := \sum_i x_i \frac{p_{ij}}{p_{\bullet j}}$$

is called the *conditional expectation* of X with respect to the event $\{Y = y_i\}$. The random variable $E(X|Y)$ is called the *conditional expectation* of X with respect to Y if $E(X|Y)$ is equal to $E(X|Y = y_i)$ on every set $\{\omega : Y = y_i\}$. In particular, for indicators $X = I_A$ and $Y = I_B$, we obtain

$$E(X|Y) = P(A|B) = \frac{P(AB)}{P(B)},$$

which is called the *conditional expectation of the event A given the event B.* We mention some properties of conditional expectations:

1. $E(X) = E\big(E(X|Y)\big)$, in particular, for $X = I_A$ and $Y = I_B$, we have $P(A) = P(B)P(A|B) + P(\Omega \setminus B)P(A|\Omega \setminus B)$;

2. If X and Y are independent, then $E(X|Y) = E(X)$;

3. Since by definition $E(X|Y)$ is a function of Y, then conditional expectation can be interpreted as a *prediction* of X given the information from the "observed" random variable Y.

Finally, for a random variable X with values in $\{0, 1, 2, \ldots\}$ we introduce the notion of a *generating function*

$$\phi_X(x) = E(x^X) = \sum_i x^i p_i.$$

We have

$$\phi(1) = 1, \quad \frac{d^k \phi}{dx^k}\Big|_{x=0} = k! p_k$$

and

$$\phi_{X_1 + \cdots + X_k}(x) = \prod_{i=1}^{k} \phi_{X_i}(x)$$

for independent random variables $X_1, \ldots, X_k$.

Example 1.1 (Trading on a stock exchange: Revisited)

Consider the following time scale: $n = 0$ (present time), $\ldots, n = N$ (can be one month, quarter, year, etc.).

An elementary outcome can be written in the form of a sequence $\omega = (\omega_1, \ldots, \omega_N)$, where ω_i is an elementary outcome representing the results of trading at time step $i = 1, \ldots, N$. Now we consider a probability space $(\Omega, \mathcal{F}_N, P)$ that contains all trading results up to time N. For any $n \leq N$, we also introduce the corresponding probability space $(\Omega, \mathcal{F}_n, P)$ with elementary outcomes $(\omega_1, \ldots, \omega_n) \in \mathcal{F}_n \subseteq \mathcal{F}_N$.

Thus, to describe the evolution of trading on a stock exchange, we need a *filtered probability space* $(\Omega, \mathcal{F}_N, \mathbb{F}, P)$ called a *stochastic basis*, where $\mathbb{F} = (\mathcal{F}_n)_{n \leq N}$ is called a *filtration* (or *information flow*):

$$\mathcal{F}_0 = \{\emptyset, \Omega\} \subseteq \mathcal{F}_1 \subseteq \ldots \subseteq \mathcal{F}_N.$$

By definition, the set $\mathcal{F}_n$ contains all trading information up to time n; therefore, if $A \in \mathcal{F}_n \in \mathbb{F}$, then $\mathcal{F}_n$ also contains the complement of A and it is closed under taking countable unions and intersections. The sets of events that satisfy such properties are σ-algebras, and we will discuss them in detail later in this chapter.

Now consider a (B, S)-market. Since asset B is non-risky, we can assume that $B(\omega) \equiv B_n$ for all $\omega \in \Omega$. For a risky asset S, it is natural to assume that prices $S_1, \ldots, S_N$ are random variables on stochastic basis $(\Omega, \mathcal{F}_N, \mathbb{F}, P)$. Each of S_n is completely determined by the trading results up to time $n \leq N$ or, in other words, by the set of events $\mathcal{F}_n$. We also assume that the sources of trading randomness are exhausted by the stock prices; that is, $\mathcal{F}_n = \sigma(S_1, \ldots, S_n)$ is a set generated by random variables $S_1, \ldots, S_n$.

Let us consider a specific example of a (B, S)-market. Let $\rho_1, \ldots, \rho_N$ be independent random variables taking values a and b ($a < b$) with probabilities $P(\{\omega : \rho_k = b\}) = p$ and $P(\{\omega : \rho_k = a\}) = 1 - p \equiv q$, respectively. Define the stochastic basis: $\Omega = \{a, b\}^N$ is the space of sequences of length N whose elements are equal to either a or b; $\mathcal{F} = 2^\Omega$ is the set of all subsets of Ω. The filtration $\mathbb{F}$ is generated by the prices (S_n) or equivalently by the sequence (ρ_n):

$$\mathcal{F}_n = \sigma(S_1, \ldots, S_n) = \sigma(\rho_1, \ldots, \rho_n),$$

which means that every random variable on the probability space $(\Omega, \mathcal{F}_n, P)$ is a function of $S_1, \ldots, S_n$ or, equivalently, of $\rho_1, \ldots, \rho_n$ due to relations

$$\frac{\Delta S_k}{S_{k-1}} = \rho_k, \quad k = 1, 2, \ldots.$$

A financial (B, S)-market defined on this stochastic basis is called *binomial*.

Consider a contingent claim f_N. Since its maturity date is N, then, in general, $f_N = f(S_1, \ldots, S_N)$ is a function of all "history" $S_1, \ldots, S_N$. The key problem now is to estimate (or predict) f_N at any time $n \leq N$

given the available market information $\mathcal{F}_n$. We would like these predictions $E(f_N|\mathcal{F}_n)$, $n = 0, 1, \ldots, N$, to have the following intuitively natural properties:

1. $E(f_N|\mathcal{F}_n)$ is a function of $S_1, \ldots, S_n$, but not of future prices $S_{n+1}, \ldots, S_N$;

2. A prediction based on the *trivial* information $\mathcal{F}_0 = \{\emptyset, \Omega\}$ should coincide with the mean value of a contingent claim: $E(f_N|\mathcal{F}_0) = E(f_N)$;

3. Predictions must be compatible:

$$E(f_N|\mathcal{F}_n) = E\Big(E(f_N|\mathcal{F}_{n+1})\Big|\mathcal{F}_n\Big),$$

in particular

$$E\Big(E(f_N|\mathcal{F}_n)\Big) = E\Big(E(f_N|\mathcal{F}_n)\Big|\mathcal{F}_0\Big) = E(f_N);$$

4. A prediction based on all possible information $\mathcal{F}_N$ should coincide with the contingent claim: $E(f_N|\mathcal{F}_N) = f_N$;

5. Linearity:

$$E(\phi f_N + \psi g_N|\mathcal{F}_n) = \phi E(f_N|\mathcal{F}_n) + \psi E(g_N|\mathcal{F}_n)$$

for ϕ and ψ defined by the information in $\mathcal{F}_n$;

6. If f_N does not depend on the information in $\mathcal{F}_n$, then a prediction based on this information should coincide with the mean value

$$E(f_N|\mathcal{F}_n) = E(f_N);$$

7. Denote $f_n = E(f_N|\mathcal{F}_n)$; then from Property 3, we obtain

$$E(f_{n+1}|\mathcal{F}_n) = E\Big(E(f_N|\mathcal{F}_{n+1})\Big|\mathcal{F}_n\Big) = E(f_N|\mathcal{F}_n) = f_n$$

for all $n \leq N$. Such stochastic sequences are called *martingales*.

Now we consider how to calculate predictions. Comparing the notions of a conditional expectation and a prediction, we see that a prediction of f_N based on $\mathcal{F}_n = \sigma(S_1, \ldots, S_n)$ is equal to the conditional expectation of a random variable f_N with respect to random variables $S_1, \ldots, S_n$.

Worked Example 1.2 *Suppose that the monthly price evolution of stock S is given by*

$$S_n = S_{n-1}(1 + \rho_n), \quad n = 1, 2, \ldots,$$

where profitabilities ρ_n are independent random variables taking values 0.2 and -0.1 with probabilities 0.4 and 0.6, respectively. Given that the current price $S_0 = 200$ (\$), find the predicted mean price of S for the next two months.

Solution Since

$$E(\rho_1) = E(\rho_2) = 0.2 \cdot 0.4 - 0.1 \cdot 0.6 = 0.02,$$

then

$$
\begin{aligned}
E\left(\frac{S_1 + S_2}{2}\bigg| S_0 = 200\right) &= E\left(\frac{S_0(1 + \rho_1) + S_0(1 + \rho_1)(1 + \rho_2)}{2}\bigg| S_0 = 200\right) \\
&= \frac{S_0}{2}\Big[E(1 + \rho_1) + E(1 + \rho_1)E(1 + \rho_2)\Big] \\
&= 100\big[1.02 + 1.02 \cdot 1.02\big] = 206.4. \;\square
\end{aligned}
$$

We finish this section noting that there are various indications that the use of discrete probability spaces can significantly limit the class of stochastic experiments available for stochastic modeling. Below, we discuss one of most illustrative considerations of this nature.

Let function $f(x)$, $x \in \mathbb{R}$, be non-negative with $\int_{-\infty}^{\infty} f(x)\,dx = 1$. Then function

$$F(x) = \int_{-\infty}^{x} f(y)\,dy$$

satisfies properties (D1)–(D2) of a distribution function: F is a non-decreasing function of x such that $F(x) \searrow 0$ as $x \to -\infty$ and $F(x) \nearrow 1$ as $x \to +\infty$. In this case, function f is referred to as *density* of the distribution function F. For example, function

$$f(x) = \frac{1}{\sigma\sqrt{2\pi}}\, e^{-\frac{(x-\mu)^2}{2\sigma^2}}$$

defines a *normal* distribution function F with parameters $\mu \in \mathbb{R}$ and $\sigma \in \mathbb{R}_+$. In particular, a *standard normal distribution* is a normal distribution with $\mu = 0$ and $\sigma^2 = 1$. Usually, it is denoted Φ or $\mathcal{N}(0, 1)$.

One can pose a natural question: given an arbitrary function F that satisfies all properties of a distribution function, is there a probability space and a random variable X such that F is the distribution function of X?

The following example gives a positive answer to this question and illustrates that non-discrete probability spaces and non-discrete random variables (that are of particular importance in stochastic modeling) exist.

Consider the *Borel* algebra $\mathcal{B}([0, 1])$. It is the set of all Borel subsets of the interval $[0, 1]$, which contains all possible subintervals of $[0, 1]$, their complements and countable unions and intersections, and therefore it is a σ-algebra. For elements of $\mathcal{B}([0, 1])$ we can introduce a (unique) Lebesgue measure m, that, for intervals, is equal to their length. Let $\Omega = [0, 1]$, $\mathcal{F} = \mathcal{B}([0, 1])$, and $P = m$, then $(\Omega, \mathcal{F}, P)$ is a probability space. We now define a random variable $X : [0, 1] \to \mathbb{R}$ in the following way: $X(0) = X(1) = 0$; further for each $x \in (0, 1)$, let $\omega = F(x)$ and define $X(\omega) = x$. Then the probability of $\{\omega : X(\omega) \le x\}$ is equal to the length of the interval $[0, F(x)] = F(x)$.

In the next section, we discuss in detail the fundamental general notion of a probability space that facilitates the quantitative description of numerous statistically stable stochastic experiments.

1.3 Elements of probability theory and stochastic analysis

One of the pivotal foundations of modern probability theory is the notion of a *probability space* $(\Omega, \mathcal{F}, P)$, where

- a set Ω is the *space of elementary outcomes* ω;

- the *space of all events* is represented by $\mathcal{F}$, the set of all subsets of Ω, including $\emptyset$ and Ω, which also contains all their complements and countable unions and intersections (σ-algebra);

- a *probability measure* (or simply, *probability*) P is a function $P : \mathcal{F} \to [0, 1]$ that satisfies the following properties:

  1. $P(\Omega) = 1$, $P(\emptyset) = 0$;

  2. $P(\Omega \setminus A) = 1 - P(A)$ for all $A \in \mathcal{F}$;

  3. $P\left(\bigcup_k A_k\right) = \sum_k P(A_k)$ for any sequence of pairwise disjoint sets $A_k \in \mathcal{F}$.

This construction is often referred to as *Kolmogorov's axiomatic formulation* of probability theory. For detailed accounts of its various aspects, we refer the reader to standard probability textbooks.

Any pair $(\Omega, \mathcal{F})$ in this construction is called a *measurable space*. In particular, the pair $(\mathbb{R}, \mathcal{B}(\mathbb{R}))$ is called the *Borel space*.

We say that function $X : \Omega \to \mathbb{R}$ is *measurable* if the inverse image of any Borel set $B \in \mathcal{B}(\mathbb{R})$ is in $\mathcal{F}$:

$$X^{-1}(B) = \{\omega \,:\, X(\omega) \in B\} \in \mathcal{F}.$$

Such functions are called *random variables*. The *indicator*

$$X(\omega) = I_A(\omega) = \begin{cases} 1, & \omega \in A \\ 0, & \omega \notin A \end{cases}, \quad A \in \mathcal{F},$$

and linear combinations of indicators

$$X(\omega) = \sum_{k \geq 1} x_k \, I_{A_k}(\omega), \quad A_k \in \mathcal{F}, \; x_k \in \mathbb{R}$$

are the simplest examples of random variables.

A random variable X induces a probability measure on the Borel space $(\mathbb{R}, \mathcal{B}(\mathbb{R}))$ defined by

$$P_X(B) := P\{\omega \,:\, X(\omega) \in B\}, \quad B \in \mathcal{B}(\mathbb{R}).$$

For sets $B = (-\infty, x]$, $x \in \mathbb{R}$, then we can introduce a function

$$F(x) = F_X(x) = P_X((-\infty, x]), \quad x \in \mathbb{R}, \tag{1.4}$$

which is called the *distribution function* of random variable X. This definition generalizes the notion of the distribution function for a discrete random variable introduced by formula (1.1), and if X is discrete, then function (1.4) coincides with (1.2). Note that function F satisfies properties (D1)–(D2) of a distribution function and that it is right continuous.

A large class of non-discrete random variables can be characterized by absolutely continuous distribution functions of the integral form:

$$F(x) = \int_{-\infty}^{x} f(y)\, dy,$$

where non-negative function f is called the *density* of a random variable. Here, we mention some useful examples of such distributions:

1. *Uniform* on $[a, b]$ distribution:

$$f(y) = \begin{cases} 1/(b-a), & y \in [a, b] \\ 0, & y \notin [a, b] \end{cases} ;$$

2. *Normal* (Gaussian) distribution $\mathcal{N}(\mu, \sigma^2)$ with $\mu \in \mathbb{R}$ and $\sigma \in \mathbb{R}_+$:

$$f(y) = \frac{1}{\sigma\sqrt{2\pi}} e^{-\frac{(y-\mu)^2}{2\sigma^2}}, \quad y \in \mathbb{R};$$

3. *Gamma* distribution with parameters $\alpha > 0$ and $\beta > 0$:

$$f(y) = \frac{y^{\alpha-1} e^{-y/\beta}}{\Gamma(\alpha)\, \beta^{\alpha}}, \quad y \geq 0,$$

where Γ is the gamma function. In particular, if $\alpha = 1$ and $\beta = 1/\lambda$, then this distribution is referred to as *exponential* distribution with parameter λ. Its density has the form

$$f(y) = \lambda e^{-\lambda y}, \quad y \geq 0, \ \lambda > 0;$$

4. *Student* distribution:

$$f(y) = \frac{\Gamma((n+1)/2)}{\sqrt{n\pi}\,\Gamma(n/2)} \left(1 + \frac{y^2}{n}\right)^{-\frac{n+1}{2}}, \quad y \in \mathbb{R},$$

with degrees of freedom $n = 1, 2, \ldots$;

5. χ^2 distribution:

$$f(y) = \frac{y^{n/2-1} e^{-y/2}}{\Gamma(n/2)\, 2^{n/2}}, \quad y \geq 0, \ n = 1, 2, \ldots,$$

which is a gamma distribution with parameters $\alpha = n/2$ and $\beta = 2$.

Given a random variable X, one can introduce various random variables associated with it in the following way. Consider an arbitrary measurable (Borel) function $\varphi : (\mathbb{R}, \mathcal{B}(\mathbb{R})) \to (\mathbb{R}, \mathcal{B}(\mathbb{R}))$; then the composition of functions X and φ,

$$Y = \varphi \circ X, \tag{1.5}$$

is a random variable, since, for any $B \in \mathcal{B}(\mathbb{R})$, we have

$$
\begin{aligned}
\{\omega : Y(\omega) \in B\} &= \{\omega : \varphi(X(\omega)) \in B\} \\
&= \{\omega : X(\omega) \in \varphi^{-1}(B)\} \in \mathcal{F}.
\end{aligned}
$$

In particular, we can introduce random variables $X^+ := \max\{0, X\}$, $X^- := \max\{0, -X\}$, $|X| = X^+ + X^-$, X^k, and so forth.

Denote $\mathcal{F}^X$ the set of all random events of the type $\{\omega : X(\omega) \in B\}$ for all possible $B \in \mathcal{B}(\mathbb{R})$. It is not difficult to show that $\mathcal{F}^X$ is a σ-algebra. This referred to as σ-*algebra generated by random variable* X. We also note that, if Y as a random variable that is measurable with respect to the σ-algebra $\mathcal{F}^X$, then there exists a Borel function φ such that the representation (1.5) holds for Y. Later, we will discuss martingale representations as dynamic versions of this fact.

In the previous section, in the case of discrete random variables on a discrete probability space, we saw that distribution functions and expectations are the key quantitative characteristics of random variables. We now introduce the notion of expectation in the non-discrete case. First, we consider a discrete random variable X on probability space $(\Omega, \mathcal{F}, P)$. Suppose it can attain the finite number of values: $x_1, \ldots, x_n$, then we can write

$$X(\omega) = \sum_{k=1}^{n} x_k I_{A_k}(\omega),$$

where $A_k = \{\omega : X(\omega) = x_k\}$, $k = 1, 2, \ldots, n$, and $\bigcup_{k=1}^{n} A_k = \Omega$.

As in the case of a discrete probability space, we define the *expectation* of X by

$$E(X) = \sum_{k=1}^{n} x_k P(A_k).$$

Now let X be an arbitrary non-negative random variable on $(\Omega, \mathcal{F}, P)$. We define a sequence $(X_n)_{n=1,2,\ldots}$ of discrete random variables with finite numbers of values by

$$X_n(\omega) = \sum_{k=1}^{n\,2^n} \frac{k-1}{2^n} I_{\left\{\omega \,:\, \frac{k-1}{2^n} \le X(\omega) < \frac{k}{2^n}\right\}}(\omega) + n\, I_{\left\{\omega \,:\, X(\omega) \ge n\right\}}(\omega).$$

Clearly, $X_n(\omega) \nearrow X(\omega)$ for each $\omega \in \Omega$. Since sequence $(E(X_n))_{n=1,2,\ldots}$ is non-decreasing, then there exists the limit

$$E(X) := \lim_{n \to \infty} E(X_n),$$

which is called the expectation of X. Note that this limit is not necessarily finite.

An arbitrary random variable X on $(\Omega, \mathcal{F}, P)$ can be written in the form $X = X^+ - X^-$, where X^+ and X^- are non-negative random variables. If at least one of the expectations is finite, we define

$$E(X) = E(X^+) - E(X^-).$$

If both $E(X^+) < \infty$ and $E(X^-) < \infty$, then

$$E(|X|) = E(X^+) + E(X^-) < \infty,$$

and we say that X is a random variable with *finite* expectation.

This construction of the expectation is identical to the definition of the Lebesgue integral for a measurable function X, and the following integral notation is also common in the probability theory:

$$E(X) = \int_\Omega X \, dP = \int_\Omega X(\omega) \, P(d\omega).$$

Because of construction, expectations are linear and monotonic. Change of variables in the Lebesgue integral allows the following representation of expectation in terms of the distribution function of X:

$$E(\varphi(X)) = \int_\Omega \varphi(X) \, dP = \int_{\mathbb{R}} \varphi(x) \, P_X(dx) = \int_{\mathbb{R}} \varphi(x) \, dF_X(x),$$

where φ is a Borel function that is integrable with respect to P_X.

For $\varphi(x) = x^k$, $k = 1, 2, \ldots$, the expectation $E(X^k)$ is called the k-th *moment* of random variable X. Suppose $E(X) = \mu$ and let $\varphi(x) = (x - \mu)^k$, then the corresponding moments are called *centered* moments. The second centered moment is called the *variance* of X: $Var(X) = E((X - \mu)^2)$, and it is one of the key measures of the dispersion of values of X about the mean value μ. The common additional measures are *skewness*:

$$S = \frac{E((X - \mu)^3)}{(Var(X))^{3/2}},$$

and *kurtosis*:

$$K = \frac{E((X - \mu)^4)}{(Var(X))^2}.$$

If a probability space $(\Omega, \mathcal{F}, P)$ is non-discrete, then some further important properties of expectations involve the notion of P-*almost surely* or simply *almost surely*. We say that a property holds almost surely (we write a.s.) if there is a set $N \in \mathcal{F}$ such that the probability (measure) of N is zero and the property holds for all $\omega \in \Omega \setminus N$. The following properties can be readily derived:

1. If $X = 0$ (a.s.), then $E(X) = 0$;

2. If $X = Y$ (a.s.) and $E|X| < \infty$, then $E|Y| < \infty$ and $E(X) = E(Y)$;

3. If $X \geq 0$ and $E(X) = 0$, then $X = 0$ (a.s.);

4. If $E|X| < \infty$, $E|Y| < \infty$, and $E(X\,I_A) \leq E(Y\,I_A)$ for all $A \in \mathcal{F}$, then $X \leq Y$ (a.s.).

Let $(X_n)_{n=1,2,\dots}$ be a sequence of arbitrary random variables on a non-discrete probability space $(\Omega, \mathcal{F}, P)$. Suppose that $X_n \to X$ (a.s.) as $n \to \infty$, where X is some random variable. We are now interested in conditions that can guarantee that $E(X_n)$ converges to $E(X)$. The most general condition of this nature is the *uniform integrability* of the sequence $(X_n)_{n=1,2,\dots}$:

$$\lim_{c \to \infty} \sup_n E|X_n| \cdot I_{\{\omega \,:\, |X_n(\omega)| > c\}} = 0 \,.$$

The following result is fundamental.

Theorem 1.1 *Let* $(X_n)_{n=1,2,\dots}$ *be a uniformly integrable sequence of random variables such that* $X_n \to X$ *(a.s.) as* $n \to \infty$. *Then the limit random variable* X *has finite expectation. Furthermore,* $\lim_{n\to\infty} E(X_n) = E(X)$ *and* $\lim_{n\to\infty} E|X_n - X| = 0$ *(convergence in* L^1*).*

Often there is a need for modeling with a n-dimensional random variable $X = (X_1, \dots, X_n)$. Denote $\mathcal{B}(\mathbb{R}^n)$ the Borel algebra on $\mathbb{R}^n$, then the *random vector* X can be defined as a measurable mapping from a measurable space $(\Omega, \mathcal{F})$ into the Borel space $(\mathbb{R}^n, \mathcal{B}(\mathbb{R}^n))$. A natural quantitative probabilistic characteristic of the random vector X is the n-dimensional distribution function

$$F_X(x_1, \dots, x_n) = P(\omega \,:\, X_1(\omega) \leq x_1, \dots, X_n(\omega) \leq x_n), \quad (x_1, \dots, x_n) \in \mathbb{R}^n \,,$$

which is also referred to as a *joint distribution function* of random variables $X_1, \dots, X_n$.

Random variables $X_1, \dots, X_n$ are said to be *independent* if $F_X(x_1, \dots, x_n) = F_{X_1}(x_1) \times \dots \times F_{X_n}(x_n)$, where $F_{X_1}, \dots, F_{X_n}$ are the corresponding one-dimensional distribution functions. Similarly to the one-dimensional case, a n-dimensional distribution function F_X generates a probability measure P_X on $(\mathbb{R}^n, \mathcal{B}(\mathbb{R}^n))$ such that

$$P_X((-\infty, x_1] \times \dots \times (-\infty, x_n]) = F_X(x_1, \dots, x_n) \,,$$

which is called a n-*dimensional distribution*. We denote it P_n, and we note its *consistency* property:

$$P_n(B) = P_{n+1}(B \times \mathbb{R}), \quad n = 1, 2, \dots \tag{1.6}$$

for all $B \in \mathcal{B}(\mathbb{R}^n)$.

Many important facts in probability theory are related to infinite families of random variables. A sequence of random variables $(X_n)_{n=1,2,\dots}$ is one of the typical examples of such families. Very often it is natural to assume that finite dimensional distributions P_k of $X_1, \dots, X_k$ are known for all $k = 1, 2, \dots$. In this case, it is important to consider the question of the existence of a probability space that will accommodate sequence $(X_n)_{n=1,2,\dots}$. The affirmative answer to this question is given by the celebrated *Kolmogorov's consistency theorem*, where the consistency property of finite dimensional distributions becomes the consistency condition on finite dimensional distributions. Note that the example that we discussed at the end of the previous section, which motivates the study of non-discrete probability spaces and non-discrete random variables, is a simple consequence of Kolmogorov's theorem.

The notion of a joint distribution of random variables $X_1, \dots, X_n$ allows one to compute quantitative characteristics of random variables that are functions of $X_1, \dots, X_n$. We illustrate this in the case of two random variables.

Theorem 1.2 *Let X_1 and X_2 be independent random variables with finite expectations. Then*

$$E(X_1 X_2) = E(X_1) E(X_2). \tag{1.7}$$

Proof We need to prove this statement only for non-negative random variables since the general case then follows from the following relations:

$$X_i = X_i^+ - X_i^- , \ i = 1, 2,$$

and

$$X_1 X_2 = X_1^+ X_2^+ - X_1^- X_2^+ - X_1^+ X_2^- + X_1^- X_2^- .$$

Let

$$X_{i,n} := \sum_{k=0}^{\infty} \frac{k}{n} I_{\left\{\omega : \frac{k}{n} \leq X_i(\omega) < \frac{k+1}{n}\right\}} , \ i = 1, 2,$$

and note that by construction we have

$$X_{i,n} \leq X_i \quad \text{and} \quad |X_{i,n} - X_i| \leq \frac{1}{n}, \ i = 1, 2.$$

Since discrete random variables $X_{1,n}$ and $X_{2,n}$ are independent, we obtain

$$
\begin{aligned}
E(X_{1,n} X_{2,n}) &= \sum_{k,l=0}^{\infty} \frac{k\,l}{n^2} E\left(I_{\left\{\omega : \frac{k}{n} \leq X_1(\omega) < \frac{k+1}{n}\right\}} I_{\left\{\omega : \frac{l}{n} \leq X_2(\omega) < \frac{l+1}{n}\right\}} \right) \\
&= \sum_{k,l=0}^{\infty} \frac{k\,l}{n^2} E\left(I_{\left\{\omega : \frac{k}{n} \leq X_1(\omega) < \frac{k+1}{n}\right\}} \right) E\left(I_{\left\{\omega : \frac{l}{n} \leq X_2(\omega) < \frac{l+1}{n}\right\}} \right) \\
&= E(X_{1,n}) E(X_{2,n}).
\end{aligned}
$$

By Theorem 1.1, we have

$$E(X_{i,n}) \to E(X_i) \quad \text{as } n \to \infty, \ i = 1, 2,$$

and therefore

$$
\begin{aligned}
\left| E(X_1 X_2) - E(X_{1,n} X_{2,n}) \right| &\leq E\left(\left| X_1 X_2 - X_{2,n} X_{2,n} \right| \right) \\
&\leq E\left(|X_1| \, |X_2 - X_{2,n}| \right) + E\left(|X_{2,n}| \, |X_1 - X_{1,n}| \right) \\
&\leq \frac{1}{n} E\left(|X_1| \right) + \frac{1}{n} E\left(|X_2 + \frac{1}{n}| \right) \to 0 \quad \text{as } n \to \infty.
\end{aligned}
$$

Thus,

$$
\begin{aligned}
E(X_1 X_2) &= \lim_{n \to \infty} E(X_{1,n} X_{2,n}) = \lim_{n \to \infty} E(X_{1,n}) \lim_{n \to \infty} E(X_{2,n}) \\
&= E(X_1) E(X_2). \ \square
\end{aligned}
$$

In the context of this result, it is useful to introduce the notion of *covariance* of random variables X_1 and X_2 with expectations μ_1 and μ_2, respectively:

$$Cov(X_1 X_2) = E\left((X_1 - \mu_1)(X_2 - \mu_2) \right),$$

and the notion of the *correlation coefficient* of X_1 and X_2:

$$Cor(X_1 X_2) = \frac{Cov(X_1 X_2)}{\sqrt{Var(X_1)}\sqrt{Var(X_2)}}.$$

If random variables X_1 and X_2 are independent, then Theorem 1.2 implies that $Cov(X_1 X_2) = Cor(X_1 X_2) = 0$. In this case, we say that X_1 and X_2 are *uncorrelated*. Note that the inverse is not true: the uncorrelated random variables are not necessarily independent. We also note here that for random variables that are not necessarily independent, equality (1.7) can be replaced with an appropriate inequality. The following *Cauchy-Bunyakovskyi inequality* is a classical example of such inequalities.

Theorem 1.3 *Let X_1 and X_2 be random variables with finite second moments. Then*

$$E\left(|X_1 X_2| \right) \leq \sqrt{E(X_1^2) E(X_2^2)}.$$

Note that the Cauchy-Bunyakovskyi inequality immediately implies that $|Cor(X_1 X_2)| \leq 1$.

The next property of expectations is formulated for random variables that are defined via formula (1.5): $Y = \varphi \circ X$. Let the Borel function φ be convex downward; that is, suppose that for each x_0 there exists a number $\lambda = \lambda(x_0)$ such that

$$\varphi(x) \geq \varphi(x_0) + (x - x_0)\lambda(x_0) \tag{1.8}$$

for all $x \in \mathbb{R}$. We arrive at the following *Jensen's inequality*.

Theorem 1.4 *Suppose that φ is a convex downward Borel function and a random variable X is such that the expectations $E(|X|)$ and $E(\varphi(X))$ are finite. Then*

$$\varphi(E(X)) \leq E(\varphi(X)).$$

Proof Let $x_0 = E(X)$, then inequality (1.8) with $x = X$ implies

$$\varphi(X) \geq \varphi(E(X)) + (X - E(X))\,\lambda(E(X)).$$

Using linearity and monotonicity of expectations, we conclude

$$
\begin{aligned}
E(\varphi(X)) &\geq \varphi(E(X)) + E(X - E(X))\,\lambda(E(X)) \\
&= \varphi(E(X)) + \lambda(E(X))\,E(X - E(X)) = \varphi(E(X)).\ \square
\end{aligned}
$$

The last property of expectations that we wish to mention in this section is called the *Chebyshev inequality*.

Theorem 1.5 *Let X be a non-negative random variable with finite expectation. Then for any $\varepsilon > 0$,*

$$P(X \geq \varepsilon) \leq \frac{E(X)}{\varepsilon}. \tag{1.9}$$

If X is an arbitrary random variable with finite variance, then

$$P(|X - E(X)| \geq \varepsilon) \leq \frac{Var(X)}{\varepsilon^2}. \tag{1.10}$$

Proof Inequality (1.9) holds true since

$$
\begin{aligned}
P(X \geq \varepsilon) &= E\left(I_{\left\{\omega\,:\,X(\omega) \geq \varepsilon\right\}}\right) = \frac{\varepsilon\,E\left(I_{\left\{\omega\,:\,X(\omega) \geq \varepsilon\right\}}\right)}{\varepsilon} \\
&\leq \frac{E\left(X\,I_{\left\{\omega\,:\,X(\omega) \geq \varepsilon\right\}}\right)}{\varepsilon} \leq \frac{E(X)}{\varepsilon}.
\end{aligned}
$$

Inequality (1.10) is in fact the inequality (1.9) when applied to the non-negative random variable $(X - E(X))^2$ with $\varepsilon^2 > 0$. $\square$

In order to formulate one of the most important corollaries of inequalities (1.9)–(1.10), we need to introduce the following type of convergence. We say that sequence $(X_n)_{n=1,2,\dots}$ *converges in probability* to a random variable X if for any $\varepsilon > 0$

$$P(|X_n - X| \geq \varepsilon) \to 0 \quad \text{as } n \to \infty.$$

Consider a sequence $(X_n)_{n=1,2,\dots}$ of independent identically distributed random variables with expectations μ and variances σ^2. Define a new sequence

$$\overline{X}_n = \frac{1}{n}\sum_{k=1}^{n} X_k,$$

where, clearly, $E(\overline{X}_n) = \mu$ and $Var(\overline{X}_n) = \sigma^2/n$. Then for any fixed $\varepsilon > 0$, inequality (1.10) implies

$$P\big(|\overline{X}_n - \mu| \geq \varepsilon\big) \leq \frac{1}{n}\frac{\sigma^2}{\varepsilon^2} \to 0 \quad \text{as } n \to \infty.$$

Thus, the sequence of the *empirical means* $\overline{X}_n$ converges in probability to the *theoretical mean* μ. This result is usually referred to as the *law of large numbers* (LLN).

We also mention here the following *statistical application* of the law of large numbers. Suppose that parameters μ and σ^2 of sequence $\big(X_k\big)_{k=1,2,...}$ are unknown and we wish to estimate them from a finite subsequence $X_1, \ldots, X_n$. It turns out that $\overline{X}_n$ and

$$s_n^2 = \frac{1}{n-1}\sum_{k=1}^{n}(X_k - \overline{X}_n)^2$$

are adequate approximations for μ and σ^2, respectively. These approximations are not biased since $E(\overline{X}_n) = \mu$ and $E(s_n^2) = \sigma^2$, and the law of large numbers guarantees that $\overline{X}_n$ and s_n^2 converge in probability to μ and σ^2, respectively.

Recall, that so far we have discussed three types of convergence of a sequence $\big(X_n\big)_{n=1,2,...}$:

1. convergence almost surely;

2. convergence in mean; and

3. convergence in probability.

Note that these three types of convergence do not involve distributions of random variables X_n. Is it possible to introduce a type of convergence that would involve only distributions of random variables X_n? The answer to this question is affirmative, and it reflects one of the key essential ideas of probability theory.

We say that a sequence of random variables $\big(Y_n\big)_{n=1,2,...}$ with distribution functions $\big(F_{Y_n}\big)_{n=1,2,...}$ *converges in distribution* to a random variable Y with a continuous distribution function F_Y, if for all $x \in \mathbb{R}$

$$F_{Y_n}(x) \to F_Y(x) \quad \text{as } n \to \infty.$$

We now consider again a sequence $\big(X_n\big)_{n=1,2,...}$ of independent identically distributed random variables with expectations μ and variances σ^2. Define

$$Y_n = \frac{\sum_{k=1}^{n} X_k - \mu n}{\sigma \sqrt{n}}, \quad n = 1, 2, \ldots.$$

Sequence $\left(Y_n\right)_{n=1,2,\ldots}$ converges in distribution to a standard normal random variable $Y = \mathcal{N}(0,1)$; that is, for all $x \in \mathbb{R}$,

$$F_{Y_n}(x) \to \Phi(x) = \frac{1}{\sqrt{2\pi}} \int_{-\infty}^{\infty} e^{-y^2/2}\, dy \quad \text{as } n \to \infty.$$

Because of the extreme importance of this result in probability theory and statistics, it (and its various modifications) is referred to as the *central limit theorem* (CLT).

Independent *Bernoulli random variables*

$$X_n = \begin{cases} 1 & \text{with probability } p \in [0,1] \\ 0 & \text{with probability } q = 1-p \end{cases}, \quad n = 1,2,\ldots,$$

are often used in our discussions throughout this book. Denote $S_n = X_1 + \ldots + X_n$, then since for all $k = 1,2,\ldots$, $E(X_k) = p$ and $Var(X_k) = p \cdot q$ for such variables, the central limit theorem takes the following form

$$\frac{S_n - np}{\sqrt{npq}} \to \mathcal{N}(0,1) \quad \text{(in distribution)}.$$

This result is usually referred to as *integral De Moivre-Laplace theorem* and it follows readily from the following *local De Moivre-Laplace theorem*:

$$B_n(k) = P(S_n = k) = \binom{n}{k} p^k q^{n-k} \sim \frac{1}{\sqrt{2\pi npq}} e^{-\frac{(k-np)^2}{2npq}}$$

for $k = 0,1,\ldots,n \to \infty$, or denoting $k = np + x\sqrt{npq} = 0,1,\ldots,n$, we can write[1]

$$B_n(k) \sim \frac{1}{\sqrt{2\pi npq}} e^{-\frac{x^2}{2}}.$$

Note that the above approximations with normal distribution and normal density are at least of the order $n^{-1/2}$.

The general notion of *conditional expectation* is essential in modern probability theory. Its rigorous introduction involves the use of absolutely continuous measures. Let $(\Omega, \mathcal{F}, P)$ be a probability space and consider a σ-algebra $\mathcal{G} \subseteq \mathcal{F}$ which is a sub-algebra of the original σ-algebra $\mathcal{F}$.

Consider a non-negative random variable X. Denote $E(X|\mathcal{G})$ a random variable on $(\Omega, \mathcal{F}, P)$ that is measurable with respect to the smaller σ-algebra $\mathcal{G}$ and that

$$E(X\, I_G) = E\big(E(X|\mathcal{G}) \cdot I_G\big) \quad \text{for any } G \in \mathcal{G}. \tag{1.11}$$

Random variable $E(X|\mathcal{G})$ is called the *conditional expectation* of random variable X with respect to the σ-algebra $\mathcal{G}$. It is *well defined* in the following sense:

---

[1] We say that functions f and g are *equivalent*: $f \sim g$, if the limit of their ratio is equal to one.

if Y' and Y'' are two σ-measurable random variables that satisfy equality (1.11), then they are almost surely equal. Indeed, from (1.11) we have

$$E\big((Y' - Y'') \cdot I_G\big) = 0 \quad \text{for any } G \in \mathcal{G}\,.$$

Consider sets

$$G' = \{\omega : Y' - Y'' > 0\} \quad \text{and} \quad G' = \{\omega : Y' - Y'' < 0\}\,,$$

then

$$E\big((Y' - Y'') \cdot I_{G'}\big) = 0 = E\big((Y' - Y'') \cdot I_{G''}\big)$$

or $E|Y' - Y''| = 0$, and therefore

$$P\big(\{\omega : Y'(\omega) \neq Y''(\omega)\}\big) = 0\,.$$

In order to explore the existence of a conditional expectation, we now introduce the notion of absolutely continuous measures. We say that measure Q is *absolutely continuous* with respect to measure P (we write $Q \ll P$) if for any set $A \in \mathcal{F}$, $P(A) = 0$ implies that $Q(A) = 0$. For example, if Z is a non-negative random variable on $(\Omega, \mathcal{F}, P)$ with finite expectation, then measure Q, defined by the equality $Q(A) = E(ZI_A)$, is absolutely continuous with respect to measure P. It turns out the inverse statement holds true, and it is known as the *Radon-Nikodým* theorem: for any measure $Q \ll P$, there exists a measurable function $Z = \frac{dQ}{dP}$ (the density), such that

$$Q(A) = \int_A Z\,dP\,. \tag{1.12}$$

Applying this result to measure Q with $Q(G) = E(XI_G)$ for all $G \in \mathcal{G}$, we conclude that there exists a $\mathcal{G}$-measurable density Z, which is indeed the conditional expectation $E(X|\mathcal{G})$.

This construction can be extended to an arbitrary random variable X in the following standard way. We use the decomposition $X = X^+ - X^-$, and assuming that at least one of the expectations $E(X^\pm|\mathcal{G})$ is finite, we define

$$E(X|\mathcal{G}) = E(X^+|\mathcal{G}) - E(X^-|\mathcal{G})\,.$$

The classical notion of conditional expectation of a random variable X with respect to a random variable Y also can be obtained from the construction above:

$$E(X|Y) := E(X|\mathcal{F}^Y)\,,$$

where $\mathcal{F}^Y$ is a σ-algebra generated by Y. Furthermore, if we take $X = I_A$ and $Y = I_B$, then we arrive at the definition of conditional probability $P(A\,|\,B) := E(I_A\,|\,I_B)$.

Let us list some useful properties of conditional expectations.

1. If random variable X is constant, then $E(X|\mathcal{G}) = X$ (a.s.);

2. If $X \leq Y$ (a.s.), then

$$E(X|\mathcal{G}) \leq E(Y|\mathcal{G}) \quad \text{(a.s.)},$$

and in particular, $|E(X|\mathcal{G})| \leq E(|X||\mathcal{G})$ (a.s.);

3. If $\mathcal{G} = \mathcal{F}$, then $E(X|\mathcal{G}) = X$ (a.s.);

4. $E\big(E(X|\mathcal{G})\big) = E(X)$ (a.s.);

5. If $\mathcal{G}_1 \subseteq \mathcal{G}_2$, then $E\big(E(X|\mathcal{G}_2)|\mathcal{G}_1\big) = E(X|\mathcal{G}_1)$ (a.s.);

6. We say that random variable X is *independent* of the σ-algebra $\mathcal{G}$, if all events of the type

$$\{\omega \ : \ X(\omega) \in B\}_{B \in \mathcal{B}(\mathbb{R})}$$

are independent of any event from the σ-algebra $\mathcal{G}$. In this case, $E(X|\mathcal{G}) = X$ (a.s.);

7. If random variables φ_1 and φ_2 are measurable with respect to $\mathcal{G}$, then

$$E(\varphi_1 X_1 + \varphi_2 X_2|\mathcal{G}) = \varphi_1 E(X_1|\mathcal{G}) + \varphi_2 E(X_2|\mathcal{G}) \ (a.s.),$$

where X_1 and X_2 are given random variables. In particular, if φ_1 and φ_2 are constant, then this property generalizes the linearity of expectations;

8. If a sequence $\big(X_n\big)_{n=1,2,\dots}$ of random variables is such that $|X_n| \leq Y$ (a.s.), $E(Y) < \infty$, and $X_n \to X$ (a.s.), then the dominated convergence theorem implies

$$E(X_n|\mathcal{G}) \longrightarrow E(X|\mathcal{G}) \ (a.s.) \quad \text{as } n \to \infty.$$

Further insights about probabilistic properties of sequence $\big(X_n\big)_{n=1,2,\dots}$ of random variables on probability space $(\Omega, \mathcal{F}, P)$ can be obtained by introducing the corresponding sequence of σ-algebras $\big(\mathcal{F}_n^X\big)_{n=1,2,\dots}$, where σ-algebra $\mathcal{F}_n^X$ is generated by the first n random variables: $X_1, \dots, X_n$. It is customary to interpret each $\mathcal{F}_n^X$ as information associated with the given sequence up to time n. Since more information becomes available as n increases, the sequence $\big(\mathcal{F}_n^X\big)$ is usually referred to as *natural information flow* or as *natural filtration*. As we mentioned earlier, such an approach to studying sequences of random variables corresponds well to the nature of financial markets.

Thus, when modeling financial markets on probability space $(\Omega, \mathcal{F}, P)$, it is natural to introduce a filtration $\mathbb{F} = \big(\mathcal{F}_n\big)_{n=0,1,2,\dots}$. It is customary to assume that $\mathcal{F}_0 = \emptyset$ and $\mathcal{F}_{n-1} \subseteq \mathcal{F}_n$, $n = 1, 2, \dots$. The probability space with filtration $(\Omega, \mathcal{F}, \mathbb{F}, P)$ is referred to as *stochastic basis*. One of the advantages of this general approach is an opportunity to accommodate a situation when the observed information flow is bigger or smaller than the filtration generated by a specific sequence $\big(X_n\big)$. Financial interpretations of such situations include, for example, the case of obtaining some additional information (e.g., insider information) and the case when it is impossible to obtain the complete information (e.g., in the case of non-tradable assets). We say that a sequence $(X_n)_{n=1,2,\dots}$ of random variables is a *stochastic sequence* if it is adapted to a filtration $\mathbb{F}$; that is, $\mathcal{F}_n^X \subseteq \mathcal{F}_n$ for all n. Motivated by financial applications, we focus our attention on studying such sequences. We say that an integrable stochastic sequence $(X_n)_{n=1,2,\dots}$ defined on a stochastic basis $(\Omega, \mathcal{F}, \mathbb{F}, P)$ is a *martingale* if

$$E(X_n|\mathcal{F}_{n-1}) = X_{n-1} \quad \text{(a.s.)}$$

for all $n \geq 1$.

If

$$E(X_n|\mathcal{F}_{n-1}) \geq X_{n-1} \quad \text{a.s. or} \quad E(X_n|\mathcal{F}_{n-1}) \leq X_{n-1} \quad \text{(a.s.)}$$

for all $n \geq 1$, then X is called a *submartingale* or a *supermartingale*, respectively.

Example 1.2

Let $(Y_n)_{n=0,1,\dots}$ be a sequence of independent random variables with zero expectations $E(Y_n)$. Then the sequence of the partial sums $X_n = \sum_{k=0}^{n} Y_k$ is a martingale with respect to the natural filtration $(\mathcal{F}_n^X)_{n=0,1,\dots}$. If expectations $E(Y_n)$ are non-negative (non-positive), then (X_n) is a submartingale (supermartingale), respectively. □

Example 1.3

Let $(Y_n)_{n=0,1,\dots}$ be a sequence of independent random variables with expectations $E(Y_n) = 1$. Then the sequence of the partial products $X_n = \prod_{k=0}^{n} Y_k$ is a martingale with respect to the natural filtration $(\mathcal{F}_n^X)_{n=0,1,\dots}$. If expectations $E(Y_n) \geq 1$ (or ≤ 1), then (X_n) is a submartingale (supermartingale), respectively. □

Remark Given a martingale $(X_n)_{n=0,1,\dots}$, there is a simple way of constructing submartingales (and therefore supermartingales, due to the symmetric relationship between submartingales and supermartingales). Suppose that φ is a convex downward Borel function such that $E|\varphi(X_n)| < \infty$ for all $n = 0, 1, \dots$. Then Jensen's inequality implies that $(\varphi(X_n))_{n=0,1,\dots}$ is a submartingale.

The following notion of a *stopping time* (or *Markov time*) is closely related to the introduced notion of a stochastic sequence. A random variable $\tau : \Omega \longrightarrow \mathbb{Z}_+ \equiv \{0, 1, \dots\}$ is a *stopping time* if

$$\{\omega : \tau(\omega) \leq n\} \in \mathcal{F}_n \quad \text{for all } n = 0, 1, \dots,$$

or equivalently, $\{\omega : \tau(\omega) = n\} \in \mathcal{F}_n$ for all $n = 0, 1, \dots$. We can interpret stopping times as random times where randomness does not depend on the future (beyond time n). Thus, we arrive at the following definition. Suppose (for technical reasons) that σ-algebra $\mathcal{F}$ is the minimal σ-algebra that contains the filtration $(\mathcal{F}_n)_{n=0,1,...}$. Let τ be a stopping time, then σ-algebra

$$\mathcal{F}_\tau = \{A \in \mathcal{F} : A \cap \{\tau = n\} \in \mathcal{F}_n \quad \text{for all } n = 0, 1, \dots\}$$

is referred to as the *information that is available up to the stopping time* τ. Clearly, if $\tau_1 \le \tau_2$ (a.s), then $\mathcal{F}_{\tau_1} \subseteq \mathcal{F}_{\tau_2}$.

Now let us discuss some useful properties of martingales and stopping times.

1. Let $(X_n)_{n=0,1,...}$ be a martingale (submartingale, supermartingale) and suppose that stopping times $\tau_1 \le \tau_2 \le N$ (a.s), then

$$E(X_{\tau_2} \,|\, \mathcal{F}_{\tau_1}) = X_{\tau_1} \tag{1.13}$$

$(E(X_{\tau_2} \,|\, \mathcal{F}_{\tau_1}) \ge X_{\tau_1},\ E(X_{\tau_2} \,|\, \mathcal{F}_{\tau_1}) \le X_{\tau_1}$, respectively). In particular, $E(X_{\tau_1}) = E(X_{\tau_2})$.

Proof of this property, which generalizes the obvious property of deterministic times, readily follows from the following observations:

$$E\left(I_{A \cap \{\tau_1 = n\}} (X_{\tau_2} - X_{\tau_1})\right) = E\left(I_{A \cap \{\tau_1 = n\}} \sum_{k=n}^{N} (X_{k+1} - X_k) I_{\{\tau_2 > k\}}\right)$$

$$= \sum_{k=n}^{N} E\left(I_{A \cap \{\tau_1 = n\}} I_{\{\tau_2 > k\}} (X_{k+1} - X_k)\right)$$

$$= \sum_{k=n}^{N} E\left(E\left[(X_{k+1} - X_k) \,|\, \mathcal{F}_k\right] I_{A \cap \{\tau_1 = k\}} I_{\{\tau_2 > k\}}\right) = 0$$

for $A \in \mathcal{F}_{\tau_1}$ and $n < N$. Similarly, we can show $E\left(I_{A \cap \{\tau_1 = n\}} (X_{\tau_2} - X_{\tau_1})\right) \ge 0$ for submartingales and $E\left(I_{A \cap \{\tau_1 = n\}} (X_{\tau_2} - X_{\tau_1})\right) \le 0$ for supermartingales.

2. Kolmogorov-Doob inequalities. Let $(X_n)_{n=0,1,...,N}$ be a submartingale. Then, for any real $a > 0$, we have

$$P\{\omega : \max_{0 \le n \le N} X_n \ge a\} \le \frac{E(X_N^+)}{a}. \tag{1.14}$$

Proof Consider the stopping time

$$\tau = \inf\{n \le N : X_n \ge a\}$$

and the set

$$A = \{\omega : \max_{0 \le n \le N} X_n \ge a\} \in \mathcal{F}_\tau.$$

Note $a \cdot I_A \le X_\tau \cdot I_A$. Calculating expectations we obtain

$$a \cdot P(A) \le E(X_\tau \cdot I_A) \le E(X_N \cdot I_A) \le E(X_N^+ \cdot I_A) \le E(X_N^+),$$

which proves (1.14).

If $(X_n)_{n=0,1,\ldots,N}$ is a martingale, then by applying (1.14) to submartingales $|X_n|$ and X_n^2, we arrive at the following inequalities for martingales:

$$P\Big\{\omega : \max_{0 \le n \le N} |X_n| \ge a\Big\} \le \frac{E(|X_N|)}{a} \qquad (1.15)$$

and

$$P\Big\{\omega : \max_{0 \le n \le N} |X_n| \ge a\Big\} \le \frac{E(|X_N|^2)}{a^2} \qquad (1.16)$$

under assumption that $E(|X_N|^2) \le \infty$. Taking limits in (1.14)–(1.16) as $N \to \infty$, we obtain the following inequalities for infinite time intervals:

$$P\Big\{\omega : \sup_n X_n \ge a\Big\} \le \frac{\sup_n E(X_n^+)}{a}, \qquad (1.17)$$

$$P\Big\{\omega : \sup_n |X_n| \ge a\Big\} \le \frac{\sup_n E(|X_n|)}{a}, \qquad (1.18)$$

and

$$P\Big\{\omega : \sup_n |X_n| \ge a\Big\} \le \frac{\sup_n E(|X_n|^2)}{a^2} \qquad (1.19)$$

under assumption that all supremums in the right-hand sides are finite.

3. Convergence of martingales and submartingales. Let $(X_n)_{n=0,1,\ldots}$ be a submartingale with $\sup_n E(|X_n|) < \infty$, then there exists

$$X_\infty = \lim_n X_n \quad \text{such that } E(|X_\infty|) < \infty.$$

Proof Suppose that this limit does not exist on a set of positive probability measure; that is,

$$P\Big\{\omega : \limsup_n X_n > \liminf_n X_n\Big\}. \qquad (1.20)$$

Note that the set $\{\omega : \limsup_n X_n > \liminf_n X_n\}$ can be written as a countable union of sets of the form $\{\omega : \limsup_n X_n > y > x > \liminf_n X_n\}$, where x and y are all possible rational numbers. Thus, (1.20) implies that there exist rational numbers $x < y$ such that

$$P\Big\{\omega : \limsup_n X_n > y > x > \liminf_n X_n\Big\}. \qquad (1.21)$$

Denote $\beta(x,y)$ the number of upcrossings of interval (x,y) by the submartingale $(X_n)_{n=0,1,\ldots}$. Kolmogorov-Doob inequalities imply (we omit this rather technical proof) the following Doob estimate for the expected number of upcrossings:

$$E(\beta(x,y)) \le \frac{\sup_n E(X_n^+) + |x|}{y - x}.$$

Since $(X_n)_{n=0,1,\dots}$ is a submartingale, we have

$$\sup_n E(|X_n|) < \infty \iff \sup_n E(X_n^+) < \infty$$

and Doob estimate implies that $E(\beta(x,y)) < \infty$. Thus, $\beta(x,y) < \infty$ (a.s.), which contradicts (1.21) and (1.20). $\square$

If in addition to almost sure convergence of the submartingale $(X_n)_{n=0,1,\dots}$ one wishes to establish convergence in L^1, then the condition of finiteness of $\sup_n E(|X_n|)$ must be replaced by a stronger condition of the uniform integrability. In particular, in the case of martingales we arrive at the following *Lévy's structural characterization theorem*: a stochastic sequence $(X_n)_{n=0,1,\dots}$ is a uniformly integrable martingale if and only if there exists a uniformly integrable random variable X_∞ such that $X_n = E(X_\infty | \mathcal{F}_n)$, $n = 0, 1, \dots$, and $X_n \to X_\infty$ both almost surely and in L^1.

Note that, in particular, a martingale $(X_n)_{n=0,1,\dots,N}$ is uniformly integrable since it is a finite family of random variables; therefore, by Lévy's characterization theorem, we have $X_n = E(X_N | \mathcal{F}_n)$, $n = 0, 1, \dots, N$. This property of martingales can be readily obtained from the first principles without Lévy's theorem, and it explains why the use of martingales is so natural in financial applications of the *dynamic programming* method.

We can use now the convergence properties of uniformly integrable martingales to extend property (1.13) from the case of bounded stopping times to the case of almost surely finite stopping times. Namely, let $(X_n)_{n=0,1,\dots}$ be a uniformly integrable martingale and consider stopping times $\sigma \le \tau < \infty$ (a.s.). Then

$$E(X_\tau \,|\, \mathcal{F}_\sigma) = X_\sigma \quad \text{(a.s.)}. \tag{1.22}$$

In particular, for a finite stopping time τ, we have

$$E(|X_\tau|) < \infty \quad \text{and} \quad E(X_\tau) = E(X_0). \tag{1.23}$$

We omit the complete proof of property (1.22); we only explain (1.23). For a fixed N, we introduce a bounded stopping time $\tau_N := \tau \wedge N = \min(\tau, N)$. Property (1.13) implies that $E(X_0) = E(X_{\tau_N})$, and we observe

$$E(|X_{\tau_N}|) = 2E(X_{\tau_N}^+) - E(X_{\tau_N}) \le 2E(X_{\tau_N}^+) - E(X_0).$$

Since $(X_n^+)_{n=0,1,\dots}$ is a submartingale, we obtain

$$
\begin{aligned}
E(X_{\tau_N}^+) &= \sum_{k=0}^{N} E(X_k^+ I_{\{\tau_N = k\}}) + E(X_N^+ I_{\{\tau > N\}}) \\
&\le \sum_{k=0}^{N} E(X_N^+ I_{\{\tau_N = k\}}) + E(X_N^+ I_{\{\tau > N\}}) \\
&= E(X_N^+) \le E(|X_N|) \le \sup_n E(|X_n|).
\end{aligned}
$$

Thus,

$$E\big(|X_{\tau_N}|\big) \le 3 \sup_n E\big(|X_n|\big),$$

and after taking limits as $N \to \infty$, we arrive at

$$E\big(|X_\tau|\big) \le \limsup_n E\big(|X_{\tau_N}|\big) \le 3 \sup_n E\big(|X_n|\big) < \infty.$$

Further, the uniform integrability of $(X_n)_{n=0,1,\ldots}$ implies

$$E\big(|X_n| \cdot I_{\{\tau>n\}}\big) \to 0 \quad \text{as } n \to \infty. \tag{1.24}$$

Note that $P(\tau > n) \to 0$ as $n \to \infty$ since τ is a bounded stopping time and $P(\tau < \infty) = 1$.

Consider the following decomposition:

$$X_\tau = X_{\tau \wedge n} + (X_\tau - X_n) \cdot I_{\{\tau>n\}}.$$

Taking expectations and using the martingale property, we obtain

$$\begin{aligned}
E\big(X_\tau\big) &= E\big(X_{\tau \wedge n}\big) + E\big(X_\tau \cdot I_{\{\tau>n\}}\big) - E\big(X_n \cdot I_{\{\tau>n\}}\big) \quad (1.25) \\
&= E\big(X_0\big) + E\big(X_\tau \cdot I_{\{\tau>n\}}\big) - E\big(X_n \cdot I_{\{\tau>n\}}\big).
\end{aligned}$$

Taking into account convergence (1.24) and finiteness of X_τ, we conclude that second and third terms in the right-hand side of this equality vanish as $n \to \infty$. Thus, equality (1.25) implies $E\big(X_\tau\big) = E\big(X_0\big)$. $\square$

4. Martingales and absolute continuity of probability measures. Let $(\Omega, \mathcal{F}, \mathbb{F}, P)$ be a stochastic basis, and consider another probability measure $\tilde{P} \ll P$ with density $d\tilde{P}/dP = Z$. Let $\tilde{P}_n$ and P_n be their restrictions on $\mathcal{F}_n$, and denote $Z_n := d\tilde{P}_n/dP_n$.

Sequence $(Z_n)_{n=0,1,\ldots}$ is called the *local density* of $\tilde{P}$ with respect to P and it is a martingale with respect to the initial measure P. Indeed, let $A \in \mathcal{F}_n$, then

$$\begin{aligned}
E\big(Z_{n+1} \cdot I_A\big) &= E\left(\frac{d\tilde{P}_{n+1}}{dP_{n+1}} \cdot I_A\right) = \tilde{P}_{n+1}(A) = \tilde{P}_n(A) = E\left(\frac{d\tilde{P}_n}{dP_n} \cdot I_A\right) \\
&= E\big(Z_n \cdot I_A\big) \tag{1.26}
\end{aligned}$$

for all $n = 0, 1, \ldots$. Lévy's structural characterization implies that $Z_n \to Z$ (a.s.) as $n \to \infty$, and $Z_n = E\big(Z \mid \mathcal{F}_n\big)$.

In a similar way, one can prove the following formula for *change of probability under the conditional expectation sign*: for any $\mathcal{F}_n$-measurable random variable Y, we have

$$\tilde{E}\big(Y \mid \mathcal{F}_{n-1}\big) = E\big(Z_n Z_{n-1}^{-1} Y \mid \mathcal{F}_{n-1}\big) \quad \text{(a.s.)}, \tag{1.27}$$

given that the conditional expectation $E\big(Z_n Z_{n-1}^{-1} Y \mid \mathcal{F}_{n-1}\big)$ is well defined.

The following property that connects the martingale property and absolute continuity of probability measures is related to formula (1.27) and is referred to as *Girsanov theorem*. Let $(M_n)_{n=0,1,\ldots}$ (with $M_0 = 0$) be a martingale with respect to the original probability P and suppose $E(|Y_n Y_{n-1}^{-1} \triangle M_n|) < \infty$ for all $n = 1, 2, \ldots$. Define $(\widetilde{M}_n)_{n=0,1,\ldots}$ (with $\widetilde{M}_0 = 0$) by

$$\triangle \widetilde{M}_n = \triangle M_n - E(Y_n Y_{n-1}^{-1} \triangle M_n \,|\, \mathcal{F}_{n-1}).$$

Using (1.27), we calculate

$$\widetilde{E}(\triangle \widetilde{M}_n \,|\, \mathcal{F}_{n-1}) = \widetilde{E}\Big(\triangle M_n - E(Y_n Y_{n-1}^{-1} \triangle M_n \,|\, \mathcal{F}_{n-1}) \,|\, \mathcal{F}_{n-1} \Big)$$
$$= \widetilde{E}(\triangle M_n \,|\, \mathcal{F}_{n-1}) - \widetilde{E}\Big(\widetilde{E}(Y_n Y_{n-1}^{-1} \triangle M_n \,|\, \mathcal{F}_{n-1}) \,|\, \mathcal{F}_{n-1} \Big) = 0,$$

which implies that $(\widetilde{M}_n)_{n=0,1,\ldots}$ is a martingale with respect to $\widetilde{P} \ll P$.

5. Doob decomposition and predictable characteristics of martingales. The notion of predictability is closely related to the notion of a martingale. We say that a stochastic sequence $(A_n)_{n=0,1,\ldots}$ is *predictable* if random variables A_n are $\mathcal{F}_{n-1}$-measurable for all n. We also say that a stochastic sequence (not necessarily predictable) $(A_n)_{n=0,1,\ldots}$ is *non-decreasing* if $\triangle A_n = A_n - A_{n-1} \geq 0$ (a.s.) for all n.

Let $(X_n)_{n=0,1,\ldots}$ be a submartingale. Then a martingale $(M_n)_{n=0,1,\ldots}$ and a non-decreasing stochastic sequence $(A_n)_{n=0,1,\ldots}$ exist such that the following *Doob decomposition*

$$X_n = M_n + A_n, \quad n = 0, 1, \ldots \tag{1.28}$$

holds.

To prove the existence, we set $M_0 = X_0$, $A_0 = 0$,

$$M_n = M_0 + \sum_{k=0}^{n-1} \{X_{k+1} - E(X_{k+1} \,|\, \mathcal{F}_k)\}, \text{ and } A_n = \sum_{k=0}^{n-1} \{E(X_{k+1} \,|\, \mathcal{F}_k) - X_k\}$$

for $n = 1, 2, \ldots$. Decomposition (1.28) is unique in the class of predictable stochastic sequences. Indeed, if another martingale $(M'_n)_{n=0,1,\ldots}$ and another non-decreasing stochastic sequence $(A'_n)_{n=0,1,\ldots}$ that satisfy (1.28) exist, then

$$\triangle A'_{n+1} = A'_{n+1} - A'_n = \triangle A_{n+1} + \triangle M_{n+1} - \triangle M'_{n+1}.$$

Since $(M_n)_{n=0,1,\ldots}$ and $(M'_n)_{n=0,1,\ldots}$ are martingales and since sequences $(A_n)_{n=0,1,\ldots}$ and $(A'_n)_{n=0,1,\ldots}$ are predictable, we have

$$\triangle A'_{n+1} = E(\triangle A'_{n+1} \,|\, \mathcal{F}_n) = E(\triangle A_{n+1} \,|\, \mathcal{F}_n) = \triangle A_{n+1} \quad (\text{a.s.}),$$

and therefore $A_n = A'_n$ and $M_n = M'_n$ (a.s.), $n = 0, 1, \ldots$. $\square$

If martingale $(M_n)_{n=0,1,\ldots}$ is such that $E(M_n) < \infty$ for all $n = 0, 1, \ldots$, then it is called a *square integrable* martingale. Applying Doob decomposition (1.28) to the submartingale $(M_n^2)_{n=0,1,\ldots}$, we conclude that a martingale $(m_n)_{n=0,1,\ldots}$ and a non-decreasing predictable sequence $(\langle M \rangle_n)_{n=0,1,\ldots}$ exist such that $M_n^2 = m_n + \langle M \rangle_n$. Sequence $(\langle M \rangle_n)_{n=0,1,\ldots}$ is called the *quadratic characteristic* or the *compensator* of M, and it can be constructed in the following way:

$$\langle M \rangle_n = \sum_{k=1}^{n} E\big((\triangle M_k)^2 \,|\, \mathcal{F}_{k-1}\big), \quad \langle M \rangle_0 = 0.$$

Note that

$$E\big((M_k - M_l)^2 \,|\, \mathcal{F}_l\big) = E\big(M_k^2 - M_l^2 \,|\, \mathcal{F}_l\big) = E\big(\langle M \rangle_k - \langle M \rangle_l \,|\, \mathcal{F}_l\big), \quad l \le k,$$

and $E(M_k^2) = E(\langle M \rangle_k)$ for $k = 0, 1, \ldots$.

Recall that one of the measures of association of two random variables with finite second moments is their covariance. A similar measure of association can be introduced for two square integrable martingales M and N: the sequence $(\langle M, N \rangle_n)_{n=0,1,\ldots}$ defined by

$$\langle M, N \rangle_n = \frac{1}{4}\big\{ \langle M + N \rangle_n - \langle M - N \rangle_n \big\}, \quad n = 0, 1, \ldots,$$

is called the *mutual quadratic characteristic* of M and N. It is not difficult to show that sequence

$$\Big(M_n\, N_n - \langle M, N \rangle_n \Big)_{n=0,1,\ldots}$$

is a martingale, so the mutual quadratic characteristic of M and N is the compensator of their product $(M_n\, N_n)_{n=0,1,\ldots}$. Square integrable martingales M and N are said to be *orthogonal* if $\langle M, N \rangle_n = 0$ for all $n = 0, 1, \ldots$.

6. Discrete stochastic integrals and stochastic exponentials. Let $(H_n)_{n=0,1,\ldots}$ be a predictable stochastic sequence and $(m_n)_{n=0,1,\ldots}$ be a martingale. Stochastic sequence

$$H * m_n = \sum_{k=0}^{n} H_k \Delta m_k \tag{1.29}$$

is called a *discrete stochastic integral* of H with respect to m. If martingale m is square integrable, sequence H is predictable and $E(H_n^2 \Delta \langle m \rangle_n) < \infty$ for all $n = 0, 1, \ldots$, then stochastic integral $(H * m_n)_{n=0,1,\ldots}$ is a square integrable martingale with quadratic characteristic

$$\langle H * m \rangle_n = \sum_{k=0}^{n} H_k^2 \Delta \langle m \rangle_k.$$

Further, let $(M_n)_{n=0,1,\ldots}$ be a fixed square integrable martingale, then one can consider all square integrable martingales $(N_n)_{n=0,1,\ldots}$ that are orthogonal to $(M_n)_{n=0,1,\ldots}$ and introduce a family of square integrable martingales of the following form

$$X_n = M_n + N_n. \tag{1.30}$$

Conversely, any square integrable martingale $(X_n)_{n=0,1,\ldots}$ can be written in form (1.30), where the orthogonal term N has the form of the stochastic integral (1.29) with the martingale m that is orthogonal to the given martingale M. This version of decomposition (1.30) is usually referred to as *Kunita-Watanabe decomposition*.

Discrete stochastic integrals are naturally related to discrete stochastic differential equations (or stochastic difference equations). Solutions of stochastic difference equations are often used in modeling the dynamics of asset prices in financial markets. Consider a stochastic sequence $(U_n)_{n=0,1,\ldots}$ with $U_0 = 0$. Define new stochastic sequence $(X_n)_{n=0,1,\ldots}$ with $X_0 = 1$ by

$$\Delta X_n = X_{n-1}\Delta U_n, \quad n = 1, 2, \ldots. \tag{1.31}$$

This simple linear stochastic differential equation has an obvious solution

$$X_n = \prod_{k=1}^{n} (1 + \Delta U_k) = \varepsilon_n(U),$$

which is called a *stochastic exponential*. A non-homogeneous version of equation (1.31) has the form

$$\Delta X_n = \Delta N_n + X_{n-1}\Delta U_n, \quad X_0 = N_0, \tag{1.32}$$

where $(N_n)_{n=0,1,\ldots}$ is a given stochastic sequence. A solution of the non-homogeneous equation can be written in terms of solutions of the corresponding homogeneous equation, and it has the form

$$X_n = \varepsilon_n(U)\left[N_0 + \sum_{k=1}^{n} \frac{\Delta N_k}{\varepsilon_k(U)} \right].$$

Stochastic exponentials have the following useful properties:

(a) $\frac{1}{\varepsilon_n(U)} = \varepsilon_n(-U^*)$, where $\Delta U^* = \frac{\Delta U_n}{1+\Delta U_n}$ and $\Delta U_n \neq -1$;

(b) $(\varepsilon_n(U))_{n=0,1,\ldots}$ is a martingale if and only if $(U_n)_{n=0,1,\ldots}$ is a martingale;

(c) $\varepsilon_n(U) = 0$ for all $n \geq \tau_0 := \inf\{k : \varepsilon_k(U) = 0\}$;

(d) the multiplication rule for stochastic exponentials that correspond to $(U_n)_{n=0,1,\ldots}$ and $(V_n)_{n=0,1,\ldots}$:

$$\varepsilon_n(U)\varepsilon_n(V) = \varepsilon_n(U + V + [U, V]),$$

where

$$[U, V]_n = \sum_{k=1}^{n} \Delta U_k \Delta V_k \quad \text{and} \quad [U, V]_0 = 0.$$

Chapter 2

Financial Risk Management in the Binomial Model

2.1 The binomial model of a financial market. Absence of arbitrage, uniqueness of a risk-neutral probability measure, martingale representation.

The binomial model of a (B, S)-market was introduced in the previous chapter. Sometimes this model is also referred to as the Cox-Ross-Rubinstein model. Recall that the dynamics of the market are represented by equations

$$\begin{aligned}
\Delta B_n &= r B_{n-1}, & B_0 &= 1, & (2.1)\\
\Delta S_n &= \rho_n S_{n-1}, & S_0 &> 0,
\end{aligned}$$

where $r \geq 0$ is a constant rate of interest with $-1 < a < r < b$, and profitabilities or risky asset returns

$$\rho_n = \begin{cases} b & \text{with probability } p \in [0, 1] \\ a & \text{with probability } q = 1 - p \end{cases}, \quad n = 1, \dots, N,$$

form a sequence of independent identically distributed random variables. The stochastic basis in this model consists of $\Omega = \{a, b\}^N$, the space of sequences $\omega = (\omega_1, \dots, \omega_N)$ of length N whose elements are equal to either a or b; $\mathcal{F} = 2^\Omega$, the set of all subsets of Ω. The probability P has Bernoulli probability distribution with $p \in [0, 1]$, so that

$$P(\{\omega\}) = p^{\sum_{i=1}^N I_{\{b\}}(\omega_i)} (1 - p)^{\sum_{i=1}^N I_{\{a\}}(\omega_i)}.$$

The filtration $\mathbb{F}$ is generated by the sequence $(\rho_n)_{n \leq N} : \mathcal{F}_n = \sigma(\rho_1, \dots, \rho_n)$.

Note that many authors use different parameters u (up) and d (down) in the binomial model setting, which can be interpreted as possible values of stock returns, and they can be expressed in terms of our parameters a and b:

$$u = 1 + b \quad \text{and} \quad d = 1 + a.$$

In the framework of model (2.1), we can specify the following notions. A predictable sequence $\pi = (\pi_n)_{n \leq N} \equiv (\beta_n, \gamma_n)_{n \leq N}$ is an *investment strategy* (portfolio). A *contingent claim* f_N is a random variable on the stochastic basis $(\Omega, \mathcal{F}, \mathbb{F}, P)$. *Hedge* for a contingent claim f_N is a self-financing portfolio with the terminal value $X_n^\pi \geq f_N$. A hedge π^* with the value $X_N^{\pi^*} \leq X_N^\pi$ for any other hedge π is called the *minimal hedge*. A self-financing portfolio $\pi \in SF$ is called an *arbitrage* portfolio if

$$X_0^\pi = 0, \quad X_N^\pi \geq 0 \quad \text{and} \quad P(\{\omega : \ X_N^\pi > 0\}) > 0,$$

which can be interpreted as an opportunity of making a profit without risk.

Note that the risky nature of a (B, S)-market is associated with randomness of prices S_n. A particular choice of probability P (in terms of Bernoulli parameter p) allows one to numerically express this randomness. In general, the initial choice of P can give probabilistic properties of S such that the behavior of S is very different from the behavior of a non-risky asset B. On the other hand, it is clear that pricing of contingent claims should be *neutral to risk*. This can be achieved by introducing a new probability P^* such that the behaviors of S and B are similar under this probability: S and B are on average the same under P^*. In other words, the sequence of discounted prices $(S_n/B_n)_{n \leq N}$ must be, on average, constant with respect to probability P^*:

$$E^*\left(\frac{S_n}{B_n}\right) = E^*\left(\frac{S_0}{B_0}\right) = S_0 \quad \text{for all} \quad n = 1, \ldots, N.$$

For $n = 1$, this implies

$$E^*\left(\frac{S_1}{B_1}\right) = S_0\, E^*\left(\frac{1 + \rho_1}{1 + r}\right) = S_0\left[\frac{(1 + b)p^* + (1 + a)(1 - p^*)}{1 + r}\right] = S_0,$$

where p^* is a Bernoulli parameter that defines P^*. We have

$$p^* + bp^* + 1 + a - p^* - ap^* = 1 + r,$$

and therefore

$$p^* = \frac{r - a}{b - a},$$

which means that in the binomial model the risk-neutral probability P^* is *unique*, and for any $\omega = (\omega_1, \ldots, \omega_N) \in \Omega$,

$$P^*(\{\omega\}) = (p^*)^{\sum_{i=1}^N I_{\{b\}}(\omega_i)} (1 - p^*)^{\sum_{i=1}^N I_{\{a\}}(\omega_i)}.$$

Note that in this case we can find *density* Z_N^* of probability P^* with respect to probability P, that is, a non-negative random variable such that

$$E(Z_N^*) = 1 \quad \text{and} \quad P^*(A) = E(Z_N^* I_A) \quad \text{for all} \quad A \in \mathcal{F}_N.$$

Since Ω is discrete, we only need to compute values of Z_N^* for every elementary event $\{\omega\}$. We have

$$P^*(\{\omega\}) = E(Z_N^* I_{\{\omega\}}) = Z_N^*(\omega)\, P(\{\omega\}),$$

and hence

$$Z_N^*(\omega) = \frac{P^*(\{\omega\})}{P(\{\omega\})} = \left(\frac{p^*}{p}\right)^{\sum_{i=1}^N I_{\{b\}}(\omega_i)} \left(\frac{1-p^*}{1-p}\right)^{N-\sum_{i=1}^N I_{\{b\}}(\omega_i)}.$$

To describe the behavior of discounted prices S_n/B_n under the risk-neutral probability P^*, we compute the following conditional expectations for all $n \le N$. Using the independence of $(\rho_n)_{n \le N}$, we obtain

$$\begin{aligned}
E^*\left(\frac{S_n}{B_n}\bigg|\mathcal{F}_{n-1}\right) &= E^*\left(S_0 \prod_{k=1}^n \frac{1+\rho_k}{1+r}\bigg|\mathcal{F}_{n-1}\right) \\
&= \frac{S_0}{1+r^n} E^*\left(\prod_{k=1}^n (1+\rho_k)\bigg|\mathcal{F}_{n-1}\right) \\
&= \frac{S_0}{1+r^n} \prod_{k=1}^{n-1} (1+\rho_k) E^*(1+\rho_n) \\
&= \frac{S_{n-1}}{B_{n-1}} \frac{E^*(1+\rho_n)}{1+r} = \frac{S_{n-1}}{B_{n-1}} \frac{1+r}{1+r} \\
&= \frac{S_{n-1}}{B_{n-1}}.
\end{aligned}$$

This means that the sequence $(S_n/B_n)_{n \le N}$ is a martingale with respect to the risk-neutral probability P^*. This is the reason that P^* is also referred to as a *martingale probability (martingale measure)*.

The next important property of a binomial market is the absence of arbitrage strategies. Such a market is referred to as a *no-arbitrage market*. Consider a self-financing strategy, $\pi = (\pi_n)_{n \le N} \equiv (\beta_n, \gamma_n)_{n \le N} \in SF$ with discounted values X_n^π/B_n. Using predictability of β and γ and properties of martingale probability, we have for all $n \le N$

$$\begin{aligned}
E^*\left(\frac{X_n^\pi}{B_n}\bigg|\mathcal{F}_{n-1}\right) &= E^*\left(\beta_n + \gamma_n \frac{S_n}{B_n}\bigg|\mathcal{F}_{n-1}\right) \\
&= E^*(\beta_n|\mathcal{F}_{n-1}) + \gamma_n E^*\left(\frac{S_n}{B_n}\bigg|\mathcal{F}_{n-1}\right) \\
&= \beta_n + \gamma_n \frac{S_{n-1}}{B_{n-1}} = \frac{\beta_n B_{n-1} + \gamma_n S_{n-1}}{B_{n-1}} \\
&= \frac{X_{n-1}^\pi}{B_{n-1}},
\end{aligned}$$

which implies that the discounted value of a self-financing strategy is a martingale with respect to the risk-neutral probability P^*. This property is usually referred to as the *martingale characterization of self-financing strategies SF*.

Further, suppose there exists an arbitrage strategy $\tilde{\pi}$. By its definition, we have

$$E\left(\frac{X_N^{\tilde{\pi}}}{B_N}\right) = \frac{E(X_N^{\tilde{\pi}})}{B_N} > 0,$$

and the martingale property of X_n^π / B_n implies

$$E^*\left(\frac{X_N^{\tilde\pi}}{B_N}\right) = E^*\left(\frac{X_0^{\tilde\pi}}{B_0}\right) = E^*(X_0^{\tilde\pi}) = 0\,.$$

Now, for probabilities P and P^* there is a positive density Z^* so that $P^*(A) = E(Z_N^* I_A)$ for any event $A \in \mathcal{F}_N$. Therefore,

$$
\begin{aligned}
0 &= X_0^{\tilde\pi} = X_0^{\tilde\pi}/B_0 = E^*\left(\frac{X_N^{\tilde\pi}}{B_N}\right) = \frac{E^*(X_N^{\tilde\pi})}{B_N} = \frac{E(Z_N^* X_N^{\tilde\pi})}{B_N} \\
&\geq \frac{\min_\omega \left[Z_N^*(\omega)\right] E(X_N^{\tilde\pi})}{B_N} > 0\,,
\end{aligned}
$$

which contradicts the assumption of existence of arbitrage.

Now we prove that, in the binomial market framework, any martingale can be represented in the form of a discrete stochastic integral with respect to some basic martingale. Let $(\rho_n)_{n \leq N}$ be a sequence of independent random variables on $(\Omega, \mathcal{F}, P^*)$ defined by

$$\rho_n = \begin{cases} a & \text{with probability } p^* = \frac{r-a}{b-a} \\ b & \text{with probability } q^* = 1 - p^* \end{cases}\,,$$

where $-1 < a < r < b$. Consider filtration $\mathbb{F}$ generated by the sequence (ρ_n) : $\mathcal{F}_n = \sigma(\rho_1, \ldots, \rho_n)$. Any martingale $(M_n)_{n \leq N}$, $M_0 = 0$, can be written in the form

$$M_n = \sum_{k=1}^{n} \phi_k \Delta m_k\,, \tag{2.2}$$

where $(\phi_n)_{n \leq N}$ is a predictable sequence, and

$$\left(\sum_{k=1}^{n} \Delta m_k\right)_{n \leq N} = \left(\sum_{k=1}^{n}(\rho_k - r)\right)_{n \leq N}$$

is a martingale. We will refer to it as the Bernoulli martingale.

Since σ-algebras $\mathcal{F}_n$ are generated by $\rho_1, \ldots, \rho_n$, and M_n are completely determined by $\mathcal{F}_n$, then functions $f_n = f_n(x_1, \ldots, x_n)$ with x_k equal to either a or b exist, such that

$$M_n(\omega) = f_n(\rho_1(\omega), \ldots, \rho_n(\omega))\,, \quad n \leq N\,.$$

The required representation (2.2) can be rewritten in the form

$$\Delta M_n(\omega) = \phi_k(\omega)\Delta m_k$$

or

$$f_n(\rho_1(\omega), \ldots, \rho_{n-1}(\omega), b) - f_{n-1}(\rho_1(\omega), \ldots, \rho_{n-1}(\omega)) = \phi_n(\omega)(b - r)$$

$$f_n(\rho_1(\omega), \ldots, \rho_{n-1}(\omega), a) - f_{n-1}(\rho_1(\omega), \ldots, \rho_{n-1}(\omega)) = \phi_n(\omega)(a - r),$$

or

$$\phi_n(\omega) = \frac{f_n(\rho_1(\omega), \ldots, \rho_{n-1}(\omega), b) - f_{n-1}(\rho_1(\omega), \ldots, \rho_{n-1}(\omega))}{(b - r)}$$

$$= \frac{f_n(\rho_1(\omega), \ldots, \rho_{n-1}(\omega), a) - f_{n-1}(\rho_1(\omega), \ldots, \rho_{n-1}(\omega))}{(a - r)}.$$

The martingale property of $(M_n)_{n \leq N}$ implies

$$E^* \left(f_n(\rho_1, \ldots, \rho_n) - f_{n-1}(\rho_1, \ldots, \rho_{n-1}) \middle| \mathcal{F}_{n-1} \right) = 0,$$

or

$$p^* f_n(\rho_1, \ldots, \rho_{n-1}, b) + (1 - p^*) f_n(\rho_1, \ldots, \rho_{n-1}, a) = f_{n-1}(\rho_1, \ldots, \rho_{n-1}).$$

Rearranging this equality

$$p^* \left(f_n(\rho_1, \ldots, \rho_{n-1}, b) - f_{n-1}(\rho_1, \ldots, \rho_{n-1}) \right)$$
$$+ (1 - p^*) \left(f_n(\rho_1, \ldots, \rho_{n-1}, a) - f_{n-1}(\rho_1, \ldots, \rho_{n-1}) \right) = 0$$

and dividing both sides by $p^* (1 - p^*)$, we arrive at

$$\frac{f_n(\rho_1(\omega), \ldots, \rho_{n-1}(\omega), b) - f_{n-1}(\rho_1(\omega), \ldots, \rho_{n-1}(\omega))}{1 - p^*}$$

$$= \frac{f_n(\rho_1(\omega), \ldots, \rho_{n-1}(\omega), a) - f_{n-1}(\rho_1(\omega), \ldots, \rho_{n-1}(\omega))}{p^*},$$

which in view of choice $p^* = (r - a)/(b - a)$ proves the result.

Similarly, any martingale $(M_n)_{n \leq N}$ on $(\Omega, \mathcal{F}, \mathbb{F}, P)$ can be written in the form (2.2), where the basic martingale $(m_n)_{n \leq N}$ has the form

$$\left(\sum_{k=1}^{n} (\rho_k - \mu) \right)_{n \leq N}$$

with $\mu = E(\rho_1) = \ldots = E(\rho_N)$.

Using the established *martingale representations* we now can prove the following *representation for density* Z_N^* of the martingale probability P^* with respect to P:

$$Z_N^* = \prod_{k=1}^{N} \left(1 - \frac{\mu - r}{\sigma^2} (\rho_k - \mu) \right) = \varepsilon_N \left(- \frac{\mu - r}{\sigma^2} \sum_{k=1}^{N} (\rho_k - \mu) \right), \qquad (2.3)$$

where $\mu = E(\rho_k)$, $\sigma^2 = Var(\rho_k)$, $k = 1, \ldots, N$.

Indeed, consider $Z_n^* = E(Z_N^* | \mathcal{F}_n)$, $n = 0, 1, \ldots, N$. Properties of conditional expectations imply that $(Z_n^*)_{n \leq N}$ is a martingale with respect to probability P and filtration $\mathcal{F}_n = \sigma(\rho_1, \ldots, \rho_n)$. Therefore, Z_n^* can be written in the form

$$Z_n^* = 1 + \sum_{k=1}^{n} \phi_k \left(\rho_k - \mu \right),$$

where (ϕ_k) is a predictable sequence. Since $Z_n^* > 0$, the following stochastic differential equation

$$Z_n^* \;=\; 1 + \sum_{k=1}^{n} Z_{k-1}^* \frac{\phi_k}{Z_{k-1}^*} \left(\rho_k - \mu \right) = 1 + \sum_{k=1}^{n} Z_{k-1}^* \, \psi_k \left(\rho_k - \mu \right)$$

is satisfied with $\psi_k = \phi_k / Z_{k-1}^*$, $k = 1, \ldots, N$. Hence, it can be written in the form of a stochastic exponential

$$Z_n^* = \prod_{k=1}^{n} \left(1 + \psi_k \left(\rho_k - \mu \right) \right).$$

Taking into account that Z_N^* is the density of a martingale probability, we can compute the coefficients $\psi_k = \phi_k / Z_{k-1}^*$. Since ρ_1 is independent of $\mathcal{F}_0$, we have

$$
\begin{aligned}
0 &= E^* \left((\rho_1 - r) \big| \mathcal{F}_0 \right) = E^*(\rho_1 - r) = E \left(Z_1^* (\rho_1 - r) \right) \\
&= E \left(\left(1 + \psi_1 \left(\rho_1 - \mu \right) \right) (\rho_1 - r) \right) = (\mu - r) + \psi_1 \sigma^2 \, ;
\end{aligned}
$$

thus, $\psi_1 = -(\mu - r)/\sigma^2$.

Now suppose that $\psi_k = -(\mu - r)/\sigma^2$ for all $k = 1, \ldots, N-1$, then using independence of $\rho_1, \ldots, \rho_N$ we obtain

$$
\begin{aligned}
0 &= E^* \left((\rho_N - r) \big| \mathcal{F}_{N-1} \right) = \frac{E \left(Z_N^* (\rho_N - r) \big| \mathcal{F}_{N-1} \right)}{Z_{N-1}^*} \\
&= E \left(\left(1 + \psi_N \left(\rho_N - \mu \right) \right) (\rho_N - r) \big| \mathcal{F}_{N-1} \right) \\
&= E \left((\rho_N - r) + \psi_N \left(\rho_N - \mu \right) (\rho_N - r) \big| \mathcal{F}_{N-1} \right) \\
&= E(\rho_N - r) + \psi_N E \left((\rho_N - \mu)(\rho_N - r) \big| \mathcal{F}_{N-1} \right) \\
&= (\mu - r) + \psi_N \sigma^2 \, .
\end{aligned}
$$

Mathematical induction implies $\psi_N = -(\mu - r)/\sigma^2$, which proves the claim.

2.2 Hedging contingent claims in the binomial market model. The Cox-Ross-Rubinstein formula

In the framework of a binomial (B, S)-market, we consider a financial contract associated with a contingent claim f_N with the future repayment date N.

If f_N is deterministic, then its market risk can be trivially computed since $E(f_N|\mathcal{F}_N) \equiv f_N$. In fact, there is no risk associated with the repayment of this claim as one easily can find the present value of the discounted claim f_N/B_N.

If f_N depends on the behavior of the market during the contract period $[0, N]$, then it is a random variable. The intrinsic risk in this case is related to the ability to repay f_N. To estimate and manage this risk, one should be able to predict f_N given the current market information $\mathcal{F}_n$, $n \leq N$.

We start the discussion of a methodology of pricing contingent claims with two simple examples that illustrate the essence of *hedging*.

Worked Example 2.1 *Let* $\Omega = \{\omega_1, \omega_2\}$ *and* $\mathcal{F}_0 = \{\emptyset, \Omega\}$, $\mathcal{F}_1 = \{\emptyset, \{\omega_1\}, \{\omega_2\}, \Omega\}$. *Consider a single-period binomial* (B, S)*-market with* $B_0 = 1\,(\$)$, $S_0 = 100\,(\$)$, $B_1 = B_0(1 + r) = 1 + r = 1.2\,(\$)$ *assuming that the annual rate of interest is* $r = 0.2$, *and*

$$S_1 = \begin{cases} 150\,(\$) & \text{with probability } p = 0.4 \\ 70\,(\$) & \text{with probability } 1 - p = 0.6\,. \end{cases}$$

Find the price for a European call option $f_1 = (S_1 - K)^+ \equiv \max\{0, S_1 - K\}\,(\$)$ *with strike price* $K = 100\,(\$)$.

Solution Clearly,

$$f_1 = (S_1 - 100)^+ \equiv \max\{0, S_1 - 100\} = \begin{cases} 50\,(\$) & \text{with probability } 0.4 \\ 0\,(\$) & \text{with probability } 0.6\,. \end{cases}$$

The intuitive (heuristic) price for this option is

$$E\left(\frac{f_1}{1+r}\right) = \frac{0.4 \times 50}{1.2} = 16\,.$$

Now, using the minimal hedging approach to pricing, we construct a self-financing strategy $\pi_0 = (\beta_0, \gamma_0)$ that replicates the final value of the option: $X_1^\pi = f_1$. Since $X_1^\pi = \beta_0(1 + r) + \gamma_0 S_1$, then we have

$$\beta_0\, 1.2 + \gamma_0\, 150 = 50\,,$$
$$\beta_0\, 1.2 + \gamma_0\, 70 = 0\,,$$

which gives $\beta_0 = -36.5$ and $\gamma_0 = 5/8$. Therefore, the 'minimal hedging' price is

$$X_0^\pi = \beta_0 + \gamma_0 S_0 = -36.5 + 100 \times 5/8 \approx 26.$$

Note that this strategy of managing risk (of repayment) assumes that the writer of the option at time 0 sells this option for 26 dollars, borrows 36.5 dollars (as β_0 is negative) and invests the obtained 62.5 dollars in 5/8 ($= 62.5/100$) shares of the stock S.

Alternatively, we can find a risk-neutral probability p^* from the equation

$$100 = S_0 = E^* \left(\frac{S_1}{1+r} \right) = \frac{150\,p^* + 70\,(1 - p^*)}{1.2}.$$

So $p^* = 5/8$ and the "risk-neutral" price is

$$E^* \left(\frac{f_1}{1+r} \right) = \frac{50 \times 5/8}{1.2} \approx 26.\ \square$$

Worked Example 2.2 *On the same market, find the price of an option with the terminal payment $f_1 = \max\{S_0, S_1\} - S_1$.*

Solution Note that

$$f_1 = \begin{cases} 30\ (\$) & \text{with probability } 0.6 \\ 0\ (\$) & \text{with probability } 0.4. \end{cases}$$

The heuristic price for this option is

$$E \left(\frac{f_1}{1+r} \right) = \frac{0.6 \times 30}{1.2} = 15.$$

Using a minimal hedging self-financing strategy $\pi_0 = (\beta_0, \gamma_0)$, we have

$$\beta_0\, 1.2 + \gamma_0\, 150 = 0,$$
$$\beta_0\, 1.2 + \gamma_0\, 70 = 30;$$

hence, $\gamma_0 = -3/8$ and $\beta_0 = 3/8 \times 150/1.2 = 450/96 \approx 46.8$. Therefore, the "minimal hedging" price is

$$X_0^\pi = \beta_0 + \gamma_0 S_0 = 46.8 - 100 \times 3/8 = 9.3.$$

Finally, the "risk-neutral" price is

$$E^* \left(\frac{f_1}{1+r} \right) = \frac{30 \times 3/8}{1.2} = \frac{90}{9.6} \approx 9.3.$$

In contrast to the previous example, this strategy assumes that the writer of the option at time 0 sells this option for 9.3 dollars, borrows 3/8 shares of the stock S (worth 37.5 dollars) and invests the obtained 46.8 dollars in a bank account. $\square$

Note that in both examples the "minimal hedging" price coincides with the "risk-neutral" price and that they differ from the intuitive price for the option. This observation leads us to a more general statement: *the price of a contingent claim is equal to the expectation of its discounted value with respect to a risk-neutral probability.*

To verify this, we consider a contingent claim f_N on a binomial (B, S)-market (2.1). The conditional expectation (with respect to a risk-neutral probability) of its discounted value

$$M_n^* = E^* \left(\frac{f_N}{B_N} \Big| \mathcal{F}_n \right), \quad n = 0, \ldots, N,$$

is a martingale with the boundary values $M_0^* = E^*(f_N/B_N)$ and $M_N^* = f_N/B_N$. It admits the following martingale representation

$$M_n^* = M_0^* + \sum_{k=1}^{n} \phi_k^* (\rho_k - r),$$

where $\phi_k^* = \phi_k^*(S_1, \ldots, S_{k-1})$ are completely determined by $S_1, \ldots, S_{k-1}$. Let

$$\gamma_n^* = \phi_n^* \frac{B_n}{S_{n-1}} \quad \text{and} \quad \beta_n^* = M_{n-1}^* - \gamma_n^* \frac{S_{n-1}}{B_{n-1}},$$

then we obtain a strategy $\pi^* = (\pi_n^*) \equiv (\beta_n^*, \gamma_n^*)$ with values

$$X_n^{\pi^*} = \beta_n^* B_n + \gamma_n^* S_n \quad \text{and} \quad M_n^{\pi^*} = \frac{X_n^{\pi^*}}{B_n}, \quad n = 0, \ldots, N.$$

In particular, for $n = N$, we have the following equality

$$X_N^{\pi^*} = B_N M_N^{\pi^*} = \frac{B_N f_N}{B_N} = f_N,$$

which means that π^* is a hedge for f_N. For any other hedge π, from properties of conditional expectations we have

$$\frac{X_n^{\pi}}{B_n} = E^* \left(\frac{X_N^{\pi}}{B_N} \Big| \mathcal{F}_n \right) \geq E^* \left(\frac{f_N}{B_N} \Big| \mathcal{F}_n \right) = M_n^* = \frac{X_n^{\pi^*}}{B_n}$$

for $n \leq N$. Thus, π^* is the minimal hedge for a contingent claim f_N.

The initial value $C_N(f) = X_0^{\pi^*}$ of this minimal hedge is called the *fair price* of a contingent claim f_N. As we observed before, it is equal to $E^*(f_N/B_N)$.

Now we compute the fair price of an arbitrary European call option on a binomial (B, S)-market. In this case, $f_N = (S_N - K)^+ \equiv \max\{0, S_N - K\}$. Recall that a European call option gives its holder the right to buy shares of the stock S at a fixed strike price K (which can be distinct from the market price S_N) at time N. The writer of such an option is obliged to sell shares at this price K.

Using the methodology described above, we have

$$C_N \equiv C_N\big((S_N - K)^+\big) = E^*\left(\frac{(S_N - K)^+}{(1+r)^N}\right) = \frac{E^*\big((S_N - K)\,I_{\{\omega:\ S_N \geq K\}}\big)}{(1+r)^N}.$$

To compute the latter expectation, we use the representation (2.3) for the density Z^*:

$$
\begin{aligned}
E^*\big((S_N - K)^+\big) &= E\big(Z_N^*(S_N - K)^+\big) \\
&= E\left(\varepsilon_N\left(-\frac{\mu - r}{\sigma^2}\sum_{k=1}^{N}(\rho_k - \mu)\right)(S_N - K)\,I_{\{\omega:\ S_N \geq K\}}\right).
\end{aligned}
$$

Denote

$$k_0 := \min\left\{k \leq N \ :\ S_0(1+b)^k(1+a)^{N-k} \geq K\right\},$$

then

$$k_0 = \left[\!\left[\, \ln\frac{K}{S_0\,(1+a)^N} \Big/ \ln\frac{1+b}{1+a} \right]\!\right] + 1\,,$$

where $[\![x]\!]$ is the integer part of a real number x. Now since

$$p^* = \frac{r - a}{b - a}\,, \qquad \mu = p\,(b - a) + a\,, \qquad \sigma^2 = (b - a)^2 p\,(1 - p)\,,$$

$$1 - \frac{\mu - r}{\sigma^2}\,(b - \mu) = \frac{p^*}{p} \quad \text{and} \quad 1 - \frac{\mu - r}{\sigma^2}\,(a - \mu) = \frac{1 - p^*}{1 - p}\,,$$

we have

$$
\begin{aligned}
&E\left(\varepsilon_N\left(-\frac{\mu - r}{\sigma^2}\sum_{k=1}^{N}(\rho_k - \mu)\right)K\,I_{\{\omega:\ S_N \geq K\}}\right) \\
&= K\sum_{k=k_0}^{N}\binom{N}{k}\left[1 - \frac{\mu - r}{\sigma^2}(b - \mu)\right]^{k} p^k \left[1 - \frac{\mu - r}{\sigma^2}(a - \mu)\right]^{N-k}(1 - p)^{N-k} \\
&= K\sum_{k=k_0}^{N}\binom{N}{k}\left[\frac{p^*}{p}\right]^{k} p^k \left[\frac{1 - p^*}{1 - p}\right]^{N-k}(1 - p)^{N-k} \\
&= K\sum_{k=k_0}^{N}\binom{N}{k}(p^*)^k\,(1 - p^*)^{N-k}\,.
\end{aligned}
$$

Next, using the stochastic exponentials' representation of $S_n = S_0\,\varepsilon_N\big(\sum_{k=1}^{N}\rho_k\big)$ and observing that

$$1 - \frac{\mu - r}{\sigma^2}\,(b - \mu) + b - \frac{\mu - r}{\sigma^2}\,(b - \mu)\,b = \frac{p^*}{p}\,(1 + b)$$

and

$$1 - \frac{\mu - r}{\sigma^2}\,(a - \mu) + a - \frac{\mu - r}{\sigma^2}\,(a - \mu)\,a = \frac{1 - p^*}{1 - p}\,(1 + a)\,,$$

we obtain

$$E\left(\varepsilon_N\left(-\frac{\mu-r}{\sigma^2}\sum_{k=1}^{N}(\rho_k-\mu)\right)S_N I_{\{\omega:\ S_N\geq K\}}\right)$$

$$= S_0\, E\left(\varepsilon_N\left(-\frac{\mu-r}{\sigma^2}\sum_{k=1}^{N}(\rho_k-\mu)\right)\varepsilon_N\left(\sum_{k=1}^{N}\rho_k\right)I_{\{\omega:\ S_N\geq K\}}\right)$$

$$= S_0\, E\left(\varepsilon_N\left(-\frac{\mu-r}{\sigma^2}\sum_{k=1}^{N}(\rho_k-\mu)\right.\right.$$

$$\left.\left.+\sum_{k=1}^{N}\rho_k-\frac{\mu-r}{\sigma^2}\sum_{k=1}^{N}(\rho_k-\mu)\rho_k\right)I_{\{\omega:\ S_N\geq K\}}\right)$$

$$= S_0\sum_{k=k_0}^{N}\binom{N}{k}\left[1-\frac{\mu-r}{\sigma^2}(b-\mu)+b-\frac{\mu-r}{\sigma^2}(b-\mu)b\right]^k p^k$$

$$\times\left[1-\frac{\mu-r}{\sigma^2}(a-\mu)+a-\frac{\mu-r}{\sigma^2}(a-\mu)a\right]^{N-k}(1-p)^{N-k}$$

$$= S_0\sum_{k=k_0}^{N}\binom{N}{k}\left[\frac{p^*}{p}(1+b)\right]^k p^k\left[\frac{1-p^*}{1-p}(1+a)\right]^{N-k}(1-p)^{N-k}$$

$$= S_0\sum_{k=k_0}^{N}\binom{N}{k}\left[p^*(1+b)\right]^k\left[(1-p^*)(1+a)\right]^{N-k}$$

$$= S_0\,(1+r)^N\sum_{k=k_0}^{N}\binom{N}{k}\left[p^*\frac{1+b}{1+r}\right]^k\left[(1-p^*)\frac{1+a}{1+r}\right]^{N-k}.$$

Introducing the notation

$$\widetilde{p}:=\frac{1+b}{1+r}\,p^*,\quad\text{and}\quad B(j,N,p):=\sum_{k=j}^{N}\binom{N}{k}p^k(1-p)^{N-k},$$

we arrive at the *Cox-Ross-Rubinstein formula*

$$C_N = S_0\, B(k_0,N,\widetilde{p}) - K\,(1+r)^{-N}\,B(k_0,N,p^*).$$

The obtained formula gives the fair price of the call $(S_N - K)^+$ at time 0. More generally, the price of this call at any time $n \leq N$ is given by

$$C_{N,n} = S_n\, B(k_n,N-n,\widetilde{p}) - K\,(1+r)^{-(N-n)}\,B(k_n,N-n,p^*),\qquad(2.4)$$

where $k_n := \min\{n \leq k \leq N : S_n(1+b)^k(1+a)^{N-k} \geq K\}$.

From our earlier discussions, we know that the price $C_{N,n}$ is equal to the value of the minimal hedge at time $n \leq N$. We also observe that the risk component of the minimal hedge $\pi^* = (\beta_n^*, \gamma_n^*)_{n\leq N}$ is related to the structure of $C_{N,n}$ in formula (2.4): $\gamma_n^* = B(k_n, N - n, \tilde{p})$. The other component β_n^* is determined by the condition of self-financing and is equal to $-K (1+r)^{-(N)} B(k_n, N-n, p^*)$. Thus, the Cox-Ross-Rubinstein formula gives a complete description of risk-neutral strategies for European call options.

Next, we consider a *European put option* with contingent claim $f_N = (K - S_N)^+$, which gives its holder the right to sell shares of stock S at a fixed strike price K at time N.

Denote the price of a European put option by P_N. Taking into account the martingale property of $(S_n/B_n)_{n \leq N}$ and the equality $(K - S_N)^+ = (S_N - K)^+ - S_N + K$, we obtain

$$
\begin{aligned}
P_N &= E^* \left(\frac{(K - S_N)^+}{(1 + r)^N} \right) = E^* \left(\frac{\max\{0, K - S_N\}}{(1 + r)^N} \right) \qquad (2.5) \\
&= C_N - E^* \left(\frac{S_N}{(1 + r)^N} \right) + \frac{K}{(1 + r)^N} = C_N - E^*(S_0) + \frac{K}{(1 + r)^N} \\
&= C_N - S_0 + \frac{K}{(1 + r)^N} \, .
\end{aligned}
$$

This connection (2.5) between the prices P_N and C_N is called the *call-put parity* relation. It obviously allows one to express the price of a European put option in terms of the price of a European call option (and vice versa). Further, we note that this is possible not just for a European put option, but also for a whole class of contingent claims of the form $f_N = g(S_N)$, where $g(\cdot)$ is a smooth function on $[0, \infty)$. Indeed, from Taylor's formula

$$
g(x) = g(0) + g'(0)\, x + \int_0^\infty (x - y)^+ g''(y) dy \, ,
$$

and therefore, using the same argument as in (2.5), we have

$$
C_N(f) = C_N \big(g(S_N) \big) = \frac{g(0)}{(1 + r)^N} + S_0\, g'(0) + \int_0^\infty C_N \big((S_N - y)^+ \big) g''(y) dy \, .
$$

So one can use the Cox-Ross-Rubinstein formula for a European call option to find $C_N(f) = C_N \big(g(S_N) \big)$ for any smooth (twice continuously differentiable) function g.

Now we can summarize that the fair price $C_N(f) := E^*(S_N/B_N)$ for an arbitrary contingent claim f_N has the following properties:

1. It is "fair" both for the writer of the contract (as it is always possible to invest amount C_N in order to gain the amount f_N and to make the payment at time N) and for the holder (who pays the price that is equal to the minimal amount necessary for hedging). Note that this minimizes risk for both parties.

2. If the writer sells the contract at a price $x > C_N$, then there is an arbitrage opportunity: the amount C_N can be invested in a minimal hedge, and $x - C_N$ is a guaranteed non-risky profit.

3. Conversely, if $x < C_N$, then the holder of the contract can gain an arbitrage profit $C_N - x > 0$.

Thus, the set of all possible prices consists of two regions of arbitrage prices that are separated by C_N, which is therefore referred to as a *non-arbitrage price*.

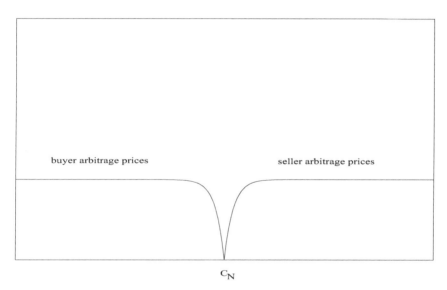

buyer arbitrage prices   seller arbitrage prices

C_N

FIGURE 2.1: Non-arbitrage price of contingent claim f_N.

In the following example, we demonstrate an elegant application of the theory of minimal hedging and of the Cox-Ross-Rubinstein formula to pricing equity-linked life insurance contracts, where terminal payment depends on the price of stock. This contract is attractive to a policyholder since stock may appreciate much faster than money held in a bank account. Additionally, this contract guarantees some minimal payment that protects the policyholder in the case of stock depreciation. However, a competitive market environment encourages insurance companies to offer innovative products of this type. Thus, they face a problem of pricing such contracts.

Worked Example 2.3 *In the framework of a binomial (B, S)-market, an insurance company issues a* **pure endowment insurance contract.** *According to this contract, the policyholder is paid*

$$f_N = \max\{S_N, K\}$$

on survival to the time N, where S_N is the stock price and K is the guaranteed minimal payment. Find the "fair" price for such an insurance policy.

Solution Let l_x be the number of policyholders of age x. Each policyholder i, $i = 1, \ldots, l_x$ can be characterized by a positive random variable T_i representing the time elapsed between age x and death. Denote $p_x(n) = P(\{\omega : T_i > n\})$, the conditional expectation for a policyholder to survive another n years from the age of x. Suppose that T_i, $i = 1, \ldots, l_x$, are both mutually independent and independent of $\rho_1, \ldots, \rho_N$.

According to the theory developed in this section, it is natural to find the required price C by equating the sum of all premiums to the average sum of all payments:

$$C \times l_x = E^* \left(\sum_{i=1}^{l_x} \frac{f_N}{B_N} I_{\{\omega : \, T_i > N\}} \right),$$

where expectation E^* is taken with respect to a martingale probability.

Taking into account that $\max\{S_N, K\} = K + (S_N - K)^+$ and independence of all T_i and ρ_k values, we use the Cox-Ross-Rubinstein formula to obtain

$$
\begin{aligned}
C &= \frac{1}{l_x} \sum_{i=1}^{l_x} E^* \left(\frac{f_N}{B_N} I_{\{\omega : \, T_i > N\}} \right) = \frac{1}{l_x} l_x \, p_x(N) \, E^* \left(\frac{K + (S_N - K)^+}{B_N} \right) \\
&= p_x(N) \frac{K}{(1+r)^N} + p_x(N) \left[S_0 \, B(k_0, N, \widetilde{p}) - \frac{K}{(1+r)^N} B(k_0, N, p^*) \right]. \; \square
\end{aligned}
$$

Next, we illustrate how arbitrage considerations can be used in pricing forward and futures contracts.

A *forward contract* is an agreement between two parties to buy or sell a specified asset S for the delivery price F at the delivery date N. Let us consider forwards as investment tools in the framework of a binomial (B, S)-market. Since such agreements can be reached at any date $n = 0, 1, \ldots, N$, it is important to determine the corresponding delivery prices $F_0, \ldots, F_N$. Note that we clearly have $F_0 = F$ and $F_N = S_N$.

Consider an investment portfolio $\pi = (\beta, \gamma)$ with values

$$X_n^\pi = \beta_n B_n + \gamma_n D_n,$$

where γ_n is the number of units of asset S, $D_k = 0$ for $n \le k \le N$, and $D_N = S_N - F_n$.

Taking into account that for a forward contract traded at time n, $\gamma_k = 0$ for $k \le n$ and $\gamma_k = \gamma_{n+1}$ for $k \ge n + 1$, we compute the discounted value of this portfolio:

$$\Delta \left(\frac{X_k^\pi}{B_k} \right) = \gamma_k \frac{\Delta D_k}{B_k},$$

$$\frac{X_N^\pi}{B_N} = \frac{X_n^\pi}{B_n} + \sum_{k=n+1}^N \gamma_k \frac{\Delta D_k}{B_k} = \frac{X_n^\pi}{B_n} + \gamma_{n+1} \frac{S_N - F_N}{B_N}.$$

Using the no-arbitrage condition for strategy π, we can now find forward price F_n:

$$0 = E^*\left(\frac{S_N - F_n}{B_N}\bigg|\mathcal{F}_n\right) = \frac{S_n}{B_n} - \frac{F_n}{B_N},$$

hence

$$F_n = B_N \frac{S_n}{B_n}.$$

Therefore, we have

$$E^*\left(\frac{X_N^\pi}{B_N}\right) = E^*\left(\frac{X_n^\pi}{B_n}\right),$$

which guarantees that π is a no-arbitrage strategy.

A *futures contract* is the same agreement, but the trading takes place on a stock exchange. The clearinghouse of the exchange opens *margin accounts* for both parties that are used for repricing the contract on a daily basis.

Let $F_0^*, \ldots, F_N^*$ be futures prices. Suppose that the parties enter a futures contract on the stock S at time n with the strike price F_n^*. At time $n+1$, the clearinghouse announces a new quoted price F_{n+1}^*. If $F_{n+1}^* > F_n^*$, then the seller of S loses and must deposit the variational margin $F_{n+1}^* - F_n^*$. Otherwise, the buyer deposits $F_n^* - F_{n+1}^*$.

Denote $\delta_0 = F_0^*$ and

$$\delta_n = F_n^* - F_{n-1}^*, \quad D_n = \delta_0 + \delta_1 + \cdots + \delta_n, \quad \Delta D_n = \delta_n$$

for $n \geq 1$. Consider an investment portfolio π with β_n representing investment in a bank account and γ_n equal to the number of shares of S traded via futures contracts. Then

$$\frac{X_N^\pi}{B_N} = \frac{X_n^\pi}{B_n} + \gamma_{n+1} \sum_{k=n+1}^{N} \frac{\Delta D_k}{B_k}.$$

From the no-arbitrage condition, we have

$$E^*\left(\sum_{k=n+1}^{N} \frac{\Delta D_k}{B_k}\bigg|\mathcal{F}_n\right) = 0,$$

which is equivalent to the fact that $(D_n)_{n \leq N}$ is a martingale with respect to P^*, and hence $D_n = E^*(D_N|\mathcal{F}_n)$. Taking into account the equalities $D_N = F_N^* = S_N$ and $D_n = \delta_0 + \delta_1 + \cdots + \delta_n = F_n^*$, we obtain

$$F_n^* = E^*(S_N|\mathcal{F}_n) = B_N E^*\left(\frac{S_N}{B_N}\bigg|\mathcal{F}_n\right) = B_N \frac{S_n}{B_n} = F_n.$$

Thus, we arrive at the following general conclusion: *on a complete no-arbitrage binomial* (B, S)*-market prices of forward and futures contracts coincide.*

2.3   Pricing and hedging American options

In a binomial (B, S)-market with the time horizon N, we consider a sequence of contingent claims $(f_n)_{n \le N}$, where each f_n has the repayment date $n = 0, 1, \ldots, N$. Managing such a collection is not difficult, as we can straightforwardly price each claim f_n:

$$C_n(f_n) = E^* \left(\frac{f_n}{(1+r)^n} \right),$$

and therefore the price of the whole collection is

$$C \left((f_n)_{n \le N} \right) = \sum_{n=0}^{N} C_n(f_n) = E^* \left(\sum_{n=0}^{N} \frac{f_n}{(1+r)^n} \right).$$

In elementary financial and actuarial mathematics, a series of deterministic payments (f_n) is called an *annuity*. Thus, using this terminology, the latter formula gives the price of a *stochastic annuity*. Note that the linear structure of the collection of contingent claims was used in the calculation of this price. In general, the structure of a series of claims can be much more complex.

Let $(f_n)_{n=0}^{N}$ be a non-negative stochastic sequence adopted to filtration $\mathbb{F} = (\mathcal{F}_n)_{n=0}^{N}$, where $\mathcal{F}_n = \sigma(S_0, \ldots, S_n)$. Suppose $\tau : \Omega \to \{0, 1, \ldots, N\}$ is a *stopping time* (or a *Markov time*); that is, it does not depend on the future. Using sequence $(f_n)_{n=0}^{N}$ and a stopping time τ, we define the following contingent claim

$$f_\tau(\omega) \equiv f_{\tau(\omega)}(\omega) = \sum_{n=0}^{N} f_n(\omega) \, I_{\{\omega: \ \tau=n\}} \, .$$

It is clear from the definition that this claim is determined by all trading information up to time N, but it is exercised at a *random time* τ, which is therefore called the *exercise time*.

According to the aforementioned methodology of managing risk associated with a contingent claim in the framework of a binomial market (B, S)-market, we can price this claim using averaging with respect to a risk-neutral probability P^*:

$$C(f_\tau) = E^* \left(\frac{f_\tau}{B^\tau} \right) = E^* \left(\frac{f_\tau}{(1+r)^\tau} \right).$$

If we denote $\mathcal{M}_0^N$ the collection of all stopping times, then we have a collection of contingent claims corresponding to these stopping times $\tau \in \mathcal{M}_0^N$, which is called an *American contingent claim* or an *American option* . Since $C(f_\tau)$ are risk-neutral predictions of future payments f_τ, then the rational price for an American claim must be

$$C_N^{\mathrm{am}} = \sup_{\tau \in \mathcal{M}_0^N} C(f_\tau) = \sup_{\tau \in \mathcal{M}_0^N} E^* \left(\frac{f_\tau}{(1+r)^\tau} \right).$$

Now, since the collection $(C(f_\tau))_{\tau \in \mathcal{M}_0^N}$ is finite, then a stopping time $\tau^* \in \mathcal{M}_0^N$ exists such that

$$C(f_{\tau^*}) = E^* \left(\frac{f_{\tau^*}}{(1+r)^{\tau^*}} \right) = \sup_{\tau \in \mathcal{M}_0^N} E^* \left(\frac{f_\tau}{(1+r)^\tau} \right) = C_N^{\mathrm{am}},$$

which must be the exercise time for an American contingent claim $(f_\tau)_{\tau \in \mathcal{M}_0^N}$.

Note that, from a mathematical point of view, the pair $(C_N^{\mathrm{am}}, \tau^*)$ solves the problem of finding an optimal stopping time for the stochastic sequence $(f_n/(1+r)^n)_{n=0}^N$. The financial interpretation of this mathematical problem is pricing an American contingent claim with an exercise time up to the maturity date N. More than 90% of options traded on exchanges are of American type.

Example 2.1 (Examples of American-type options)

1. *American call and put options* are defined by the following sequences of claims:

$$f_n = (S_n - K)^+ \quad \text{and} \quad f_n = (K - S_n)^+, \quad n \leq N,$$

respectively.

2. *Russian option* is defined by

$$f_n = \max_{k \leq n} S_k.$$

Now we describe the *methodology* for pricing such options. As in the case of European options, we use the notion of a strategy (portfolio) $\pi = (\pi_n)_{n=0}^N = (\beta_n, \gamma_n)_{n=0}^N$ with values $X_n^\pi = \beta_n B_n + \gamma_n S_n$. A self-financing strategy is called a hedge if $X_n^\pi \geq f_n$ for all $n = 0, 1, \ldots, N$. In particular, $X_\tau^\pi \geq f_\tau$ for all stopping times $\tau \in \mathcal{M}_0^N$. A hedge π^* such that $X_n^{\pi^*} \leq X_n^\pi$ for all $n \leq N$ for any other hedge π is called the *minimal hedge*.

Let $\mathcal{M}_n^N$, $0 \leq n \leq N$, be the collection of all stopping times with values in $\{n, \ldots, N\}$. Consider the stochastic sequence

$$Y_n := \sup_{\tau \in \mathcal{M}_n^N} E^* \left(\frac{f_\tau}{(1+r)^\tau} \bigg| \mathcal{F}_n \right), \quad n = 0, 1, \ldots, N,$$

which has the initial value $Y_0 = C_N^{\mathrm{am}}$ and the terminal value $Y_N = f_N/(1+r)^N$. To find the structure of sequence $(Y_n)_{n=0}^N$, we write

$$Y_N = Y_{\tau_N^*} = \frac{f_N}{(1+r)^N},$$

where $\tau_N^* \equiv N$ is the only stopping time in class $\mathcal{M}_N^N$. Now, for $n = N - 1$, we have

$$Y_{N-1} = \begin{cases} \frac{f_{N-1}}{(1+r)^{N-1}} & \text{if } \frac{f_{N-1}}{(1+r)^{N-1}} \geq E^*\left(\frac{f_N}{(1+r)^N}\big|\mathcal{F}_{N-1}\right) \\ E^*\left(\frac{f_N}{(1+r)^N}\big|\mathcal{F}_{N-1}\right) & \text{otherwise} \end{cases},$$

which is equivalent to the formula

$$Y_{N-1} = \max\left\{\frac{f_{N-1}}{(1+r)^{N-1}}, E^*\left(Y_N\big|\mathcal{F}_{N-1}\right)\right\}.$$

Setting

$$\tau_{N-1}^* = \begin{cases} N-1 & \text{if } \frac{f_{N-1}}{(1+r)^{N-1}} \geq E^*\left(\frac{f_N}{(1+r)^N}\big|\mathcal{F}_{N-1}\right) \\ N & \text{otherwise} \end{cases},$$

we obtain $Y_{\tau_{N-1}^*}$ equal either to

$$\frac{f_{N-1}}{(1+r)^{N-1}}$$

or

$$E^*\left(\frac{f_N}{(1+r)^N}\bigg|\mathcal{F}_{N-1}\right).$$

For an arbitrary $n \leq N$, we obtain expressions

$$Y_n = \max\left\{\frac{f_n}{(1+r)^n}, E^*\left(Y_{n+1}\big|\mathcal{F}_n\right)\right\}$$

and

$$\tau_n^* = \inf_{n \leq k \leq N}\left\{k : Y_k = \frac{f_k}{(1+r)^k}\right\}.$$

Finally,

$$C_N^{\text{am}} = Y_0, \qquad \tau^* = \tau_0^*.$$

Now, using sequence Y_n, we construct a hedging strategy. Since

$$Y_n \geq E^*\left(Y_{n+1}\big|\mathcal{F}_n\right) \quad \text{for all} \quad n \leq N - 1,$$

then $(Y_n)_{n \leq N}$ is a supermartingale that admits Doob decomposition:

$$Y_n = M_n - A_n,$$

where $(M_n)_{n \leq N}$ is a martingale with $M_0 = Y_0$, and $(A_n)_{n \leq N}$ is a predictable non-decreasing sequence with $A_0 = 0$. We also have the following martingale representation

$$M_n = M_0 + \sum_{k=1}^{n} \gamma_k^* \frac{S_{k-1}}{B_k}(\rho_k - r),$$

where (γ_k^*) is a predictable sequence which is defined by relation $\phi_k^* = \gamma_k^* \frac{S_{k-1}}{B_k}$ due to (2.2).

Using this sequence (γ_n^*), we define a self-financing strategy $\pi^* = (\beta_n^*, \gamma_n^*)$ with values $X_n^{\pi^*} = B_n M_n$.

This gives us the required hedge, as for all $n \le N$

$$X_n^{\pi^*} = M_n B_n = (Y_n + A_n) B_n \ge Y_n B_n = \sup_{\tau \in \mathcal{M}_n^N} E^* \left(\frac{f_\tau}{(1+r)^\tau} \middle| \mathcal{F}_n \right) B_n$$

$$= \sup_{\tau \in \mathcal{M}_n^N} E^* \left(\frac{f_\tau B_n}{B_\tau} \middle| \mathcal{F}_n \right) \ge f_n ,$$

and

$$X_0^{\pi^*} = Y_0 = \sup_{\tau \in \mathcal{M}_0^N} E^* \left(\frac{f_\tau}{(1+r)^\tau} \right) = C_N^{\mathrm{am}} .$$

Worked Example 2.4 *On a two-step (B, S)-market, price an American option with payments*

$$f_0 = (S_0 - 90)^+ \quad f_1 = (S_1 - 90)^+ \quad f_2 = (S_2 - 120)^+ ,$$

where $S_0 = 100\,(\$)$, $\Delta S_i = S_{i-1}\,\rho_i$, with

$$\rho_i = \begin{cases} 0.5 & \text{with probability } 0.4 \\ -0.3 & \text{with probability } 0.6 \end{cases} , \quad i = 1, 2,$$

and annual interest rate $r = 0.2$.

Solution It is clear that the risk-neutral probability is defined by Bernoulli's probability $p^* = 5/8$. We have that

$$Y_2 = \frac{(S_2 - 120)^+}{(1+r)^2} = \frac{(S_1(1+\rho_2) - 120)^+}{(1.2)^2} ,$$

$$Y_1 = \max \left\{ \frac{f_1}{(1+r)} , E^*(Y_2|\mathcal{F}_1) \right\} ,$$

$$Y_0 = \max \left\{ f_0 , E^*(Y_1|\mathcal{F}_0) \right\} .$$

Computing

$$E^*(Y_2|\mathcal{F}_1) = \begin{cases} \frac{p^*(225-120)}{(1+r)^2} = \frac{5/8 \times 105}{(1.2)^2} \approx 46 & \text{on the set } \{\omega : S_1 = 150\} \\ 0 & \text{on the set } \{\omega : S_1 = 70\} \end{cases} ,$$

we obtain

$$Y_1 = \begin{cases} \max \left\{ \frac{150-90}{1.2} , \frac{5/8 \times 105}{(1.2)^2} \right\} = 50 = \frac{f_1}{1+r} & \text{on the set } \{\omega : S_1 = 150\} \\ 0 & \text{on the set } \{\omega : S_1 = 70\} \end{cases} .$$

Taking into account that $E^*(Y_1|\mathcal{F}_0) = E^*(Y_1) \approx 31$ we obtain

$$Y_0 = \max\{0, 31\} = 31 \neq 10 = f_0 \,,$$

and the optimal stopping time

$$\tau^* \equiv \tau_0^* \equiv \tau_1^* \equiv 1 \,. \ \square$$

We complete this section with the following general remark regarding situations when the optimal stopping time for an American option is equal to the terminal time N. Let $f_n = g(S_n)$, where g is some non-negative convex function. Suppose for simplicity that $r = 0$. We have

$$C_N^{\mathrm{am}}(f) = \sup_{\tau \in \mathcal{M}_0^N} E^*(f_\tau) = \sup_{\tau \in \mathcal{M}_0^N} E^*(g(S_\tau)) \,.$$

Since by Jensen's inequality $(g(S_\tau))_{n \leq N}$ is a submartingale, then for any $\tau \leq N$

$$E^*(g(S_\tau)) \leq E^*(g(S_N)) \,,$$

which implies that $\tau^* \equiv N$ is the optimal stopping time.

2.4 Utility functions and St. Petersburg's paradox. The problem of optimal investment.

In the previous sections, we studied investment strategies (portfolios) from the point of view of hedging contingent claims. Another criterion for comparing investment strategies can be formulated in terms of utility functions. A continuously differentiable function $U : [0, \infty) \to \mathbb{R}$ is called a *utility function* if it is non-decreasing, concave, and

$$\lim_{x \downarrow 0} U'(x) = \infty \,, \qquad \lim_{x \to \infty} U'(x) = 0 \,.$$

An investor's aim to maximize $U(X_N^\pi)$ can lead to a difficult problem, as X_N^π is a random variable. Therefore, it is natural to compare average utilities: we say that a strategy π' is preferred to strategy π if

$$E(U(X_N^{\pi'})) \geq E(U(X_N^\pi)) \,.$$

One of the fundamental notions in this area of financial mathematics is the notion of *risk aversion*. Its mathematical description is given by the Arrow-Pratt function

$$R_A(\cdot) := -\frac{U''(\cdot)}{U'(\cdot)}$$

(in the case when U is twice continuously differentiable). This function characterizes decreasing risk aversion if $R'_A < 0$ and increasing risk aversion if $R'_A > 0$.

Thus, such utility functions allow one to introduce a *measure of investment preferences* for risk-averse participants in a market.

Historically, the theory of optimal investment with the help of utility functions grew from the famous Bernoulli's *St. Petersburg's paradox.*

Worked Example 2.5 (St.Petersburg's paradox) *Peter challenges Paul to a game of coin–toss. The game ends when the tail appears for the first time. If this happens after n tosses of a coin, then Peter pays Paul 2^{n-1} dollars. What price C should Paul pay Peter for an opportunity to enter this game?*

Solution Let X be Paul's prize money, which is a random variable. An intuitive way of finding C suggests computing the average of X:

$$E(X) = 1 \times 1/2 + 2 \times 1/4 + \ldots + 2^{n-1}/2^n + \ldots = 1/2 + \ldots + 1/2 \ldots = \infty.$$

Thus, since the average of Paul's prize money is infinite, then Paul can agree to any price offered by Peter, which is clearly paradoxical.

Bernoulli suggested that the price C can be found from the equation

$$E(\ln X) = \ln C,$$

which implies $C = 2$, as

$$
\begin{aligned}
E(\ln X) &= \sum_{n=1}^{\infty} \frac{\ln 2^{n-1}}{2^n} = \sum_{n=1}^{\infty} \frac{(n-1)\ln 2}{2^n} \\
&= \ln 2 \sum_{n=1}^{\infty} \frac{n-1}{2^n} = \ln 2 \times 1 = \ln 2. \quad \Box
\end{aligned}
$$

In general, given a utility function U, consider a problem of finding a self-financing strategy π^* such that

$$\max_{\pi \in SF} E\Big(U\big(X_N^\pi(x)\big)\Big) = U\big(X_N^{\pi^*}(x)\big), \tag{2.6}$$

where x is the initial value and $X_N^\pi(x)$ is the terminal value of strategy π on market (2.1). For simplicity, let $U(x) = \ln x$. Then

$$\ln X_N^\pi(x) = \ln \frac{X_N^\pi(x)}{B_N} + \ln B_N,$$

and therefore, the optimization problem (2.6) reduces to finding the maximum of

$$E\left(\ln \frac{X_N^\pi(x)}{B_N}\right)$$

overall $\pi \in SF$.

Let us denote $Y_n(x) := X_n^\pi(x)/B_n$ the discounted value of a self-financing portfolio π. Recall that $(Y_n)_{n \le N}$ is a positive martingale with respect to a risk-neutral probability P^*. Thus, we arrive at the problem of finding a positive martingale $Y^*(x) \equiv (Y_n^*)_{n \le N}$ with $Y_0^* = x$, such that

$$\max_Y E\big(\ln Y_N(x) \big) = E\big(\ln Y_N^*(x) \big) ,$$

where the maximum is taken over the set of all positive martingales with the initial value x.

Let $Y_N^*(x) = x/Z_N^*$, where Z_N^* is the density of the martingale probability P^*. All other values of $Y^*(x)$ are defined as the following conditional expectations with respect to P^*:

$$Y_0^* = x, \quad Y_n^*(x) = E^* \left(\frac{x}{Z_N^*} \Big| \mathcal{F}_n \right), \quad n = 1, \ldots, N .$$

For any other martingale Y, using Taylor decomposition for the logarithmic function and the martingale property of $Y(x)$ with respect to probability P^*, we have

$$
\begin{aligned}
E\big(\ln Y_N(x) \big) &= E\left(\ln \frac{x}{Z_N^*} + \left[\ln Y_N(x) - \ln \frac{x}{Z_N^*} \right] \right) \\
&\le E\left(\ln \frac{x}{Z_N^*} \right) + E\left(\frac{Z_N^*}{x} \left[Y_N(x) - \frac{x}{Z_N^*} \right] \right) \\
&= E\left(\ln \frac{x}{Z_N^*} \right) + \left[E^*(Y_N(x)) - E^*\left(\frac{x}{Z_N^*} \right) \right] \Big/ x \\
&= E\left(\ln \frac{x}{Z_N^*} \right) + \frac{x - x}{x} = E\left(\ln \frac{x}{Z_N^*} \right) \\
&= E\big(\ln Y_N^*(x) \big) .
\end{aligned}
$$

Thus, $Y^*(x)$ is an optimal martingale. Recall that, for such a martingale, $Y_N^*(x)$ necessarily coincides with the discounted value of some self-financing strategy π^*. To find this optimal portfolio $\pi^* = (\beta_n^*, \gamma_n^*)_{n \le N}$, we introduce quantities

$$\alpha_n^* := \gamma_n^* \frac{S_{n-1}}{X_{n-1}^{\pi^*}} ,$$

which represent the *proportion* of risky capital in the portfolio.

Using mathematical induction in N, we obtain

$$\frac{X_N^{\pi^*}(x)}{B_N} = x \prod_{k=1}^N \left(1 - \frac{\alpha_k^*}{1+r} (\rho_k - r) \right).$$

However, because of (2.3), we have

$$\frac{X_N^{\pi^*}(x)}{B_N} = \frac{x}{Z_N^*} = x \prod_{k=1}^N \left(1 - \frac{\mu - r}{\sigma^2} (\rho_k - r) \right)^{-1},$$

where $\mu = E(\rho_k)$. Using the two equalities above, we arrive at the following equation for α_k^*:

$$\prod_{k=1}^{N} \left(1 - \frac{\alpha_k^*}{1+r}(\rho_k - r)\right) \times \left(1 - \frac{\mu - r}{\sigma^2}(\rho_k - r)\right) = 1.$$

Let $N = 1$; then the latter equation reduces to

$$\left(1 - \frac{\alpha_1^*}{1+r}(\rho_1 - r)\right) \times \left(1 - \frac{\mu - r}{\sigma^2}(\rho_1 - r)\right) = 1,$$

and on the set $\{\omega : \rho_1(\omega) = b\}$, we have

$$\left(1 - \frac{\alpha_1^*}{1+r}(b - r)\right) \times \left(1 - \frac{\mu - r}{\sigma^2}(b - r)\right) = 1,$$

which implies that

$$\alpha_1^* = \frac{(1+r)(\mu - r)}{(r - a)(b - r)}.$$

On the set $\{\omega : \rho_1(\omega) = a\}$, the expression for α_1^* is exactly the same. Next, suppose that $\alpha_1^* \equiv \alpha_2^* \equiv \ldots \equiv \alpha_{N-1}^*$, then by induction, we obtain that α_N^* is also given by this expression.

Thus, the constant proportion of risky capital

$$\alpha^* = \frac{(1+r)(\mu - r)}{(r - a)(b - r)} \tag{2.7}$$

is a characteristic property of the optimal strategy π^* that solves the optimization problem (2.6) with the logarithmic utility function. Therefore, in this case, management of the risk associated with an investment portfolio reduces to retaining the proportion of risky capital in this portfolio at the constant level (2.7).

Note that management of this type of risk differs from hedging contingent claims. To illustrate this, we revisit Worked Examples 2.1 and 2.2. Recall that in these examples we consider a single-period binomial (B, S)-market with the annual rate of interest $r = 0.2$ and with the profitability of the risky asset

$$\rho_1 = \begin{cases} 0.5 & \text{with probability } 0.4 \\ -0.3 & \text{with probability } 0.6 \end{cases}.$$

The average profitability

$$\mu = E(\rho_1) = 0.5 \times 0.4 - 0.3 \times 0.4 = 0.02$$

is less than $r = 0.2$, and the optimal proportion of risky capital

$$\alpha^* = \frac{1.2 \times (-0.18)}{0.5 \times 0.3} \approx -1.5$$

is negative. This indicates that an investor should prefer depositing money in
a bank account.

Recall that for the contingent claim in Worked Example 2.1 we have

$$f_1 = (S_1 - 100)^+ = \begin{cases} 50 & \text{with probability } 0.4 \\ 0 & \text{with probability } 0.6 \end{cases}$$

and its minimal hedging price is 26. For the contingent claim in Worked Example 2.2, it is

$$f_1 = \max\{S_0, S_1\} - S_1 = \begin{cases} 0 & \text{with probability } 0.4 \\ 30 & \text{with probability } 0.6 \end{cases}$$

and 9.3, respectively.

Now we compute terminal values of optimal investment portfolios with
$\alpha^* = -1.5$, and the initial values $X_0^{\alpha^*} = 26$:

$$X_1^{\alpha^*} = X_0^{\alpha^*} + \Delta X_1^{\alpha^*} = X_0^{\alpha^*} + \left(r X_0^{\alpha^*} + \alpha^* X_0^{\alpha^*} (\rho_1 - r) \right) \Big|_{X_0^{\alpha^*} = 26}$$

$$\approx \begin{cases} 19.5 & \text{with probability } 0.4 \\ 50.7 & \text{with probability } 0.6 \end{cases} \neq \begin{cases} 50 & \text{with probability } 0.4 \\ 0 & \text{with probability } 0.6 \end{cases},$$

and $X_0^{\alpha^*} = 9.3$:

$$X_1^{\alpha^*} = X_0^{\alpha^*} + \left(r X_0^{\alpha^*} + \alpha^* X_0^{\alpha^*} (\rho_1 - r) \right) \Big|_{X_0^{\alpha^*} = 9.3}$$

$$\approx \begin{cases} 4 & \text{with probability } 0.4 \\ 14 & \text{with probability } 0.6 \end{cases} \neq \begin{cases} 0 & \text{with probability } 0.4 \\ 30 & \text{with probability } 0.6 \end{cases}.$$

Thus, the optimal strategy of managing the investment risk differs from both
strategies of minimal hedging.

Remark 2.1

In a single-period binomial (B, S)-market, every portfolio can be associated
with the pair (α_0, α_1) of non-negative real numbers $\alpha_0, \alpha_1 \in [0, 1]$, $\alpha_0 + \alpha_1 = 1$ that represent the proportions of the capital invested in assets B and S,
respectively. Then the profitability of a portfolio is equal to the weighted sum
of the profitabilities r and ρ_1:

$$\rho(\alpha_0, \alpha_1) = \alpha_0 r + \alpha_1 \rho_1 .$$

In this case, the optimal portfolio (α_0^*, α_1^*) can be found as a solution to the
following optimization problem

$$E\big(\rho(\alpha_0^*, \alpha_1^*)\big) = \max_{(\alpha_0, \alpha_1)} E\big(\rho(\alpha_0, \alpha_1)\big)$$

under assumption of either

$$Var\big(\rho(\alpha_0, \alpha_1)\big) \leq const$$

or

$$P\Big(\{\rho(\alpha_0, \alpha_1) \leq const\}\Big) \leq c, \quad c \in (0,1).$$

Solving this type of optimization problem leads to the introduction of a notion of optimal (effective) portfolio, to the Markovitz theory and to the capital asset pricing model. The concept of *Value at Risk* also originates from this type of problem, and it is widely used in financial practice.

2.5 The term structure of prices, hedging, and investment strategies in the Ho-Lee model

Bonds are debt securities issued by a government or a company for accumulating capital. Bonds are issued for a specified period of time $[0, N]$, where N is called the *maturity (exercise, redemption) time*, and they are characterized by their *face value (par value, nominal value)*. Payments up to redemption are called *coupons*. We consider zero-coupon bonds with face value 1. To satisfy the no-arbitrage condition, one has to assume that

$$0 < B(n, N) < 1, \quad n < N,$$

where $B(n, N)$ is the price at time n of a bond with the exercise time N. Suppose that the evolution of these prices is described by the *Ho-Lee model*:

$$B(n+1, N) = \frac{B(n, N)}{B(n, n+1)} h(\xi_{n+1}; n+1, N), \qquad (2.8)$$

where $(\xi_n)_{n \leq N}$ is a sequence of independent random variables taking values 0 or 1 with probabilities p and $1 - p$, respectively. The *perturbation function h* has the following properties: $h(\cdot; 0, 0) = h(\cdot; N, N) = 1$ and $0 < h(1; n, N) < 1 < h(0; n, N)$ for $n < N$. Note that many authors also use notation $h_d(n, N)$ and $h_{up}(n, N)$ for $h(1; n, N)$ and $h(0; n, N)$, respectively, which highlights the perturbation movements *down* and *up*.

As in the case of a binomial (B, S)-market, we can take the stochastic basis $(\Omega, \mathcal{F}, \mathbb{F}, P)$ with $\Omega = \{0, 1\}^{N^*}$, $\mathcal{F}_n = \sigma(\xi_1, \dots, \xi_n)$, $\mathcal{F} = \mathcal{F}_{N^*}$, and with probability P defined by a Bernoulli parameter $p \in [0, 1]$. The family $\big(B(n, N)\big)_{n \leq N^*}$ in (2.8) is said to be arbitrage-free if for any $n \leq N^*$ the stochastic sequence

$$\big(B_n^{-1} B(n, N)\big)_{n \leq N}, \quad \text{where} \quad B_n^{-1} = \prod_{k=0}^{k} B(k-1, k),$$

is a martingale with respect to some probability P^*. Sequence (B_n) can be interpreted here as a bank account with the interest rate $r_n = \frac{1}{B(n-1,n)} - 1$.

The no-arbitrage condition implies the existence of $p = p^*$ such that

$$p^* h(0; n, N) + (1 - p^*) h(1; n, N) = 1. \tag{2.9}$$

Further, there is a $\delta_* > 1$ such that

$$h^{-1}(0; n, N) = p^* + (1 - p^*)\delta_*^{N-n}, \tag{2.10}$$

$$h(1; n, N) = \delta_*^{N-n} \left(p^* + (1 - p^*)\delta_*^{N-n} \right)^{-1},$$

and

$$\delta_*^{N-n} = h(1; n, N) \, h^{-1}(0; n, N).$$

To verify equalities (2.9)–(2.10), we consider a portfolio π, where one unit is invested in a zero-coupon bond with the maturity time N, and $\widetilde{\gamma}$ units are invested in a zero-coupon bond with the maturity time $\widetilde{N}$. The value of this portfolio at time n is

$$
\begin{aligned}
X_n^\pi &= B(n, N) + \widetilde{\gamma} \, B(n, \widetilde{N}) \\
&= B(n - 1, N) \frac{h(\xi_n; n, N)}{B(n - 1, N)} + \widetilde{\gamma} \, B(n - 1, \widetilde{N}) \frac{h(\xi_n; n, \widetilde{N})}{B(n - 1, \widetilde{N})}.
\end{aligned}
$$

We say that portfolio π is a *risk-free* portfolio if its value X_n^π is *independent* of an increase or decrease of values of the zero-coupon bonds. In this case, we have

$$
B(n - 1, N) \frac{h(0; n, N)}{B(n - 1, N)} + \widetilde{\gamma} \, B(n - 1, \widetilde{N}) \frac{h(0; n, \widetilde{N})}{B(n - 1, \widetilde{N})}
$$

$$
= B(n - 1, N) \frac{h(1; n, N)}{B(n - 1, N)} + \widetilde{\gamma} \, B(n - 1, \widetilde{N}) \frac{h(1; n, \widetilde{N})}{B(n - 1, \widetilde{N})},
$$

and hence,

$$
\widetilde{\gamma} = \frac{B(n - 1, N)}{B(n - 1, \widetilde{N})} \times \frac{h(0; n, N) - h(1; n, N)}{h(0; n, \widetilde{N}) - h(1; n, \widetilde{N})}.
$$

With this optimal strategy, the value of portfolio π at time n is

$$
X_n^\pi = \frac{B(n - 1, N)}{B(n - 1, \widetilde{N})} \times \frac{h(1; n, N) \, h(0; n, \widetilde{N}) - h(0; n, N) \, h(1; n, \widetilde{N})}{h(0; n, \widetilde{N}) - h(1; n, \widetilde{N})}.
$$

Absence of arbitrage implies that the return of the portfolio must be equal to the risk-free rate of return r_n of the one-period zero-coupon bond $B(n-1, N)$:

$$
\frac{\triangle X_n^\pi}{X_{n-1}^\pi} = r_n = \frac{1 - B(n - 1, N)}{B(n - 1, N)}.
$$

Substituting the optimal $\widetilde{\gamma}$ we arrive at the following equality

$$
\frac{1 - h(0; n, N)}{h(1; n, N) - h(0; n, N)} = \frac{1 - h(0; n, \widetilde{N})}{h(1; n, \widetilde{N}) - h(0; n, \widetilde{N})}
$$

for an arbitrary $\widetilde{N}$. Denoting

$$1 - p^* = \frac{1 - h(0; n, N)}{h(1; n, N) - h(0; n, N)} \tag{2.11}$$

we obtain relation (2.9).

We now rewrite model (2.8) in the form

$$\triangle B(n, N) = B(n - 1, N)\,\rho_n^N, \quad B(0, N) > 0, \tag{2.12}$$

where $\left(\rho_n^N\right)_{n=1,\dots,N}$ is a sequence of random variables such that

$$\rho_n^N = \begin{cases} b_n = \frac{h(0;n,N)}{B(n-1,N)} - 1, & \text{with probability } p; \\ a_n = \frac{h(1;n,N)}{B(n-1,N)} - 1, & \text{with probability } 1 - p. \end{cases}$$

We also note that the evolution of the bank account can be written as

$$\triangle B_n = B_{n-1}\,r_n, \quad B_0 = 1. \tag{2.13}$$

Formulas (2.12)–(2.13) indicate the similarity between the Ho-Lee model for a bond market and the binomial model for a stock market.

Further, calculating

$$
\begin{aligned}
E^*\left(1 + \rho_n^N\right) &= p^* \frac{h(0; n, N)}{B(n - 1, N)} + (1 - p^*)\,\frac{h(1; n, N)}{B(n - 1, N)} \\
&= 1/B(n - 1, N) = 1 + r_n,
\end{aligned}
$$

and using (2.11), we obtain

$$p^* = 1 - \frac{1 - h(0; n, N)}{h(1; n, N) - h(0; n, N)} = \frac{h(1; n, N) - 1}{h(1; n, N) - h(0; n, N)}$$

$$= \frac{a_n\,B(n - 1, N) + B(n - 1, N) - 1}{a_n\,B(n - 1, N) + B(n - 1, N) - 1 - b_n\,B(n - 1, N) - B(n - 1, N) + 1}$$

$$= \frac{(1 + a_n)\,B(n - 1, N) - (1 + r_n)\,B(n - 1, N)}{(a_n - b_n)\,B(n - 1, N)}$$

$$= \frac{a_n - r_n}{a_n - b_n} = \frac{r_n - a_n}{b_n - a_n},$$

which is similar to the formula for the unique martingale probability p^* in the framework of the binomial stock market.

We now prove (2.10). In order to have a recombining binomial tree, we consider

$$B(n + 2, N) = B(n, N)(1 + a_{n+1})(1 + b_{n+2}) = B(n, N)(1 + b_{n+1})(1 + a_{n+2}).$$

Using (2.8) and keeping in mind that $B(n+1, n+2)$ satisfies a similar relation, we obtain

$$
\begin{aligned}
B(n+2, N) &= B(n, N) \frac{h(1; n+1, N)}{B(n, n+1)} \frac{h(0; n+2, N)}{B(n+1, n+2)} \\
&= B(n, N) \frac{h(1; n+1, N)}{B(n, n+1)} \frac{h(0; n+2, N) B(n, n+1)}{B(n, n+2) h(1; n+1, n+2)} \\
&= B(n, N) \frac{h(1; n+1, N) h(0; n+2, N)}{B(n, n+2) h(1; n+1, n+2)}
\end{aligned}
$$

and

$$
\begin{aligned}
B(n+2, N) &= B(n, N) \frac{h(0; n+1, N)}{B(n, n+1)} \frac{h(1; n+2, N)}{B(n+1, n+2)} \\
&= B(n, N) \frac{h(0; n+1, N)}{B(n, n+1)} \frac{h(1; n+2, N) B(n, n+1)}{B(n, n+2) h(0; n+1, n+2)} \\
&= B(n, N) \frac{h(0; n+1, N) h(1; n+2, N)}{B(n, n+2) h(0; n+1, n+2)} .
\end{aligned}
$$

These two equalities together with (2.9) imply

$$
\frac{1}{h(0; n+2, N)} = \frac{\left(1 - h(0; n+1, n+2)\right) p^*}{1 - p^* h(0; n+1, n+2)}
$$
$$
+ \frac{(1 - p^*) h(0; n+1, n+2)}{1 - p^* h(0; n+1, n+2)} \frac{1}{h(0; n+1, N)},
$$

which can be written in the form

$$
\frac{1}{h(0; n+1, N)} = \frac{\gamma_*}{\delta_*} + \frac{1}{\delta_*} \frac{1}{h(0; n, N)} \tag{2.14}
$$

with

$$
\delta_* = \frac{1 - p^* h(0; n, n+1)}{(1 - p^*) h(0; n, n+1)} \quad \text{and} \quad \gamma_* = \frac{\left(1 - h(0; n, n+1)\right) p^*}{(1 - p^*) h(0; n, n+1)} .
$$

Equation (2.14) has a general solution

$$
h(0; n, N) = \frac{1}{p^* + c \delta_*^{N-n}}
$$

for some constant c that depends on the initial condition $h(1; 0, 0)$. Clearly, $c = 1 - p^*$, so we arrive at

$$
h(0; n, N) = \frac{1}{p^* + (1 - p^*) \delta_*^{N-n}} .
$$

From (2.9), we obtain

$$
h(1; n, N) = \frac{\delta_*^{N-n}}{p^* + (1 - p^*) \delta_*^{N-n}} ,
$$

which proves (2.10).

Now consider the introduced family of bonds $(B(n,N))_{n \leq N*}$ and a bank account $(B_n)_{n \leq N*}$ with the rate of interest $r_n \geq 0$. For a perturbation function

$$h(\xi_j; j, N) = \delta_*^{(N-n)\xi_j} h(0; j, N),$$

we have

$$B(n, N) = \frac{B(0, N)}{B(0, n)} \prod_{j=1}^{n} h^{-1}(\xi_j; j, n) h(\xi_j; j, N).$$

Further, introducing a new parameter $\delta = \ln \delta_*$, we can rewrite the perturbation function h in the form

$$h(\xi_n; n, N) = e^{(N-n)\xi_n \delta} \left(p^* + (1 - p^*) e^{(N-n)\delta} \right)^{-1}.$$

We obtain the following term structure of bond prices in the Ho-Lee model:

$$
\begin{aligned}
B(n, N) &= B(0, N) \prod_{i=1}^{n} B(i-1, i)^{-1} \frac{\prod_{i=1}^{n} \delta_*^{(N-i)\xi_i}}{\prod_{i=1}^{n} \left(p^* + (1-p^*)\delta_*^{(N-i)} \right)} \\
&= B(0, N) B_n \left[e^{\delta \sum_{i=1}^{n} (N-i)\xi_i} \Big/ E^* \left(e^{\delta \sum_{i=1}^{n} (N-i)\xi_i} \right) \right] \\
&= \frac{B(0, N)}{B(0, n)} e^{\delta (N-n) \sum_{i=1}^{n} \xi_i} \prod_{i=1}^{n} \frac{p^* + (1-p^*)\delta_*^{(n-i)}}{p^* + (1-p^*)\delta_*^{(N-i)}}.
\end{aligned}
$$

Now let us choose a particular bond $(B(n, N^1))_{n \leq N^1}$ from the family $(B(n, N))_{n \leq N \leq N*}$. Using it as a risky asset, and a bank account $(B_n)_{n \leq N^1}$ as a non-risky asset, we can form a financial market. A portfolio π is formed by β_n units of asset B_n and $\gamma_n(N^1)$ bonds $B(n, N^1)$ with the exercise date N^1. The values of this portfolio are

$$X_n^\pi = \beta_n B_n + \gamma_n(N^1) B(n, N^1).$$

The portfolio π is self-financing if

$$\Delta X_n^\pi = \beta_n \Delta B_n + \gamma_n(N^1) \Delta B(n, N^1).$$

Thus, this $(B_n, B(n, N^1))_{n \leq N^1}$-market is analogous to the binomial (B, S)-market with the unique martingale measure P^*.

Let us consider a contingent claim

$$f_N = \left(B(N, N^1) - K \right)^+, \quad N \leq N^1,$$

which corresponds to the European call option. Its price is uniquely determined by

$$C_N = E^* \left(B_N^{-1} \left(B(N, N^1) - K \right)^+ \right).$$

Taking into account the term structure of bond prices, $B(N, N^1) \geq K$ if not less than $k_0 := k(N, N^1, B(0, N), B(0, N^1))$ quantities $\xi_1, \ldots, \xi_N$ take value 1, where

$$k(t, T, B, B')$$
$$= \inf \left\{ k \leq t : k \geq \frac{1}{(T-t)\delta} \ln \left(K \frac{B}{B'} \prod_{i=1}^{t} \frac{p^* + (1-p^*)\delta^{(T-i)}}{p^* + (1-p^*)\delta^{(t-i)}} \right) \right\}.$$

Denote

$$\mathbb{B}(k_0, t, T, p) := \sum \left\{ \frac{\prod_{i=1}^{t} \delta_*^{x_i(T-t)}}{\prod_{i=1}^{t} \left(p^* + (1-p^*)\delta_*^{(T-i)} \right)} \, p^{t - \sum x_i} (1-p)^{\sum x_i} \right\},$$

where summation is taken over all vectors $(x_1, \ldots, x_t)$, consisting of 0s and 1s and such that $\sum x_i \geq k_0$.

We obtain that

$$
\begin{aligned}
C_N &= E^* \left(B_N^{-1} \left(B(N, N^1) - K \right)^+ \right) \\
&= E^* \left(\left(B_N^{-1} B(N, N^1) - B_N^{-1} K \right)^+ \right) \\
&= E^* \left(\left[B(0, N^1) \frac{\prod_{i=1}^{N} \delta_*^{x_i(N^1-i)}}{\prod_{i=1}^{N} \left(p^* + (1-p^*)\delta_*^{(N^1-i)} \right)} \right. \right. \\
&\qquad\qquad \left. \left. -K B(0, N) \frac{\prod_{i=1}^{N} \delta_*^{x_i(N-i)}}{\prod_{i=1}^{N} \left(p^* + (1-p^*)\delta_*^{(N-i)} \right)} \right]^+ \right) \\
&= B(0, N^1) \, \mathbb{B}(k_0, N, N^1, p^*) - K \, B(0, N) \, \mathbb{B}(k_0, N, N, p^*).
\end{aligned}
$$

Now denoting $k_n := k(N-n, N^1-n, B(n, N), B(n, N^1))$ we obtain the structure of the minimal hedge π^*:

$$
\begin{aligned}
X_n^{\pi^*} &= B_n E^* \left(B_N^{-1} \left(B(N, N^1) - K \right)^+ \middle| \mathcal{F}_n \right) \\
&= B(n, N^1) \, \mathbb{B}(k_n, N-n, N^1-n, p^*) \\
&\qquad - K \, B(n, N) \, \mathbb{B}(k_n, N-n, N-n, p^*).
\end{aligned}
$$

On the same market, we now solve the optimization problem (2.6) with the logarithmic utility function. Note that the density Z_N^* of probability P^* with respect to probability P has the form

$$Z_N^* = \varepsilon_N \left(-\frac{p^* - p}{p(1-p)} \sum (\xi_n - (1-p)) \right).$$

Hence, the discounted value of the optimal strategy $\pi^* = (\beta_n^*, \gamma_n^*)$ is

$$\frac{X_N^{\pi^*}}{B_N} = \frac{x}{Z_N^*} = x \Big/ \varepsilon_N \left(-\frac{p^* - p}{p(1-p)} \sum (\xi_n - (1-p)) \right).$$

Let the proportion of risky capital be

$$\alpha_n^*(N) = \frac{\gamma_n^*(N)\,B(n-1,N)}{X_{n-1}^{\pi^*}};$$

then, since π^* is self-financing, we obtain

$$\Delta \frac{X_n^{\pi^*}}{B_n} = \gamma_n^*(N)\,\Delta \frac{B(n,N)}{B_n} = \frac{X_{n-1}^{\pi^*}\,\alpha_n^*(N)}{B(n-1,N)}\,\Delta \frac{B(n,N)}{B_n}.$$

Using structure of $\big(B(n,N)\big)$, we write

$$\Delta \frac{X_n^{\pi^*}}{B_n} = \frac{X_{n-1}^{\pi^*}\,\alpha_n^*(N)}{B_{n-1}}\left[\frac{\delta_*^{(N-n)\xi_n}}{p^* + (1-p^*)\,\delta_*^{N-n}} - 1\right]$$

and

$$\frac{X_N^{\pi^*}}{B_N} = x\,\varepsilon_N \left(\sum \alpha_n^*(N)\left[\frac{\delta_*^{(N-n)\xi_n}}{p^* + (1-p^*)\,\delta_*^{N-n}} - 1\right]\right),$$

and therefore arrive at the expression

$$\alpha_n^*(N) = \frac{p^* - p}{p\,(1-p)}\,\frac{\delta_*^{(N-n)} - 1}{p^* + (1-p^*)\,\delta_*^{N-n}}.$$

2.6 The transition from the binomial model of a financial market to a continuous model. The Black-Scholes formula and equation.

In previous sections, we dealt with discrete markets, where time horizon is described by integers $0, 1, \ldots, N$, representing some units of time (e.g., years, months). Now suppose that we wish to consider a market with time horizon $[0, T]$ for some real number $T \geq 0$. We can divide this interval into m equal parts, so that we will have a time scale with the step $\tau = T/m > 0$. Thus, it is natural to consider the following (B, S, τ)-market:

$$B_t^\tau - B_{t-\tau}^\tau = r(\tau)\,B_{t-\tau}^\tau, \quad B_0^\tau > 0,\ r(\tau) > 0,$$

$$S_t^\tau - S_{t-\tau}^\tau = \rho_t(\tau)\,S_{t-\tau}^\tau, \quad S_0^\tau > 0,$$

where $\big(\rho_t(\tau)\big)$ is a stochastic sequence of independent profitabilities with values $a(\tau)$ and $b(\tau)$ such that $-1 < a(\tau) < r(\tau) < b(\tau)$, that generates the following filtration

$$\mathcal{F}_t^\tau = \sigma\big(\rho_n(\tau),\ n \leq t\big), \quad t = 0, \tau, 2\tau, \ldots, (m-1)\tau, (T/\tau)\tau.$$

Note that this discrete market can be extended to the whole of $[0, T]$ in the following standard way: for $s \in [t, t + \tau)$, where $t = 0, \tau, \ldots, m\tau$, define

$$B_s^\tau \equiv B_t^\tau, \quad S_s^\tau \equiv S_t^\tau, \quad \mathcal{F}_s^\tau \equiv \mathcal{F}_t^\tau, \quad \rho_s^\tau \equiv \rho_t^\tau,$$

so that all stochastic sequences become *stochastic processes*, and we obtain a (formally) continuous-time model of a market.

Consider a European call option on a (B, S, τ)-market. In this case, $f_T = (S_{(T/\tau)\tau} - K)^+$, and let C_T^τ be its price. If we consider a one-parameter family of (B, S, τ)-markets with respect to $\tau > 0$, then we expect processes (B_t^τ), (S_t^τ), and prices C_T^τ to have "reasonable" limits as $\tau \to 0$.

Suppose that parameters of the (B, S, τ)-market and of the limit market satisfy the relations

$$1 + r(\tau) = e^{r\tau}, \ 1 + b(\tau) = e^{\sigma\sqrt{\tau}}, \ 1 + a(\tau) = e^{-\sigma\sqrt{\tau}}, \ r \geq 0, \ \sigma > 0.$$

We use them to find asymptotic expressions for the martingale probability

$$p_\tau^* = \frac{r(\tau) - a(\tau)}{b(\tau) - a(\tau)} \quad \text{as} \quad \tau \to 0.$$

Using Taylor expansion, we can write

$$e^{r\tau} \sim 1 + r\tau \quad \text{and} \quad e^{\pm\sigma\sqrt{\tau}} = 1 \pm \sigma\sqrt{\tau} + \frac{\sigma^2\tau}{2} \quad \text{as} \quad \tau \to 0.$$

Therefore,

$$p_\tau^* = \frac{(e^{r\tau} - 1) - (1 - e^{-\sigma\sqrt{\tau}})}{e^{\sigma\sqrt{\tau}} - e^{-\sigma\sqrt{\tau}}} \sim \frac{1}{2}\left(1 + \frac{r - \frac{\sigma^2}{2}}{\sigma}\sqrt{\tau}\right) \quad \text{as} \quad \tau \to 0.$$

Consider τ-subdivision of interval $[0, t]$ with $\tau = [\![n/\tau]\!]$ subintervals of length τ. We identify S_t^τ with $S_{n\tau}^\tau$, where

$$S_{n\tau}^\tau = S_0 \prod_{k=1}^{n} (1 + \rho_k(\tau)) = S_0 e^{-\sum_{k=1}^{n} \xi_k^\tau}$$

with independent random variables $(\xi_k^\tau)_{k=1,\ldots,n}$ such that

$$\xi_k^\tau = \begin{cases} \sigma\sqrt{\tau} & \text{with probability} \ p_\tau^* \\ -\sigma\sqrt{\tau} & \text{with probability} \ 1 - p_\tau^* \end{cases}.$$

We also identify B_t^τ with $B_{n\tau}^\tau$ and

$$B_{n\tau}^\tau = \prod_{k=1}^{n} (1 + r(\tau)) = e^{r n\tau} \sim B_t = e^{rt} \quad \text{as} \quad \tau \to \infty.$$

To analyze the limit behavior of S_t^τ as $\tau \to 0$, we need to compute $\lim \sum_{k=1}^{n} \xi_k^\tau$

as $\tau \to 0$ (or, equivalently, as $n \to \infty$). We have the following asymptotic expressions for expected values and variances of ξ_k^τ with respect to the martingale probability p_τ^*:

$$E^{*,\tau}(\xi_k^\tau) = \sigma \sqrt{\tau} (2p_\tau^* - 1) \sim \left(r - \frac{\sigma^2}{2}\right)\tau \quad \text{as } \tau \to 0$$

and

$$
\begin{aligned}
Var^{*,\tau}(\xi_k^\tau) &= E^{*,\tau}((\xi_k^\tau)^2) - \left(E^{*,\tau}(\xi_k^\tau)\right)^2 \\
&= \sigma^2\tau - \sigma^2\tau(2p_\tau^* - 1)^2 \sim \sigma^2\tau \quad \text{as } \tau \to 0.
\end{aligned}
$$

Using the cental limit theorem, we arrive at the following asymptotic expression:

$$\sum_{k=1}^{n} \xi_k^\tau \sim \mathcal{N}\left(\left(r - \frac{\sigma^2}{2}\right)t, \sigma^2 t\right);$$

that is, the limit distribution of $\sum_{k=1}^{n} \xi_k^\tau$ is normal with mean $\left(r - \frac{\sigma^2}{2}\right)t$ and variance $\sigma^2 t$. Hence, under a martingale probability, we have that S_t^τ converges in distribution to

$$S_t = S_0 \, e^{(r - \frac{\sigma^2}{2})t + \sigma \sqrt{\tau} \, \xi}$$

as $\tau \to 0$ with $\xi \sim \mathcal{N}(0, 1)$.

We can now repeat this argument for the case of the real-world probability p_τ assuming that $E(\rho_k^\tau) = e^{\mu\tau - 1}$ for some $\mu \in \mathbb{R}$. Then S_t^τ converges in distribution to

$$S_t = S_0 \, e^{(\mu - \frac{\sigma^2}{2})t + \sigma W_t}$$

as $\tau \to 0$, where $(W_t)_{t \geq 0}$ is a family of Gaussian random variables that satisfy the following properties:

1. $W_0 = 0$;

2. $W_t - W_s \sim \mathcal{N}(0, t - s)$; and

3. $W_{t_2} - W_{t_1}$ and $W_{s_2} - W_{s_1}$ are independent for all $t_2 > t_1 > s_2 > s_1$.

This family $(W_t)_{t \geq 0}$ is called *Brownian motion* or *Wiener process*, and the family $(S_t)_{t \geq 0}$ is usually referred to as *geometrical Brownian motion*. The corresponding continuous-time (B, S)-market is called the *Black-Scholes model*.

Parameters r, μ, and σ are usually referred to as *interest rate*, *drift*, and *volatility*, respectively. Consider a European call option on this continuous market, with the claim $f_T = (S_T - K)^+$. We will find its price using the passage to the limit:

$$C_T = \lim_{\tau \to 0} C_T^\tau.$$

Using the Cox-Ross-Rubinstein formula, we obtain

$$C_T^\tau = S_0\, B\big(k_0(\tau),\, m,\, \widetilde{p}_\tau\big) - K\left(1 + r(\tau)\right)^{-m} B\big(k_0(\tau),\, m,\, p_\tau^*\big),$$

where

$$m = \frac{T}{\tau}, \qquad k_0(\tau) = 1 + \frac{\ln\left(K/S_0\left(1 + a(\tau)\right)^m\right)}{\ln\left([1 + a(\tau)]/[1 + b(\tau)]\right)},$$

and

$$p_\tau^* = \frac{r(\tau) - a(\tau)}{b(\tau) - a(\tau)}, \qquad \widetilde{p}_\tau = \frac{1 + b(\tau)}{1 + a(\tau)}\, p_\tau^*.$$

By the De Moivre-Laplace limit theorem, we have

$$B\big(k_0(\tau),\, m,\, p_\tau^*\big) \sim \Phi\left(\frac{m\, p_\tau^* - k_0(\tau)}{\sqrt{m\, p_\tau^*\left(1 - p_\tau^*\right)}}\right) = \Phi\left(y_\tau^*\right),$$

$$B\big(k_0(\tau),\, m,\, \widetilde{p}_\tau\big) \sim \Phi\left(\frac{m\, \widetilde{p}_\tau - k_0(\tau)}{\sqrt{m\, \widetilde{p}_\tau\left(1 - \widetilde{p}_\tau\right)}}\right) = \Phi\left(\widetilde{y}_\tau\right).$$

Also, for $\tau \to 0$

$$k_0(\tau) \sim \frac{\ln\left(K/S_0\right) + m\,\sigma\,\sqrt{\tau}}{2\,\sigma\,\sqrt{\tau}}, \qquad \left(1 + r(\tau)\right)^{-m} \sim e^{r\,T}.$$

Finally, taking into account relations

$$m\, p_\tau^* \sim \frac{T\,\tau\left(r - \sigma^2/2\right) + T\,\sigma\,\sqrt{\tau}}{2\,\sigma\,\tau^{3/2}},$$

$$m\, \widetilde{p}_\tau - k_0(\tau) \sim \frac{T\,\tau\left(r - \sigma^2/2\right) + \tau\,\ln(S_0/K)}{2\,\sigma\,\tau^{3/2}},$$

and

$$\sqrt{m\, p_\tau^*\left(1 - p_\tau^*\right)} \sim \sqrt{T/4\tau},$$

we obtain

$$\lim_{\tau \to 0} \frac{m\, p_\tau^* - k_0(\tau)}{\sqrt{m\, p_\tau^*\left(1 - p_\tau^*\right)}} = \frac{\ln\left(S_0/K\right) + T\left(r - \sigma^2/2\right)}{\sigma\,\sqrt{T}} = y^*,$$

$$\lim_{\tau \to 0} \frac{m\, \widetilde{p}_\tau - k_0(\tau)}{\sqrt{m\, \widetilde{p}_\tau\left(1 - \widetilde{p}_\tau\right)}} = \frac{\ln\left(S_0/K\right) + T\left(r + \sigma^2/2\right)}{\sigma\,\sqrt{T}} = \widetilde{y}.$$

Thus, we arrive at the celebrated Black-Scholes formula:

$$\lim_{\tau \to 0} C_T^\tau = C_T = S_0\, \Phi(\widetilde{y}) - K\, e^{-r\,T}\, \Phi(y^*). \tag{2.15}$$

In general, one can replace interval $[0, T]$ with $[t, T]$, where $0 \le t \le T$.

In this case, we consider a contract written at time t with time to expiry $T - t$. Replacing T by $T - t$ and S_0 by S_t in formula (2.15), we obtain the corresponding version of the Black-Scholes formula:

$$C(x,t) = S_t \Phi \left(\frac{\ln(S_t/K) + (T-t)(r + \sigma^2/2)}{\sigma \sqrt{T-t}} \right)$$

$$- K e^{-r(T-t)} \Phi \left(\frac{\ln(S_t/K) + (T-t)(r - \sigma^2/2)}{\sigma \sqrt{T-t}} \right),$$

which also indicates that price C is a function of time and price of the asset $S_t = x$.

Proposition 2.1 *Suppose that function $C(\cdot, \cdot)$ is continuously differentiable in t and twice continuously differentiable in x. Then it satisfies the Black-Scholes differential equation*

$$\frac{\partial C}{\partial t} + r x \frac{\partial C}{\partial x} + \frac{1}{2} \sigma^2 x^2 \frac{\partial^2 C}{\partial x^2} - r C = 0. \tag{2.16}$$

Proof Consider a (B, S, τ)-market with parameters

$$1 + r(\tau) = e^{r\tau}, \quad 1 + b(\tau) = e^{\sigma\sqrt{\tau}}, \quad 1 + a(\tau) = e^{-\sigma\sqrt{\tau}}, \quad \sigma > 0.$$

Using Taylor expansions for $e^{r\tau}$ and $e^{\pm \sigma \sqrt{\tau}}$, we obtain the following asymptotic expression for the martingale probability p_τ^*:

$$p_\tau^* = \frac{(e^{r\tau} - 1) - (1 - e^{-\sigma\sqrt{\tau}})}{e^{\sigma\sqrt{\tau}} - e^{-\sigma\sqrt{\tau}}} \sim \frac{1}{2} \left(1 + \frac{r}{\sigma} \sqrt{\tau} \right) \quad \text{as} \quad \tau \to 0.$$

Since prices of asset S can take only two possible values on this (B, S, τ)-market, then

$$e^{r\tau} C(x,t) = p_\tau^* C(x e^{\sigma\sqrt{\tau}}, t + \tau) + (1 - p_\tau^*) C(x e^{-\sigma\sqrt{\tau}}, t + \tau).$$

Using Taylor's formula, we can write

$$e^{r\tau} C(x,t) = (1 + r\tau) C(x,t) + o(\tau),$$
$$C(x e^{\sigma\sqrt{\tau}}, t + \tau)$$
$$= C(x,t) + \frac{\partial C}{\partial t} + x \sigma \sqrt{\tau} \frac{\partial C}{\partial x} + \frac{1}{2} \sigma^2 x^2 \tau \frac{\partial^2 C}{\partial x^2} + o(\tau),$$
$$C(x e^{-\sigma\sqrt{\tau}}, t + \tau)$$
$$= C(x,t) + \frac{\partial C}{\partial t} - x \sigma \sqrt{\tau} \frac{\partial C}{\partial x} + \frac{1}{2} \sigma^2 x^2 \tau \frac{\partial^2 C}{\partial x^2} + o(\tau)$$

for $\tau \to 0$, and hence,

$$(1 + r\tau) C(x,t) = C(x,t) + \tau \frac{\partial C}{\partial t} + x r \tau \frac{\partial C}{\partial x} + \frac{1}{2} \sigma^2 x^2 \tau \frac{\partial^2 C}{\partial x^2} + o(\tau),$$

which implies equation (2.16). $\square$

Chapter 3

Advanced Analysis of Financial Risks: Discrete Time Models

3.1 Fundamental theorems on arbitrage and completeness. Pricing and hedging contingent claims in complete and incomplete markets.

Let $(\Omega, \mathcal{F}_N, \mathbb{F}, P)$ be a stochastic basis with filtration $\mathbb{F} = (\mathcal{F}_n)_{n \leq N}$:

$$\mathcal{F}_0 = \{\emptyset, \Omega\} \subseteq \mathcal{F}_1 \subseteq \ldots \subseteq \mathcal{F}_N \,.$$

Consider a (B, S)-market with a non-risky asset B defined by a deterministic (or predictable) sequence of its prices $(B_n)_{n=0}^N$, $B_0 = 1$. A risky asset S is defined by a stochastic sequence (of prices) $(S_n)_{n=0}^N$ adopted to filtration $\mathbb{F}$.

Further, consider the sequence $(S_n/B_n)_{n=0}^N$. We say that a probability $\widetilde{P}$ is a *martingale probability* if $(S_n/B_n)_{n=0}^N$ is a martingale with respect to $\widetilde{P}$. The collection of all such probabilities is denoted $\mathcal{M}(S/B)$.

As in the case of a binomial (B, S)-market, one can consider the notions of a self-financing strategy, a portfolio, and so forth. Recall that there is an *arbitrage opportunity* in this market a self-financing strategy $\widetilde{\pi} \in SF$ exists such that $X_0^{\widetilde{\pi}} = 0$ (a.s.), $X_n^{\widetilde{\pi}} \geq 0$ (a.s.), $n = 1, \ldots, N$, and $P(\{\omega : X_N^{\widetilde{\pi}} > 0\}) > 0$.

In Sections 3.1 and 3.2, we will assume that our probability space is discrete and finite. This case is rich enough to cover the situation of multinomial markets, and it is also relatively simple from a technical point of view.

Theorem 3.1 (1st Fundamental Theorem of Financial Mathematics)
A (B, S)-market is arbitrage-free if and only if $\mathcal{M}(S/B) \neq \emptyset$.

Proof We will prove the "if" part of this statement. For simplicity, suppose that $B_n \equiv 1$ for all n. Let $\widetilde{P} \in \mathcal{M}(S/B)$, then for any self-financing strategy $\pi = (\beta, \gamma)$, its discounted value

$$X_n^\pi = X_0^\pi + \sum_{k=1}^n \gamma_k \Delta S_k$$

is a martingale with respect to $\widetilde{P}$. Recall that, in the case of binomial markets, this fact is referred to as the martingale characterization of the class SF of self-financing strategies. Now suppose that π^* is an arbitrage strategy. By definition, we have $E(X_N^{\pi^*}) > 0$. However, the martingale property of $(X_n^{\pi^*})$ implies $\widetilde{E}(X_N^{\pi^*}) = \widetilde{E}(X_0^{\pi^*}) = X_0^{\pi^*} = 0$, which contradicts the assumption that π^* is an arbitrage strategy.

The proof of the converse is technically far more complex. It can be found in various technically advanced monographs. $\square$

We say that a (B, S)-market is *complete* if every contingent claim f_N can be *replicated* by some self-financing strategy; that is, there exist $\pi \in SF$ and $x \geq 0$ such that

$$X_0^\pi = x \qquad \text{and} \qquad X_N^\pi = f_N \quad \text{(a.s.)}.$$

The sequence of discounted prices $(S_n/B_n)_{n=0,\ldots,N}$ is a martingale with respect to any probability $\widetilde{P} \in \mathcal{M}(S/B)$. In the case of a complete market, it forms a basis for the space of all martingales with respect to $\widetilde{P}$: any martingale can be written in the form of a discrete stochastic integral with respect to $(S_n/B_n)_{n=0,\ldots,N}$. This property of a market is called the *martingale representation* property.

Proposition 3.1 *A (B, S)-market is complete if and only if it possesses the martingale representation property.*

Proof For simplicity, suppose that $B_n \equiv 1$ for all n. Consider a complete (B, S)-market. Let $(X_n)_{n=0}^N$ be an arbitrary martingale and define a contingent claim by $f_N \equiv X_N$. The completeness of the market implies that $\pi \in SF$ and $x \geq 0$ exist such that

$$X_N^\pi = f_N = X_N \qquad \text{and} \qquad X_n^\pi = x + \sum_{k=1}^n \gamma_k \, \Delta S_k \quad \text{(a.s.)}.$$

Since π is a self-financing strategy, the later equality means that $(X_n)_{n=0}^N$ is a martingale with respect to any probability $\widetilde{P} \in \mathcal{M}(S/B)$. Thus, we have two martingales with the same terminal value f_N, and therefore, for all $n = 0, 1, \ldots, N$

$$X_n^\pi = \widetilde{E}(f_N|\mathcal{F}_n) = X_n,$$

which gives us a representation of X in terms of the basis martingale (S/B).

Conversely, consider a contingent claim f_N and a stochastic sequence $(X_n)_{n=0}^N$, where $X_n = \widetilde{E}(f_N|\mathcal{F}_n)$ for any fixed probability $\widetilde{P} \in \mathcal{M}(S/B)$. Then we can represent this martingale in the form

$$X_n = X_0 + \sum_{k=1}^n \phi_k \, \Delta S_k,$$

where $(\phi_k)_{k=1}^N$ is a predictable sequence. Now let

$$\gamma_n^* = \phi_n\,, \qquad \beta_n^* = X_n - \gamma_n^* S_n\,, \quad n \leq N\,.$$

Note that

$$
\begin{aligned}
\beta_n^* &= X_n - \gamma_n^* S_n = X_0 + \sum_{k=1}^{n-1} \gamma_k^* S_k + \gamma_n^* (S_n - S_{n-1}) - \gamma_n^* S_n \\
&= X_0 + \sum_{k=1}^{n-1} \gamma_k^* S_k - \gamma_n^* S_{n-1}
\end{aligned}
$$

is completely determined by the information contained in $\mathcal{F}_{n-1}$; that is, $(\beta_n^*)_{n=0}^N$ is a predictable sequence. This implies that $\pi^* = (\beta_n^*, \gamma_n^*)_{n=0}^N$ is a self-financing strategy such that for all $n = 0, 1, \ldots, N$

$$X_n^{\pi^*} = X_n \quad (a.s.)\,.$$

In particular, we obtain $X_N^{\pi^*} = X_N = f_N$; that is, an arbitrary contingent claim f_N can be replicated and the market is complete. $\square$

The essential property of complete markets is characterized in the following result.

Theorem 3.2 (2nd Fundamental Theorem of Financial Mathematics)
A (B, S)-market is complete if and only if the set $\mathcal{M}(S_n/B_n) \neq \emptyset$ consists of a unique element P^*.

Proof Consider an arbitrary event $A \in \mathcal{F}_N$ and let $f_N = I_A$. This contingent claim can be replicated: there are $x > 0$ and $\pi \in SF$ such that $X_0^\pi = x$, $X_N^\pi = f_N$, and $X_n^\pi = x + \sum_{k=1}^n \gamma_k \Delta S_k$ for all $n = 0, \ldots, N$.

If P_1, $P_2 \in \mathcal{M}(S/B)$, then $(X_n^\pi)_{n=0,\ldots,N}$ form martingales with respect to both these probabilities. Therefore,

$$x = X_0^\pi = E_i(X_N^\pi | \mathcal{F}_0) = E_i(X_N^\pi) = E_i(I_A) = P_i(A)\,, \quad i = 1, 2.$$

Hence, $P_1 = P_2$.

Now we sketch the proof of the converse. Let P^* be the unique martingale measure. Using mathematical induction, we will show that $\mathcal{F}_n = \mathcal{F}_n^S = \sigma(S_0, \ldots, S_N)$. Suppose $\mathcal{F}_{n-1} = \mathcal{F}_{n-1}^S$. Let $A \in \mathcal{F}_n$ and define a random variable

$$Z = 1 + \frac{1}{2}\left[I_A - E(I_A | \mathcal{F}_n^S)\right] > 0\,.$$

Clearly, $E^*(Z) = 1$ and $E^*(Z | \mathcal{F}_n^S) = 1$. Now define a new probability $P'(C) := E^*(Z\,I_C)$. We have

$$
\begin{aligned}
E'(\Delta S_n | \mathcal{F}_{n-1}) &= E^*(Z\Delta S_n | \mathcal{F}_{n-1}) = E^*(Z\Delta S_n | \mathcal{F}_{n-1}^S) \\
&= E^*\left(E^*(Z\Delta S_n | \mathcal{F}_{n-1}) \Big| \mathcal{F}_{n-1}\right) \\
&= E^*\left(\Delta S_n E^*(Z | \mathcal{F}_{n-1}) \Big| \mathcal{F}_{n-1}\right) \\
&= E^*(\Delta S_n | \mathcal{F}_{n-1}^S) = 0\,,
\end{aligned}
$$

which implies that P' is a martingale measure. Using the uniqueness of the martingale measure P^*, we conclude that $Z = 1$ (a.s.). Hence, $I_A = E(I_A|\mathcal{F}_n^S)$, and therefore, $\mathcal{F}_n = \mathcal{F}_n^S$.

Next consider the following conditional distributions

$$P^*(\{\omega: \rho_n \in dx\}|\mathcal{F}_{n-1}), \qquad \text{where} \qquad \rho_n = \frac{\Delta S_n}{S_{n-1}}, \qquad n = 1, \ldots, N.$$

It turns out that these distributions have the following structure: a non-positive predictable sequence $(a_n)_{n\leq N}$ and a non-negative predictable sequence $(b_n)_{n\leq N}$ exist such that

$$P^*(\{\omega: \rho_n = a_n\}|\mathcal{F}_{n-1}) + P^*(\{\omega: \rho_n = b_n\}|\mathcal{F}_{n-1}) = 1, \quad n \leq N.$$

The latter equality is implied by the following result from the general probability theory: the set of all distributions F (on real line) with the properties

$$\int_{-\infty}^{\infty} |x|\, dF(x) < \infty \quad \text{and} \quad \int_{-\infty}^{\infty} x\, dF(x) = 0,$$

consists of a unique distribution F^* if and only if there exist $a \leq 0$ and $b \geq 0$ such that $F^*(\{a\}) + F^*(\{b\}) = 1$.

Now let

$$p_n^* \quad := \quad P^*(\{\omega: \rho_n = b_n\}|\mathcal{F}_{n-1}),$$
$$1 - p_n^* \quad := \quad P^*(\{\omega: \rho_n = a_n\}|\mathcal{F}_{n-1}).$$

We have $E^*(\rho_n|\mathcal{F}_{n-1}) = 0$, or equivalently

$$b_n(\omega)\, p_n^* + a_n(\omega)\, (1 - p_n^*) = 0.$$

Thus,

$$p_n^* \quad := \quad \frac{-a_n(\omega)}{b_n(\omega) - a_n(\omega)},$$
$$1 - p_n^* \quad := \quad \frac{b_n(\omega)}{b_n(\omega) - a_n(\omega)}.$$

Now, if $(X_n, \mathcal{F}_n^S)_{n\leq N}$ is a martingale with respect to P^*, then functions exist $f_n(x_1, \ldots, x_n)$ such that

$$X_n(\omega) = f_n(\rho_1(\omega), \ldots, \rho_n(\omega)), \quad n \leq N.$$

As in the case of the binomial market, we then arrive at the following martingale representation

$$X_n = X_0 + \sum_{k=1}^{n} \gamma_k\, \Delta S_k,$$

where $(\gamma_k)_{k\leq N}$ is a predictable sequence. Since this is equivalent to the completeness of the market, the proof is completed. $\square$

Now let us discuss *general methodologies of pricing contingent claims* in complete and incomplete markets. We start with a complete (B, S)-market that admits a unique martingale measure P^*. Let f_N be a contingent claim. Note that, if the probability space $(\Omega, \mathcal{F}, P)$ is not finite, then one has to assume that $E^*(f_N/B_N) < \infty$. Consider the martingale

$$M_n^* := E^*\left(\frac{f_N}{B_N} \,\Big|\, \mathcal{F}_n\right), \quad n = 0, 1, \ldots, N,$$

which has the initial and terminal values

$$M_0^* = E^*(f_N/B_N) \quad \text{and} \quad M_N^* = f_N/B_N,$$

respectively. By Theorem 3.2, M^* has the following representation

$$M_n^* = M_0^* + \sum_{k=1}^{n} \gamma_k^* \Delta \frac{S_k}{B_k}, \quad n = 0, 1, \ldots, N,$$

where $(\gamma_k)_{k\leq N}$ is a predictable sequence.

Define a strategy $\pi^* = (\beta_n^*, \gamma_n^*)_{n\leq N}$ with $\beta_n^* = M_n^* - \gamma_n^* S_n/B_n$. Then

$$\beta_n^* = M_0^* + \sum_{k=1}^{n} \gamma_k^* \Delta \frac{S_k}{B_k} - \gamma_n^* \frac{S_n}{B_n} = M_0^* + \sum_{k=1}^{n-1} \gamma_k^* \Delta \frac{S_k}{B_k}$$

is a predictable sequence. Hence, we have constructed a self-financed strategy $\pi^* \in SF$ with values given by

$$\frac{X_0^{\pi^*}}{B_0} = M_0^*,$$

$$\Delta \frac{X_n^{\pi^*}}{B_n} = \gamma_n^* \frac{S_n}{B_n} = \Delta M_n^*, \quad n \leq N,$$

$$\frac{X_n^{\pi^*}}{B_n} = M_n^* = E^*\left(\frac{f_N}{B_N} \,\Big|\, \mathcal{F}_n\right),$$

and, in particular, $f_N = X_N^{\pi^*}$ (a.s.).

Thus, the following result is proved.

Theorem 3.3 (Pricing Contingent Claims in Complete Markets)
Let f_N be a contingent claim in a complete (B, S)-market. Then there exists a self-financing strategy $\pi^* = (\beta^*, \gamma^*)$, which is a minimal hedge with the values

$$X_n^{\pi^*} = B_n E^*\left(\frac{f_N}{B_N} \,\Big|\, \mathcal{F}_n\right),$$

and β^*, γ^* are defined by relations

$$E^*\left(\frac{f_N}{B_N}\Big|\mathcal{F}_n\right) = E^*\left(\frac{f_N}{B_N}\right) + \sum_{k=1}^{n} \gamma_k^* \Delta \frac{S_k}{B_k},$$

$$X_n^{\pi^*} = \beta_n^* B_n + \gamma_n^* S_n.$$

In particular, the price of f_N is

$$C_N = E^*\left(\frac{f_N}{B_N}\right).$$

Note that the fact that π^* is the *minimal* hedge follows from the following inequalities:

$$\frac{X_n^\pi}{B_n} = E^*\left(\frac{X_N^\pi}{B_N}\Big|\mathcal{F}_n\right) \geq E^*\left(\frac{f_N}{B_N}\Big|\mathcal{F}_n\right) = \frac{X_n^{\pi^*}}{B_n} \quad \text{(a.s.)}, \quad n = 0, 1, \dots, N,$$

for any other $\pi \in SF$ hedging f_N.

Now let us consider incomplete markets. In this case, not every contingent claim can be replicated by self-financing strategies. Consider a strategy $\pi = (\beta_n, \gamma_n)_{n \leq N}$ that is not necessarily self-financing. We can write

$$\begin{aligned}
\Delta X_n^\pi &= \beta_n \Delta B_n + \gamma_n \Delta S_n + B_{n-1} \Delta \beta_n + S_{n-1} \Delta \gamma_n \\
&= \beta_n \Delta B_n + \gamma_n \Delta S_n - \Delta c_n,
\end{aligned}$$

where

$$-\Delta c_n := B_{n-1} \Delta \beta_n + S_{n-1} \Delta \gamma_n \quad n = 1, \dots, N; \quad c_0 = 0.$$

Let $c = (c_n)_{n \leq N}$ be a non-decreasing stochastic sequence (consumption process). A class of strategies (π, c) is called *consumption strategies*. Clearly, we have

$$\begin{aligned}
X_n^{\pi,c} &= X_0^{\pi,c} + \sum_{k=1}^{n} \left(\beta_k \Delta B_k + \gamma_k \Delta S_k\right) - c_n \\
&= X_0^{\pi,c} + \sum_{k=1}^{n} \left(\beta_k' \Delta B_k + \gamma_k \Delta S_k\right),
\end{aligned}$$

where

$$\beta_k' := \beta_k - \frac{\Delta c_k}{\Delta B_k}$$

is not necessarily predictable since the consumption c_n is determined by the information in $\mathcal{F}_n \supseteq \mathcal{F}_{n-1}$.

The discounted value of a consumption strategy (π, c) has the following dynamics:

$$\Delta \frac{X_n^{\pi,c}}{B_n} = \gamma_n \Delta \frac{S_n}{B_n} + \frac{\Delta c_n}{B_{n-1}},$$

which follows from the fact that

$$X_n^{\pi,c} = \beta_n \, B_n + \gamma_n \, S_n \quad \text{and} \quad X_{n-1}^{\pi,c} = \beta_n \, B_{n-1} + \gamma_n \, S_{n-1} + \Delta c_n \,.$$

Now let f_N be a contingent claim in an incomplete (B, S)-market. If it is possible to find a consumption strategy (π^*, c^*) that replicates f_N, then the value of this strategy will be the natural choice for the price of f_N.

Consider the following stochastic sequence

$$Y_n \;=\; \sup_{\widetilde{P} \in \mathcal{M}(S/B)} \widetilde{E}\!\left(\frac{f_N}{B_N}\Big|\mathcal{F}_n\right), \quad n = 1,\ldots,N, \tag{3.1}$$

$$Y_0 \;=\; \sup_{\widetilde{P} \in \mathcal{M}(S/B)} \widetilde{E}\!\left(\frac{f_N}{B_N}\right),$$

$$Y_N \;=\; \frac{f_N}{B_N} \quad \text{(a.s.)}.$$

Note that one has to assume that

$$\sup_{\widetilde{P} \in \mathcal{M}(S/B)} \widetilde{E}\!\left(\frac{f_N}{B_N}\right) < \infty,$$

which is obviously satisfied in the case of discrete markets.

It turns out that this sequence $\left(Y_n\right)_{n \leq N}$ is a positive supermartingale with respect to any probability $\widetilde{P} \in \mathcal{M}(S/B)$. Therefore, for a fixed $\widetilde{P}$ we can write the Doob decomposition:

$$Y_n = Y_0 + \widetilde{M}_n - \widetilde{A}_n \,, \quad n \leq N,$$

where $\widetilde{M}$ is a martingale with respect to $\widetilde{P}$, and $\widetilde{A}$ is a non-decreasing predictable sequence. Clearly, this decomposition depends on the choice of $\widetilde{P}$. In this case, one can find the following *uniform* or *optional* decomposition (which is analogous to Doob decomposition), which is invariant on the class of martingale measures $\mathcal{M}(S/B)$:

$$Y_n = Y_0 + M'_n - c'_n \,, \quad n = 0,\ldots,N; \ M'_0 = c'_0 = 0,$$

where M' is a martingale with respect to any probability from $\mathcal{M}(S/B)$, and c' is a non-decreasing (but not necessarily predictable) stochastic sequence.

Furthermore, M' has the following representation:

$$M'_n = \sum_{k=1}^{n} \gamma'_k \, \Delta \frac{S_k}{B_k} \,,$$

where (γ'_k) is a predictable sequence.

Now we define a consumption strategy:

$$\gamma_n^* = \gamma'_n, \quad \beta_n^* = Y_n - \gamma_n^* \frac{S_n}{B_n}, \quad c_n^* = \sum_{k=1}^{n} B_{k-1} \Delta c'_k \,.$$

We have

$$X_0^{\pi^*,c^*} = Y_0 = \sup_{\tilde{P} \in \mathcal{M}(S/B)} \tilde{E}\left(\frac{f_N}{B_N}\right),$$

$$\Delta \frac{X_n^{\pi^*,c^*}}{B_n} = \gamma_n^* \Delta \frac{S_n}{B_n} - \frac{\Delta c_n^*}{B_{n-1}} = \Delta Y_n, \quad n \leq N.$$

Thus,

$$\frac{X_N^{\pi^*,c^*}}{B_N} = Y_N = \frac{f_N}{B_N} \quad \text{(a.s.)},$$

which means that f_N is replicated by the consumption strategy (π^*, c^*).

We have almost proved the following result.

Theorem 3.4 (Pricing Contingent Claims in Incomplete Markets)
Let f_N be a contingent claim in an incomplete (B, S)-market. Then there exists a consumption strategy (π^*, c^*), which is a minimal hedge with the values

$$X_n^{\pi^*,c^*} = B_n \sup_{\tilde{P} \in \mathcal{M}(S/B)} \tilde{E}\left(\frac{f_N}{B_N} \,\Big|\, \mathcal{F}_n\right),$$

where β^*, γ^*, and c^* are defined from the optional decomposition of the positive supermartingale Y (3.1):

$$Y_n = \sup_{\tilde{P} \in \mathcal{M}(S/B)} \tilde{E}\left(\frac{f_N}{B_N}\right) + \sum_{k=1}^{n} \gamma_k^* \Delta \frac{S_k}{B_k} - \sum_{k=1}^{n} \frac{\Delta c_k^*}{B_{k-1}},$$

$$\beta_n^* = \frac{X_n^{\pi^*,c^*} - \gamma_n^* S_n}{B_n}.$$

In particular, the initial (upper) price of f_N can be defined as

$$C_N^* = \sup_{\tilde{P} \in \mathcal{M}(S/B)} \tilde{E}\left(\frac{f_N}{B_N}\right).$$

Proof We only need to show that the hedge (π^*, c^*) is the minimal hedge. Let (π, c) be an arbitrary consumption strategy hedging f_N. Then, for any $\tilde{P} \in \mathcal{M}(S/B)$, we have

$$\frac{X_n^{\pi,c}}{B_n} \geq \tilde{E}\left(\frac{X_N^{\pi,c}}{B_N} \,\Big|\, \mathcal{F}_n\right) \geq \tilde{E}\left(\frac{f_N}{B_N} \,\Big|\, \mathcal{F}_n\right), \quad n \leq N;$$

therefore, for all $n \leq N$,

$$\frac{X_n^{\pi,c}}{B_n} \geq \sup_{\tilde{P} \in \mathcal{M}(S/B)} \tilde{E}\left(\frac{f_N}{B_N} \,\Big|\, \mathcal{F}_n\right) = \frac{X_n^{\pi^*,c^*}}{B_n} \quad \text{(a.s.)},$$

which proves the claim. $\Box$

3.2   The structure of options prices in incomplete markets and in markets with constraints.

Consider an incomplete (B, S)-market. As we noted in the previous section, there may be more than one risk-neutral probability $\widetilde{P}$, and therefore the quantity $\widetilde{E}(f_N/B_N)$ is not unique. In this section, we discuss arbitrage-free pricing of a contingent claim f_N in the context of an incomplete market.

It is intuitively clear that any number from the interval

$$\left[\min_{\widetilde{P}} \widetilde{E}(f_N/B_N) , \ \max_{\widetilde{P}} \widetilde{E}(f_N/B_N) \right]$$

can be considered as an arbitrage-free price of a contingent claim f_N.

However, if we denote $X_N^\pi(x)$ the terminal value of a strategy with the initial value x, then we can define quantities

$$C^* = \min\{x : X_N^\pi(x) \geq f_N \ \text{ for some } \ \pi \in SF\},$$
$$C_* = \max\{x : X_N^\pi(x) \leq f_N \ \text{ for some } \ \pi \in SF\}.$$

If the set of all risk-neutral measures $\widetilde{P}$ consists of only one measure P^*, then a hedge π^* exists, where the initial value is $C_N = E^*(f_N/B_N)$ and the terminal value is $X_N^{\pi^*}(C_N) = f_N$. In this case, $C^* = C_* = C$.

In general, $C_* \leq C^*$ and the interval $[C_*, C^*]$ contains all possible arbitrage-free prices for f_N, that is, prices that are not risk-free for both parties involved in the contract. Note that intervals $(0, C_*)$ and (C^*, ∞) represent an option's arbitrage prices for the buyer and the seller, respectively (see Figure 3.1).

For example, if $x > C^*$, then the seller of the option can use $y \in (C^*, x)$ for building a strategy π^* with values $X_0^{\pi^*} = y$ and $X_N^{\pi^*} \geq f_N$, which is possible by the definition of C^*. Then

$$(x - f_N) + (X_N^{\pi^*} - y) = (x - y) + (X_N^{\pi^*} - f_N) \geq x - y > 0$$

is a risk-free profit for the seller.

It turns out that the intervals $[C_*, C^*]$ and

$$\left[\min_{\widetilde{P}} \widetilde{E}(f_N/B_N) , \ \max_{\widetilde{P}} \widetilde{E}(f_N/B_N) \right]$$

are the same, which gives a method of managing risks associated with a contingent claim f_N even in the case of incomplete markets.

Now let us describe *super-hedging*, which is an effective methodology for deriving upper and lower prices C^* and C_*. Given a contingent claim f_N of a possibly rather complex structure, one can consider a dominating claim

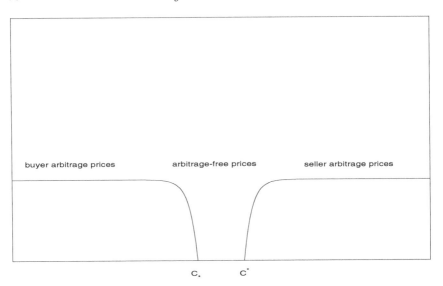

FIGURE 3.1: Arbitrage-free prices.

$\tilde{f}_N \geq f_N$ (a.s.) that can be replicated by a self-financing strategy. Then the initial value of such a strategy can be taken as a *super-price* of f_N, which naturally may be higher than required. Next, for any martingale probability $\tilde{P}$, we have $\tilde{E}(\tilde{f}_N) \geq \tilde{E}(f_N)$, which, by definition of C^* and C_*, implies that the quantities in the latter inequality coincide with the upper and lower super-prices, respectively.

We now use the European call option to illustrate this result. We have $f_N = (S_N - K)^+$. Since S_N is non-negative, then $(S_N - K)^+ \leq S_N$. Using Jensen's inequality and the fact that $(S_n/B_n)_{n=0,\ldots,N}$ is a martingale with respect to any martingale probability $\tilde{P}$, we obtain

$$\left(S_0 - \frac{K}{B_N}\right)^+ = \left(\tilde{E}\left(\frac{S_N}{B_N}\right) - \frac{K}{B_N}\right)^+ \leq \tilde{E}\left(\frac{S_N - K}{B_N}\right)^+ \leq \tilde{E}\left(\frac{S_N}{B_N}\right)$$

$$= \tilde{E}(S_0) = S_0 .$$

Thus,

$$\left(S_0 - \frac{K}{B_N}\right)^+ \leq C_* \leq C^* \leq S_0 ,$$

where because of the properties of the market, the first and the last inequalities become equalities and give us the lower and the upper prices of the option. The quantity $C^* - C_*$, called the *spread*, is a measure of market incompleteness.

Note that complete (B, S)-markets give an idealistic model of real financial markets. Incomplete markets can be regarded as a step toward more realistic models. A further step consists of introducing *markets with constraints*.

Now we consider one of the simplest models of this type. We refer to it as a (B^1, B^2, S)-market:

$$\Delta B_n^i = r^i\, B_{n-1}^i, \quad B_0^i = 1, \quad i = 1, 2,$$
$$\Delta S_n = \rho_n\, S_{n-1}, \quad S_0 \geq 0,$$
$$-1 < a < r^1 \leq r^2 < b,$$

where (ρ_n) is a sequence of independent random variables (representing profitability or return of asset S) that take values b and a with probabilities p and $1 - p$, respectively.

Assets B^1 and B^2 can be interpreted as *saving* and *credit* accounts and S represents shares. It is natural to assume that $r^1 \leq r^2$ in order to avoid the obvious arbitrage opportunity in the market. If $r^1 = r^2$, then $B^1 = B^2$ and we arrive to a (B, S)-market.

A strategy (portfolio) $\pi = (\pi_n)_{n \leq N}$ in a (B^1, B^2, S)-market is defined by three predictable sequences $(\beta_n^1, \beta_n^2, \gamma_n)_{n \leq N}$. The values of this strategy are

$$X_n^\pi = \beta_n^1\, B_n^1 + \beta_n^2\, B_n^2 + \gamma_n\, S_n.$$

A strategy π is *self-financing* if

$$\Delta X_n^\pi = \beta_n^1\, \Delta B_n^1 + \beta_n^2\, \Delta B_n^2 + \gamma_n\, \Delta S_n.$$

A strategy is *admissible* if its values are always non-negative. If credit and saving accounts have different rates of interest, then this creates an arbitrage opportunity. To avoid this, we assume that $\beta_n^1 \geq 0$ and $\beta_n^2 \leq 0$.

A strategy π will be identified with the corresponding proportion of risky capital $\alpha_n = \gamma_n\, S_{n-1}/X_{n-1}^\pi$. Let $(1 - \alpha_n)^+$ and $-(1 - \alpha_n)^-$ represent an investor's deposits in savings and credit accounts, respectively. Then the dynamics of the values of such an admissible strategy are described by

$$\Delta X_n^\pi(x) = X_{n-1}^\pi(x)\big[(1 - \alpha_n)^+\, r^1 - (1 - \alpha_n)^-\, r^2 + \alpha_n\, \rho_n\big],$$
$$X_0^\pi = x > 0.$$

Recall that, in a complete arbitrage-free (B, S)-market, contingent claims can be priced uniquely. In the case of incomplete markets, one can find an interval $[C_*, C^*]$ of arbitrage-free prices. The following methodology demonstrates that a similar result holds true in the case of (B^1, B^2, S)-markets.

Suppose that f_N is a contingent claim in a (B^1, B^2, S)-market. We will introduce an auxiliary complete market and find conditions that will guarantee that strategies with the same proportion of risky capital will have equal values in both markets. Let $d \in [0, r^2 - r^1]$. Define a (B^d, S)-market:

$$\Delta B_n^d = r^d\, B_{n-1}^d, \quad B_0^d = 1,$$
$$\Delta S_n = \rho_n\, S_{n-1}, \quad S_0 \geq 0,$$

with an interest rate $r^d = r^1 + d$.

Since the (B^d, S)-market is complete, the price of a contingent claim f_N is uniquely determined by the initial value of the minimal hedge:

$$C_N(f, r^d) = E^d \left(\frac{f_N}{B_N^d} \right),$$

where expectation is taken with respect to a martingale probability P^d in the market with the interest rate r^d.

Now let $\alpha = (\alpha_n)_{n \le N}$ be the proportion of risky capital, and $\pi(\alpha)$ and $\pi(\alpha, d)$ be the corresponding strategies in the (B^1, B^2, S)-market and (B^d, S)-market, respectively.

Lemma 3.1 *Suppose* $X_0^{\pi(\alpha)} = X_0^{\pi(\alpha,d)}$. *Then* $X_n^{\pi(\alpha)} = X_n^{\pi(\alpha,d)}$ *for all* $n \le N$ *if and only if*

$$(r^2 - r^1 - d)(1 - \alpha_n)^- + d(1 - \alpha_n)^+ = 0 \qquad (3.2)$$

for all $n \le N$.

Proof The dynamics of $X_n^{\pi(\alpha,d)}$ in the (B^d, S)-market are given by the following recurrence relation

$$
\begin{aligned}
\Delta X_n^{\pi(\alpha,d)} &= \beta_n^d \, \Delta B_n^d + \gamma_n \, \Delta S_n = \beta_n^d \, r^d \, B_{n-1}^d + \gamma_n \, \rho_n \, S_{n-1} \\
&= r^d \left(\beta_n^d \, B_{n-1}^d + \gamma_n \, S_{n-1} \right) + \gamma_n \left(\rho_n - r^d \right) S_{n-1} \\
&= r^d \, X_{n-1}^{\pi(\alpha,d)} + \gamma_n \left(\rho_n - r^d \right) S_{n-1} \\
&= r^d \, X_{n-1}^{\pi(\alpha,d)} + \alpha_n \left(\rho_n - r^d \right) X_{n-1}^{\pi(\alpha,d)} \\
&= X_{n-1}^{\pi(\alpha,d)} \left((1 - \alpha_n) \, r^d + \alpha_n \, \rho_n \right).
\end{aligned}
$$

Similarly, in the (B^1, B^2, S)-market, we obtain

$$\Delta X_n^{\pi(\alpha)} = X_{n-1}^{\pi(\alpha)} \left((1 - \alpha_n)^+ \, r^1 - (1 - \alpha_n)^- \, r^2 + \alpha_n \, \rho_n \right),$$

which proves the claim. $\square$

This result suggests the following methodology of pricing contingent claims f_N in a (B^1, B^2, S)-market. For any $d \in [0, r^2 - r^1]$, we consider a (B^d, S)-market, where one can use the initial value of the minimal hedge as a price $C_N(f, r^d)$ for f_N. Then quantities

$$\min_d C_N(f, r^d) \qquad \text{and} \qquad \max_d C_N(f, r^d)$$

are obvious natural candidates for lower and upper prices of f_N in the (B^1, B^2, S)-market.

Now we apply this methodology for pricing a European call option, that is, $f_N = (S_N - K)^+$. The Cox-Ross-Rubinstein formula gives us prices $C_N(f, r^d)$ for $r^d \in [r^1, r^2]$. Also, it is clear that the function $C_N(f, \cdot)$ is increasing on

$[r^1, r^2]$. Thus, the lower and upper prices in the (B^1, B^2, S)-market can be computed by applying the Cox-Ross-Rubinstein formula in (B^d, S)-markets with interest rates r^1 (in this case $d = 0$) and r^2 $(d = r^2 - r^1)$, respectively:

$$C_N(r^i) = S_0 \, B(k_0, N, \tilde{p}_i) - K \, (1 + r^i)^{-N} \, B(k_0, N, p_i^*),$$

$$p_i^* = \frac{r^i - a}{b - a}, \quad \tilde{p}_i = \frac{1 + b}{1 + r^i} \, p_i^*, \quad i = 1, 2.$$

Note that prices $C_N(r^1)$ and $C_N(r^2)$ illustrate the difference of interests of a buyer and a seller in a (B^1, B^2, S)-market. Price $C_N(r^2)$ is attractive to a buyer because it is the minimal price of the option that guarantees the terminal payment. Price $C_N(r^1)$ reflects the intention of a seller to keep the option as an attractive investment instrument for a buyer.

Worked Example 3.1 *Consider a (B^1, B^2, S)-market with $r^1 = 0$ and $r^2 = 0.2$. Suppose $S_0 = 100(\$)$ and*

$$S_1 = \begin{cases} 150 \, (\$) & \text{with probability } p = 0.4 \\ 70 \, (\$) & \text{with probability } 1 - p = 0.6. \end{cases}$$

Find the upper and lower prices for a European call option $f_1 = (S_1 - K)^+ \equiv \max\{0, S_1 - K\}$ with strike price $K = 100 \, (\$)$.

Solution From the Cox-Ross-Rubinstein formula, we have

$$
\begin{aligned}
C_1(0) &= S_0 \frac{r^1 - a}{b - a} \frac{1 + b}{1 + r^1} - K \, (1 + r^1)^{-1} \frac{r^1 - a}{b - a} \\
&= 100 \frac{0.3}{0.8} \frac{1.5}{1} - 100 \, (1)^{-1} \frac{0.3}{0.8} \approx 19,
\end{aligned}
$$

$$
C_1(0.2) = 100 \frac{0.5}{0.8} \frac{1.5}{1.2} - 100 \, (1.2)^{-1} \frac{0.5}{0.8} \approx 26.
$$

Thus, the *spread* in such (B^1, B^2, S)-market is equal to $C_1(0.2) - C_1(0) = 26 - 19 = 7$.

Now consider the same market with $r^1 = 0.1$ and $r^2 = 0.2$. Then compute

$$
C_1(0.1) = 100 \frac{0.3}{0.8} \frac{1.5}{1.1} - 100 \, (1.1)^{-1} \frac{0.7}{0.8} \approx 22,
$$

and the spread in this case is $C_1(0.2) - C_1(0.1) = 26 - 22 = 4$. $\square$

Note that in this example the condition (3.2) from Lemma 3.1 is satisfied, and the example illustrates that, if the gap $r^2 - r^1$ between the interest rates on saving and credit accounts becomes smaller, then the spread decreases. Spread can be regarded as a measure of proximity of (B^1, B^2, S)-market to an ideal complete market.

Next, let us consider the problem of finding an optimal strategy (a strategy that maximizes the logarithmic utility function) in a (B^1, B^2, S)-market. According to Lemma 3.1, it is equivalent to solving an optimization problem in a complete (B^d, S)-market. The optimal proportion is given by

$$\alpha_n \equiv \frac{(1 + r^d)(\mu - r^d)}{(r^d - a)(b - r^d)}, \qquad d \in [0, r^2 - r^1].$$

For the boundary values $d = 0$ and $d = r^2 - r^1$, we have

$$\alpha^{(i)} = \frac{(1 + r^{(i)})(\mu - r^{(i)})}{(r^{(i)} - a)(b - r^{(i)})}, \qquad i = 1, 2,$$

given that $\alpha^{(1)} \leq 1$ and $\alpha^{(2)} \geq 1$.

In Worked Example 3.1 with $r^1 = 0$ and $r^2 = 0.2$, we compute

$$\alpha^{(1)} = \frac{0.02}{0.3 \cdot 0.5} \approx 0.13 \leq 1, \qquad \alpha^{(2)} = \frac{-1.02 \cdot 0.18}{0.5 \cdot 0.3} < 1.$$

Thus, the optimal proportion is 0.13, and the rest of the capital must be invested in a savings account with interest rate r^1.

As another step in studying more realistic market models, let us now consider the notion of a *transaction cost*. Consider the binomial model of a (B, S)-market:

$$\begin{aligned} \Delta B_n &= r B_{n-1}, & B_0 &= 1, \\ \Delta S_n &= \rho_n S_{n-1}, & S_0 &> 0, \end{aligned}$$

where $r \geq 0$ is a constant rate of interest with $-1 < a < r < b$, and $1 \leq n \leq N$.

Now we suppose that any transaction of capital from one asset to another attracts a fee or a *transaction cost* (with a fixed parameter $\lambda \in [0, 1]$): a buyer of asset S pays $S_n(1 + \lambda)$ at time n, and a seller receives $S_n(1 - \lambda)$ accordingly.

Recall that a writer of a European call option is obliged to sell at time N one unit of asset S at a fixed price K. After receiving a premium x, the writer hedges the corresponding contingent claim by redistributing the capital between assets B and S in proportion (β, γ).

Suppose that, at terminal time N, both sell and buy prices are equal to S_N. Then the contingent claim corresponding to this option can be represented in an appropriate two-component form:

$$f = (f^1, f^2) = \begin{cases} (-K/B_N, 1) & \text{if } S_N > K \\ (0, 0) & \text{if } S_N \leq K, \end{cases}$$

where f^1 and f^2 represent the number of bonds and shares, respectively, necessary for making the repayment.

We claim that this model admits a unique "fair" arbitrage-free price C_N for such an option. First, we describe a transition from portfolio $\pi = (\beta, \gamma)$ to portfolio $\pi' = (\beta', \gamma')$ at some time $n \leq N$. Clearly, there are two cases in a situation when buying and selling shares attract transaction costs:

1. If $\gamma > \gamma'$, then we have to sell $\gamma - \gamma'$ shares and use the received capital for buying the corresponding number of bonds. This leads to the following condition:

$$(\beta' - \beta)\, B_n = (\gamma - \gamma')\, S_n\, (1 - \lambda)\,.$$

2. If $\gamma < \gamma'$, then we arrive at condition

$$(\beta - \beta')\, B_n = (\gamma' - \gamma)\, S_n\, (1 + \lambda)\,.$$

Combining these conditions results in

$$(\beta' - \beta)\, B_n + (\gamma' - \gamma)\, S_n = -\lambda\, |\gamma' - \gamma|\, S_n\,.$$

Our claim follows from the following theorem.

Theorem 3.5 (Boyle-Vorst) *In the framework of a binomial (B, S)-market with transaction costs, for any European call option, there exists a unique replicating strategy. This strategy coincides with the strategy that replicates the same option in a (complete) binomial market without transaction costs, where values $\bar{b}$ and $\bar{a}$ of the profitability sequence ρ are defined by*

$$1 + \bar{b} = (1 + b)\,(1 + \lambda) \qquad and \qquad 1 + \bar{a} = (1 + a)\,(1 - \lambda)\,.$$

Proof We use the method of backward induction. First, let us introduce the following useful notations for values of $\mathcal{F}_n$-measurable quantities β_{n+1} and γ_{n+1} on sets $\{\omega :\ \rho_n = b\}$ and $\{\omega :\ \rho_n = a\}$:

$$\beta^b_{n+1}(\rho_1, \ldots, \rho_{n-1}) = \beta_{n+1}(\rho_1, \ldots, \rho_{n-1}, b)\,,$$
$$\beta^a_{n+1}(\rho_1, \ldots, \rho_{n-1}) = \beta_{n+1}(\rho_1, \ldots, \rho_{n-1}, a)\,,$$
$$\gamma^b_{n+1}(\rho_1, \ldots, \rho_{n-1}) = \gamma_{n+1}(\rho_1, \ldots, \rho_{n-1}, b)\,,$$
$$\gamma^a_{n+1}(\rho_1, \ldots, \rho_{n-1}) = \gamma_{n+1}(\rho_1, \ldots, \rho_{n-1}, a)\,.$$

Then redistribution of capital can be expressed in the form:

$$(\beta^b_{n+1} - \beta_n)\, B_n + (\gamma^b_{n+1} - \gamma_n)\, S_{n-1}\,(1 + b) = -\lambda\, |\gamma^b_{n+1} - \gamma_n|\, S_{n-1}\,(1 + b)\,,$$
$$\tag{3.3}$$
$$(\beta^a_{n+1} - \beta_n)\, B_n + (\gamma^a_{n+1} - \gamma_n)\, S_{n-1}\,(1 + a) = -\lambda\, |\gamma^a_{n+1} - \gamma_n|\, S_{n-1}\,(1 + a)\,.$$

Subtracting the second equation from the first, we define function

$$
\begin{aligned}
g(\gamma_n) \;=\;\; & \gamma_n\, S_{n-1}\,(b - a) - \gamma^b_{n+1}\, S_{n-1}\,(1 + b) + \gamma^a_{n+1}\, S_{n-1}\,(1 + a) \\
& - \beta^b_{n+1}\, B_n + \beta^a_{n+1}\, B_n \\
& - \lambda\, |\gamma_n - \gamma^b_{n+1}|\, S_{n-1}\,(1 + b) + \lambda\, |\gamma_n - \gamma^a_{n+1}|\, S_{n-1}\,(1 + a)\,.
\end{aligned}
$$

Thus, the problem of finding β_n and γ_n given values of β_{n+1} and γ_{n+1}, becomes a question of the solvability of system (3.3), or equivalently, of finding the number of zeros of function $g(\gamma_n)$. Note that this function is continuous and linear on intervals $(-\infty, \gamma^a_{n+1})$, $(\gamma^a_{n+1}, \gamma^b_{n+1})$, and (γ^b_{n+1}, ∞) with positive derivatives equal to

$$\left[(1+\lambda)(1+b) - (1+\lambda)(1+a) \right] S_{n-1},$$

$$\left[(1+\lambda)(1+b) - (1-\lambda)(1+a) \right] S_{n-1},$$

and

$$\left[(1-\lambda)(1+b) - (1-\lambda)(1+a) \right] S_{n-1},$$

respectively.

Hence, $g(\gamma_n)$ is a strictly monotone, continuous, piece-wise linear function, which implies that there is a unique solution to equation $g(\gamma_n) = 0$. Now we have to show that $\gamma_n \in [\gamma^a_{n+1}, \gamma^b_{n+1}]$, or $g(\gamma^a_{n+1}) \le 0$ and $g(\gamma^b_{n+1}) \ge 0$. It is clear that

$$g(\gamma^a_{n+1}) = \left(\gamma^a_{n+1} - \gamma^b_{n+1} \right) S_{n-1}(1+b)(1+\lambda) - B_n\,\beta^b_{n+1} + B_n\,\beta^a_{n+1}.$$

Taking into account

$$\gamma^{ba}_{n+2} \le \gamma^b_{n+1} \le \gamma^{bb}_{n+2} \quad \text{and} \quad \gamma^{aa}_{n+2} \le \gamma^a_{n+1} \le \gamma^{ab}_{n+2} = \gamma^{ba}_{n+2},$$

we can rewrite equations (3.3) in the form

$$\left(\beta^{ba}_{n+2} - \beta^b_{n+1} \right) B_n(1+r) \; + \; \left(\gamma^{ba}_{n+2} - \gamma^b_{n+1} \right) S_{n-1}(1+b)(1+a)$$
$$= \lambda \left(\gamma^b_{n+1} - \gamma^{ba}_{n+2} \right) S_{n-1}(1+b)(1+a),$$

and

$$\left(\beta^{ba}_{n+2} - \beta^a_{n+1} \right) B_n(1+r) \; + \; \left(\gamma^{ba}_{n+2} - \gamma^a_{n+1} \right) S_{n-1}(1+b)(1+a)$$
$$= \lambda \left(\gamma^{ba}_{n+2} - \gamma^b_{n+1} \right) S_{n-1}(1+b)(1+a),$$

respectively. Subtracting the second equation from the first, we obtain

$$\left(\beta^a_{n+1} - \beta^b_{n+1} \right) B_n(1+r) = \left(\gamma^b_{n+1} - \gamma^a_{n+1} \right) S_{n-1}(1+a)(1+b)(1-\lambda),$$

and hence,

$$g(\gamma^a_{n+1}) = \left(\gamma^a_{n+1} - \gamma^b_{n+1} \right) S_{n-1}(1+b) \left[(1+\lambda) - \frac{(1+a)(1-r)}{1+r} \right] \le 0,$$

since $\gamma^b_{n+1} \ge \gamma^a_{n+1}$ and $a \le r$.

Similarly, one can prove that $g(\gamma^b_{n+1}) \ge 0$. To complete the proof, we need to check the base of induction. At the terminal time, there are two possible types of portfolios. First, $\gamma_{N+1} = 1$, $\beta_{N+1} = -K/B_n$; second, $\gamma_{N+1} = \beta_{N+1} = 0$. Note that in both cases $\gamma^b_{N+1} \ge \gamma^a_{N+1}$. Suppose $\gamma^b_{N+1} = \gamma^a_{N+1}$, then $\gamma_N = \gamma_{N+1}^b$, $\beta_N = -K/B_N$ is a unique solution of system (3.3), and $\gamma_{N+1}^a \leq \gamma_N \leq \gamma_{N+1}^b$.

If $\gamma_{N+1}^b = 1$ and $\gamma_{N+1}^a = 0$, then the unique solution has the form

$$\gamma_N = \frac{S_{N-1}(1+\bar{b}) - K}{S_{N-1}(1+\bar{b}) - S_{N-1}(1+\bar{a})}$$

with $\gamma_{N+1}^a = 0 < \gamma_N < 1 = \gamma_{N+1}^b$.

The case when $\gamma_{N+1}^b = \gamma_{N+1}^a = 0$ is trivial:

$$\gamma_N = \beta_N = 0, \qquad \gamma_{N+1}^a \leq \gamma_N \leq \gamma_{N+1}^b. \quad \square$$

3.3 Hedging contingent claims in mean square

Consider an incomplete (B, S)-market with the time horizon N. As we discussed above, for perfect hedging of contingent claims on such markets, one has to consider strategies with consumption.

An alternative approach to hedging of contingent claims was suggested by Föllmer and Sondermann. It is a combination of the ideas of hedging and of investment portfolio with the quadratic utility function.

First, we consider a one-step model. Let H be the discounted value of a contingent claim f_1. At time $n = 0$, the seller of the claim forms a portfolio $\pi_0 = (\beta_0, \gamma_0)$ with value

$$X_0^\pi = \beta_0 B_0 + \gamma_0 S_0,$$

and the discounted value

$$V_0^\pi = \frac{X_0^\pi}{B_0} = \beta_0 + \gamma_0 \frac{S_0}{B_0} = \beta_0 + \gamma_0 X_0,$$

where $X = S/B$.

At time $n = 1$, we replace β_0 with β_1, so that the value of the portfolio becomes

$$V_1^\pi = \beta_1 + \gamma_0 X_1,$$

where β_1 is determined by the replication condition:

$$V_1^\pi = H \quad \text{or} \quad X_1^\pi = f_1.$$

Thus, for finding an "optimal" strategy $\hat{\pi}$, one has to determine $\gamma_0 = \gamma$. We define the following *price sequence* C^π:

$$C_0^\pi = V_0^\pi, \quad C_1^\pi = V_1^\pi - \gamma(X_1 - X_0) = H - \gamma \Delta X_1.$$

This choice has an obvious interpretation: the amount $V_1^\pi = H$ must be paid to the holder of an option, and $\gamma \Delta X_1$ is the gain-loss implied by strategy γ. To determine an optimal γ, one has to solve the following optimization problem: find $\hat{\pi}$ such that

$$Var(C_1^{\hat{\pi}}) = \inf_\pi Var(C_1^\pi).$$

Further, if $Var(\Delta X_1) > 0$, then the variance

$$Var(C_1^{\hat{\pi}}) = Var(H) - 2\gamma\, Cov(H, \Delta X_1) + \gamma^2\, Var(\Delta X_1)$$

attains its unique minimum at the point

$$\hat{\gamma} = \frac{Cov(H, \Delta X_1)}{Var(\Delta X_1)},$$

and therefore,

$$
\begin{aligned}
Var(C_1^{\hat{\pi}}) &= Var(H - \hat{\gamma}\,\Delta X_1) = Var(H) - \frac{Cov^2(H, \Delta X_1)}{Var(\Delta X_1)} \\
&= Var(H)\left[1 - Cov^2(H, \Delta X_1)\right].
\end{aligned}
$$

Another natural optimization problem of finding

$$\inf_\pi E\left(C_1^\pi - C_0^\pi\right)^2$$

is obviously solved by

$$C_0^\pi = E\left(C_1^\pi\right).$$

Now we consider an arbitrary time horizon $N \geq 1$. It is clear from the one-step case that strategies $\pi = (\beta_n, \gamma_n)_{n \leq N}$ must be such that γ_n are predictable (i.e., determined by information $\mathcal{F}_{n-1}$) and β_n are adapted (i.e., determined by information $\mathcal{F}_n$).

We have the following discounted values:

$$\Delta V_n^\pi = V_n^\pi - V_{n-1}^\pi = X_{n-1}\Delta\gamma_n + \gamma_n\,\Delta X_n + \Delta\beta_n.$$

We say that a strategy π is *admissible* if

$$V_N^\pi = H \qquad \text{or} \qquad X_N^\pi = f_N.$$

The price of such a strategy at time n is

$$C_n^\pi = V_n^\pi - \sum_{k=1}^{n} \gamma_k\,\Delta X_k, \qquad n \leq N.$$

For simplicity, we assume that the original probability P is a martingale probability; that is, $P = P^*$. Now we define the following *risk sequence*:

$$R_n^\pi = E^*\left(\left(C_N^\pi - C_n^\pi\right)^2 \middle| \mathcal{F}_n\right).$$

Its initial value

$$R_0^\pi = E^* \left(H - \sum_{k=1}^N \gamma_k \, \Delta X_k - C_0^\pi \right)^2$$

is referred to as *risk* of strategy π.

Note that discounted values of a self-financing strategy π have the form

$$V_n^\pi = V_0^\pi - \sum_{k=1}^n \gamma_k \, \Delta X_k,$$

which implies that it has a constant price sequence: $C_n^\pi = C_0^\pi$ for all $n \le N$.

Now we solve the *minimization problem* in the class of all admissible strategies. Suppose that X is a square integrable martingale and $E^*(H^2) < \infty$ (note that in the case of a discrete probability space $(\Omega, \mathcal{F}, P^*)$, these integrability assumptions are trivially satisfied).

Consider another martingale

$$V_n^* = E^*(H|\mathcal{F}_n), \quad n \le N.$$

The following *Kunita-Watanabe decomposition* is a key technical tool for solving our problem.

Lemma 3.2 *The martingale* $(V_n^*)_{n \le N}$ *admits the decomposition*

$$V_n^* = V_0^* + \sum_{k=1}^n \gamma_k^H \, \Delta X_k + L_n^H,$$

where $(\gamma_n^H)_{n \le N}$ *is a predictable sequence and* L^H *is a martingale orthogonal to* X:

$$\langle X, L^H \rangle_n = 0.$$

Proof Define sequence $(\gamma_n^H)_{n \le N}$ by

$$\gamma_n^H = \frac{E^*(H \, \Delta X_n | \mathcal{F}_{n-1})}{E^*((\Delta X_n)^2 | \mathcal{F}_{n-1})}, \quad n \le N, \tag{3.4}$$

which is clearly predictable. Also let

$$L_n^H := V_n^* - V_0^* - \sum_{k=1}^n \gamma_k^H \, \Delta X_k,$$

which is a martingale being a difference of two martingales.

Using Cauchy-Schwartz inequality, we obtain

$$
\begin{aligned}
E^*(\gamma_n^H \, \Delta X_n)^2 &= E^* \left(E^* \left(\left[\frac{E^*(H \, \Delta X_n | \mathcal{F}_{n-1})}{E^*((\Delta X_n)^2 | \mathcal{F}_{n-1})} \, \Delta X_n \right]^2 \Big| \mathcal{F}_{n-1} \right) \right) \\
&= E^* \left(\frac{(E^*(H \, \Delta X_n | \mathcal{F}_{n-1}))^2}{E^*((\Delta X_n)^2 | \mathcal{F}_{n-1})} \right) \\
&\le E^* \left(E^*(H^2 | \mathcal{F}_{n-1}) \right) = E^*(H^2) < \infty,
\end{aligned}
$$

which implies that L^H is a square integrable martingale.

Now we show that the product

$$L_n^H X_n = \left(E^*(H|\mathcal{F}_n) - E^*(H) - \sum_{k=1}^{n} \gamma_k^H \triangle X_k \right) X_n$$

forms a martingale. This follows from the definition (3.4) of γ_n^H and the equality

$$E^* \left(E^*(H|\mathcal{F}_n) X_n | \mathcal{F}_{n-1} \right) - E^* \left(\gamma_n^H \triangle X_n X_n | \mathcal{F}_{n-1} \right) = E^*(H|\mathcal{F}_{n-1}) X_{n-1} \,.$$
$$(3.5)$$

To verify (3.5), we compute the expectations in the left-hand side. First, from properties of conditional expectations, we have

$$E^* \left(E^*(H|\mathcal{F}_n) X_n | \mathcal{F}_{n-1} \right) = E^* \left(E^*(H|\mathcal{F}_n) (X_{n-1} + \triangle X_n) | \mathcal{F}_{n-1} \right)$$
$$= E^* \left(E^*(H|\mathcal{F}_n) X_{n-1} | \mathcal{F}_{n-1} \right) + E^* \left(E^*(H|\mathcal{F}_n) \triangle X_n | \mathcal{F}_{n-1} \right)$$
$$= X_{n-1} E^*(H|\mathcal{F}_{n-1}) + E^*(H \triangle X_n | \mathcal{F}_{n-1}) \,.$$

Second, using (3.4), we obtain

$$E^* \left(\gamma_n^H \triangle X_n X_n | \mathcal{F}_{n-1} \right) = \frac{E^*(H \triangle X_n | \mathcal{F}_{n-1})}{E^*((\triangle X_n)^2 | \mathcal{F}_{n-1})} E^* \left(\triangle X_n (X_{n-1} + \triangle X_n) | \mathcal{F}_{n-1} \right)$$
$$= \frac{E^*(H \triangle X_n | \mathcal{F}_{n-1})}{E^*((\triangle X_n)^2 | \mathcal{F}_{n-1})} \left\{ X_{n-1} E^*(\triangle X_n | \mathcal{F}_{n-1}) + E^*((\triangle X_n)^2 | \mathcal{F}_{n-1}) \right\}$$
$$= E^*(H \triangle X_n | \mathcal{F}_{n-1}) \,,$$

which proves (3.5). Since sequence γ^H is predictable, then $L^H X$ is a martingale. Indeed,

$$E^* \left(L_n^H X_n | \mathcal{F}_{n-1} \right) = E^* \left([V_n^* - V_0^* - \sum_{k=1}^{n} \gamma_k^H \triangle X_k] X_n | \mathcal{F}_{n-1} \right)$$
$$= E^* \left([E^*(H|\mathcal{F}_n) - E^*(H) - \sum_{k=1}^{n} \gamma_k^H \triangle X_k] X_n | \mathcal{F}_{n-1} \right)$$
$$= X_{n-1} E^*(H|\mathcal{F}_{n-1}) + E^*(H \triangle X_n | \mathcal{F}_{n-1}) - X_{n-1} E^*(H)$$
$$- E^* \left(X_n \sum_{k=1}^{n} \gamma_k^H \triangle X_k | \mathcal{F}_{n-1} \right)$$
$$= X_{n-1} E^*(H|\mathcal{F}_{n-1}) + E^*(H \triangle X_n | \mathcal{F}_{n-1}) - X_{n-1} E^*(H)$$
$$- X_{n-1} \sum_{k=1}^{n} \gamma_k^H \triangle X_k - E^*(H \triangle X_n | \mathcal{F}_{n-1})$$
$$= X_{n-1} \left[E^*(H|\mathcal{F}_{n-1}) - E^*(H) - \sum_{k=1}^{n} \gamma_k^H \triangle X_k \right]$$
$$= X_{n-1} L_{n-1}^H \,.$$

Therefore,

$$E^* \left(L_n^H \sum_{k=1}^n \gamma_k^H \, \Delta X_k \Big| \mathcal{F}_{n-1} \right)$$

$$= E^* \left(L_n^H | \mathcal{F}_{n-1} \right) \sum_{k=1}^{n-1} \gamma_k^H \, \Delta X_k + \gamma_n^H \, E^* \left(L_n^H \, \Delta X_n | \mathcal{F}_{n-1} \right)$$

$$= L_{n-1}^H \sum_{k=1}^{n-1} \gamma_k^H \, \Delta X_k .$$

Hence, L^H is orthogonal to X and $\left(\sum_{k=1}^n \gamma_k^H \, \Delta X_k \right)_{n \leq N}$. Proof of the *uniqueness* of Kunita-Watanabe decomposition is straightforward. $\square$

Now, since $\sum_{k=1}^n \gamma_k \, \Delta X_k$ is a martingale, then to minimize risk R_0^π we must have $C_0^\pi = E^*(H)$. Furthermore, R_0^π does not depend on changes of β, the non-risky component of strategy π. We can rewrite R_0^π in the form

$$R_0^\pi = E^* \left(H - \sum_{k=1}^N \gamma_k \, \Delta X_k - E^*(H) \right)^2 = E^* \left(\sum_{k=1}^N (\gamma_k^H - \gamma_k) \, \Delta X_k + L_N^H \right)^2$$

$$= E^* \left(L_N^H \right)^2 + \sum_{k=1}^N E^* \left((\gamma_k^H - \gamma_k)^2 (\Delta X_k)^2 \right),$$

and therefore the required risk-minimizing strategy is uniquely determined by

$$\gamma_n = \gamma_n^H, \quad n = 1, 2, \ldots, N.$$

Similarly, we obtain the risk sequence

$$R_n^\pi = E^* \left((L_N^H - L_n^H)^2 | \mathcal{F}_n \right).$$

Thus, we obtain the following formulas for the optimal strategy $\hat{\pi} = (\hat{\beta}, \hat{\gamma})$:

$$\hat{\gamma}_n = \gamma_n^H, \quad \hat{\beta}_n = V_n^* - \hat{\gamma}_n \, X_n, \quad n \leq N.$$

The price of this strategy

$$C_n^{\hat{\pi}} = V_n^{\hat{\pi}} - \sum_{k=1}^n \hat{\gamma}_k \, \Delta X_k = V_n^* - \sum_{k=1}^n \gamma_k^H \, \Delta X_k$$

$$= E^*(H) + \sum_{k=1}^n \gamma_k^H \, \Delta X_k + L_n^H - \sum_{k=1}^n \gamma_k^H \, \Delta X_k = E^*(H) + L_n^H$$

is a martingale. Such strategies $\hat{\pi}$ are referred to as *self-financing in average* (or *in mean*).

Worked Example 3.2 *Consider a one-step* (B, S)-*market with the rate of interest* $r = 0.1$ *and profitability*

$$\rho = \rho_1 = \begin{cases} 0.2 & \text{with probability } 0.7 \\ -0.1 & \text{with probability } 0.3 \end{cases}.$$

Consider a pure endowment assurance with the claim

$$f_1 = \max\{1 + r, \, 1 + \rho\},$$

which is paid to the policyholder on survival to time $N = 1$ *(year). Suppose that the probability of death during this year is* 0.004, *and let* $B_0 = S_0 = 1 \, (\$)$. *Find* γ *and policy's initial price* C_0.

Solution Denote $C(\gamma) \equiv C^\pi$, where $\pi = (\beta, \gamma)$. We need to minimize $V\big(C_1(\gamma)\big)$ and $E\big(C_1(\gamma) - C_0(\gamma)\big)^2$. We have

$$E\big(\max\{1 + r, \, 1 + \rho\}\big) = 1.1 \cdot 0.3 + 1.2 \cdot 0.7 = 1.17$$

and

$$V(\rho - r) = V(\rho) = 0.0189.$$

If H is the discounted value of the payoff and $I_1 = I_A$, where A is the event of policyholder's survival for at least one year, then

$$H = I_1 \, \frac{\max\{1 + r, \, 1 + \rho\}}{1 + r}.$$

Further, for discounted prices of S, we have

$$\frac{S_1}{B_1} - \frac{S_0}{B_0} = \frac{1 + \rho}{1 + r} - 1 = \frac{\rho - r}{1 + r},$$

and therefore

$$\gamma = \frac{Cov\big(H, (\rho - r)/(1 + r)\big)}{V\big((\rho - r)/(1 + r)\big)} = \frac{Cov\big(I_1 \max\{1 + r, \, 1 + \rho\}, \, \rho - r\big)}{V(\rho - r)}$$

$$= 0.996 \left[0.7 \, \frac{(1.2 - 1.17)\,(0.1 - 0.01)}{0.0189} + 0.3 \, \frac{(1.1 - 1.17)\,(-0.2 - 0.01)}{0.0189} \right]$$

$$\approx 0.382.$$

Hence,

$$C_0 = E\big(C_1(\gamma)\big) = \frac{0.996 \cdot 1.17 - 0.332 \cdot 0.01}{1.1} = 1.0624.$$

Note that, if there is no additional source of risk related to the survival of a policyholder (i.e., if probability of policyholder's death is 0), then a replicating self-financing strategy can be easily found from the system

$$\begin{cases} 1.1 \cdot C_0 + 0.1 \cdot \gamma = 1.2 \\ 1.1 \cdot C_0 - 0.2 \cdot \gamma = 1.1 \end{cases},$$

which gives $\gamma = 0.333$ and $C_0 = 1.06$. $\square$

3.4 Gaussian model of a financial market in discrete time. Insurance appreciation and discrete version of the Black-Scholes formula.

Since prices S_n are always positive, we can write them in the *exponential* form

$$S_n = S_0 \, E^{W_n}, \quad S_0 > 0,$$

where $W_n = \sum_{k=1}^{n} w_k$, $W_0 = 0$, and

$$w_n = \Delta W_n = \ln \frac{S_n}{S_{n-1}} = \ln \left(1 + \frac{\Delta S_n}{S_{n-1}}\right), \quad n = 1, \ldots, N.$$

However, from the definition of stochastic exponentials and from the recurrence relations $\Delta S_n = \rho_n \, S_{n-1}$, we obtain

$$S_n = S_0 \, \varepsilon_n \left(\sum_{k=1}^{n} \rho_k\right) = S_0 \, \varepsilon_n(V),$$

where we introduced the notation $\sum_{k=1}^{n} \rho_k = V_n = \sum_{k=1}^{n} \Delta V_k = \sum_{k=1}^{n} v_k$ with $v_k = \rho_k = \Delta V_k > -1$.

Hence, we obtain the following connection between the two stochastic sequences W and V:

$$\begin{aligned}
S_0 \, e^{W_n} &= S_n = S_0 \, \varepsilon_n(V) = S_0 \, e^{W_n} e^{-W_n} \prod_{k=1}^{n} (1 + \Delta V_k) \\
&= S_0 \, e^{W_n} \prod_{k=1}^{n} (1 + \Delta V_k) \, e^{-\Delta W_k},
\end{aligned}$$

so

$$W_n = \sum_{k=1}^{n} \ln (1 + \Delta V_k)$$

and

$$V = \sum_{k=1}^{n} \left(e^{\Delta W_k} - 1\right) = W_n + \sum_{k=1}^{n} \left(e^{\Delta W_k} - \Delta W_k - 1\right).$$

Now, using Doob decomposition, we can write

$$W_n = M_n + A_n, \quad n = 1, \ldots, N,$$

where $A_0 = M_0 = W_0 = 0$, $E(|\Delta W_n|) < \infty$, sequence $\Delta A_n = E(\Delta W_n | \mathcal{F}_{n-1})$ is predictable, and $\Delta M_n = \Delta W_n - E(\Delta W_n | \mathcal{F}_{n-1})$ forms a martingale.

Thus, prices S_n can be written in the form

$$S_n = S_0 \, \exp \{M_n + A_n\}$$

on the stochastic basis $(\Omega, \mathcal{F}, \mathbb{F}, P)$, where $\mathbb{F} = (\mathcal{F}_n)_{n \leq N}$ is a filtration with $\mathcal{F}_n = \sigma(S_1, \dots, S_n)$.

Suppose that sequence $(w_k)_{k \leq N}$ consists of Gaussian random variables with means μ_k and variances σ_k^2:

$$w_k \sim \mathcal{N}(\mu_k, \sigma_k^2), \quad k = 1, \dots, N.$$

We can then write

$$w_k = \mu_k + \sigma_k \, \epsilon_k, \quad k = 1, \dots, N,$$

where $\epsilon_k \sim \mathcal{N}(0, 1)$ are standard Gaussian random variables. In this case, deterministic sequence A_n and Gaussian martingale M_n have the form

$$A_n = \sum_{k=1}^{n} \mu_k \quad \text{and} \quad M_n = \sum_{k=1}^{n} \sigma_k \, \epsilon_k,$$

and quadratic characteristic of M is deterministic: $\langle M, M \rangle_n = \sum_{k=1}^{n} \sigma_k^2, n \leq N$.

We assume that $\mathcal{F} = \mathcal{F}_N = \sigma(\epsilon_1, \dots, \epsilon_N)$. Define a stochastic sequence

$$Z_n = \exp\left\{ -\sum_{k=1}^{n} \frac{\mu_k}{\sigma_k} \epsilon_k - \frac{1}{2} \sum_{k=1}^{n} \left(\frac{\mu_k}{\sigma_k}\right)^2 \right\}, \quad Z_0 = 1, \quad n = 1, \dots, N.$$

We show that $(Z_n, \mathcal{F}_n)_{n \leq N}$ is a martingale with respect to the initial probability P. Indeed, taking into account independence of $\epsilon_1, \dots, \epsilon_N$, we have

$$E\left(\exp\left\{ -\frac{\mu_k}{\sigma_k} \epsilon_k - \frac{1}{2} \left(\frac{\mu_k}{\sigma_k}\right)^2 \right\} \right) = 1,$$

and for all $n = 1, \dots, N$,

$$
\begin{aligned}
E(Z_n | \mathcal{F}_{n-1}) &= E\left(Z_{n-1} \exp\left\{ -\frac{\mu_k}{\sigma_k} \epsilon_k - \frac{1}{2} \left(\frac{\mu_k}{\sigma_k}\right)^2 \right\} \bigg| \mathcal{F}_{n-1} \right) \\
&= Z_{n-1} E\left(\exp\left\{ -\frac{\mu_k}{\sigma_k} \epsilon_k - \frac{1}{2} \left(\frac{\mu_k}{\sigma_k}\right)^2 \right\} \right) \\
&= Z_{n-1} \quad \text{(a.s.)}.
\end{aligned}
$$

Now, since $Z_N > 0$ and $E(Z_N) = 1$, then the following probability

$$P^*(A) := E(Z_N \, I_A), \quad A \in \mathcal{F}$$

is well defined. Computing

$$
E^* \left(e^{i\lambda w_n} \right) = E \left(\exp \left\{ \left(i\lambda\sigma_n - \frac{\mu_n}{\sigma_n} \right) \epsilon_n + i\lambda\mu_n - \frac{1}{2} \left(\frac{\mu_n}{\sigma_n} \right)^2 \right\} \right)
$$

$$
= E \left(\left(i\lambda\sigma_n - \frac{\mu_n}{\sigma_n} \right) \epsilon_n - \frac{1}{2} \left(i\lambda\sigma_n - \frac{\mu_n}{\sigma_n} \right)^2 \right.
$$

$$
\times \exp \left\{ \frac{1}{2} \left(i\lambda\sigma_n - \frac{\mu_n}{\sigma_n} \right)^2 + i\lambda\mu_n - \frac{1}{2} \left(\frac{\mu_n}{\sigma_n} \right)^2 \right\}
$$

$$
= \exp \left\{ -\frac{\lambda^2 \sigma_n^2}{2} \right\}, \quad \lambda \geq 0 \quad n = 1, \dots, N,
$$

we conclude that $(w_k)_{k \leq N}$ is a sequence of Gaussian random variables with respect to P^*, with mean zero and variance σ_n^2. Note that independence of $(w_k)_{k \leq N}$ follows from the equality

$$
E^* \left(\exp \left\{ i \sum_{k=1}^{N} \lambda_k w_k \right\} \right) = E^* \left(\exp \left\{ i \sum_{k=1}^{N-1} \lambda_k w_k \right\} E^* \left(e^{i\lambda_N w_N} \middle| \mathcal{F}_{N-1} \right) \right)
$$

$$
= E^* \left(\exp \left\{ i \sum_{k=1}^{N-1} \lambda_k w_k \right\} \right) e^{-\lambda_N^2 w_N^2} = \dots
$$

$$
= \exp \left\{ -\frac{1}{2} \sum_{k=1}^{N} \lambda_k^2 w_k^2 \right\}.
$$

As a corollary, we obtain the following version of *Girsanov theorem*.

Proposition 3.2 *If under probability P random variables*

$$
w_k \sim \mathcal{N}(\mu_k, \sigma_k^2), \quad k = 1, \dots, N
$$

are independent, then they are also independent and normally distributed under probability P^* with zero mean:

$$
w_k \sim \mathcal{N}(0, \sigma_k^2), \quad k = 1, \dots, N.
$$

Next, we consider the following discrete Gaussian (B, S)-market:

$$
B_n = \prod_{k=1}^{n} (1 + r_k) = \exp \left\{ \sum_{k=1}^{n} \delta_k \right\},
$$

$$
S_n = S_0 \exp \left\{ \sum_{k=1}^{n} \mu_k + \sum_{k=1}^{n} \sigma_k \epsilon_k \right\}, \quad S_0 > 0,
$$

where non-negative deterministic sequences (r_k) and (δ_k) represent the rate of interest and such that

$$
1 + r_k = e^{\delta_k}, \quad k = 1, \dots, N.
$$

Now our aim is to construct a martingale probability P^* for this market. We are looking for a probability of the form of *Essher transform*:

$$P^*(A) = E(Z_N I_A),$$

where

$$Z_N = \prod_{n=1}^{N} z_n, \quad \text{with} \quad z_n = \frac{\exp\left\{a_n(w_n - \delta_n)\right\}}{E\left(\exp\left\{a_n(w_n - \delta_n)\right\}\right)},$$

and $(a_n)_{n\leq N}$ is some deterministic sequence. To find $(a_n)_{n\leq N}$, we use the martingale property of $(S_n/B_n)_{n\leq N}$:

$$E^*\left(\frac{S_n}{B_n}\Big|\mathcal{F}_{n-1}\right) = \frac{S_{n-1}}{B_{n-1}}, \quad n = 1,\ldots,N,$$

which is equivalent to

$$E^*\left(\exp\{\tilde{\mu}_n + \sigma_n \,\epsilon_n\}\right) = 1,$$

where $\tilde{\mu}_n = \mu_n - \delta_n$, $n = 1,\ldots,N$.

Taking into account the expression for Z_n, we obtain

$$E\left(\exp\{a_n(\tilde{\mu}_n + \sigma_n \,\epsilon_n) + \tilde{\mu}_n + \sigma_n \,\epsilon_n\}\right) = E\left(\exp\{a_n(\tilde{\mu}_n + \sigma_n \,\epsilon_n)\}\right)$$

and

$$E\left(\exp\{(a_n + 1)(\tilde{\mu}_n + \sigma_n \,\epsilon_n)\}\right) = E\left(\exp\{a_n(\tilde{\mu}_n + \sigma_n \,\epsilon_n)\}\right).$$

Since $\epsilon_n \sim \mathcal{N}(0,1)$, then

$$E\left(\exp\{a_n \sigma_n \,\epsilon_n\}\right) = \exp\{(a_n \sigma_n)^2/2\},$$

which implies

$$\exp\{(a_n + 1)\tilde{\mu}_n\} \exp\{(a_n + 1)^2 \sigma_n^2/2\} = \exp\{a_n\tilde{\mu}_n\} \exp\{a_n^2 \sigma_n^2/2\},$$

$$\tilde{\mu}_n + \frac{\sigma_n^2}{2} = -a_n \sigma_n^2.$$

Thus,

$$a_n = -\frac{\tilde{\mu}_n}{\sigma_n^2} - \frac{1}{2} = -\frac{\mu_n - \delta_n}{\sigma_n^2} - \frac{1}{2}.$$

Now using this a_n we compute

$$E\left(\exp\{a_n(w_n - \delta_n)\}\right)$$

$$= E\left(\exp\left\{-\left(\frac{\tilde{\mu}_n}{\sigma_n^2} + \frac{1}{2}\right)(\tilde{\mu}_n + \sigma_n \epsilon_n)\right\}\right)$$

$$= \exp\left\{-\frac{\tilde{\mu}_n^2}{\sigma_n^2} - \frac{\tilde{\mu}_n}{2}\right\} E\left(\exp\left\{-\sigma_n\left(\frac{\tilde{\mu}_n}{\sigma_n^2} + \frac{1}{2}\right)\epsilon_n\right\}\right)$$

$$= \exp\left\{-\frac{\tilde{\mu}_n^2}{\sigma_n^2} - \frac{\tilde{\mu}_n}{2}\right\} \exp\left\{\frac{\sigma_n^2}{2}\left(\frac{\tilde{\mu}_n}{\sigma_n^2} + \frac{1}{2}\right)^2\right\}$$

$$= \exp\left\{-\frac{\tilde{\mu}_n^2}{\sigma_n^2} - \frac{\tilde{\mu}_n}{2}\right\} \exp\left\{\frac{\sigma_n^2}{2}\left(\frac{\tilde{\mu}_n^2}{\sigma_n^4} + \frac{\tilde{\mu}_n}{\sigma_n^2} + \frac{1}{4}\right)\right\}$$

$$= \exp\left\{-\frac{\tilde{\mu}_n^2}{2\,\sigma_n^2} + \frac{\sigma_n^2}{8}\right\},$$

which gives us

$$z_n = \exp\left\{-\left(\frac{\tilde{\mu}_n}{\sigma_n} + \frac{\sigma_n}{2}\right)\epsilon_n - \frac{1}{2}\left(\frac{\tilde{\mu}_n}{\sigma_n} + \frac{\sigma_n}{2}\right)^2\right\}$$

and

$$Z_N = \exp\left\{-\sum_{n=1}^{N}\left[\left(\frac{\tilde{\mu}_n}{\sigma_n} + \frac{\sigma_n}{2}\right)\epsilon_n + \frac{1}{2}\left(\frac{\tilde{\mu}_n}{\sigma_n} + \frac{\sigma_n}{2}\right)^2\right]\right\}.$$

Now for simplicity we consider a special case of our Gaussian (B, S)-market. Let

$$B_n = (1+r)^n = e^{\delta n} \quad \text{with} \quad \delta = \ln(1+r), \quad r \geq 0,$$

$$S_n = S_0\, e^{W_n}, \quad S_0 > 0.$$

Here $W_n = \sum_{k=1}^{n} w_k$, $w_k = \mu + \sigma \epsilon_k$, and $\epsilon_n \sim \mathcal{N}(0,1)$ are independent random variables on the stochastic basis

$$(\Omega_1, \mathcal{F}^1, \mathbb{F}_1, P_1),$$

where $\mathbb{F}_1 = \left(\mathcal{F}_n^1\right)_{n \leq N}$ is a filtration with $\mathcal{F}_0^1 = \{\emptyset, \Omega\}$, $\mathcal{F}_n^1 = \sigma(\epsilon_1, \ldots, \epsilon_n)$, and $\mathcal{F}^1 = \mathcal{F}_N^1$.

Worked Example 3.3 *As in Worked Example 2.3, we consider a pure endowment assurance issued by an insurance company. According to this contract, the policyholder is paid*

$$f_N = g(S_N)$$

on survival to the time N, where S_N is the stock price and g is some function specified by the contract. Suppose $E(g^2(S_N)) < \infty$. Find the "fair" price for such insurance policy.

Solution Recall that if l_x is the number of policyholders of age x, then each policyholder i, $i = 1, \ldots, l_x$ can be characterized by a positive random variable T_i representing the time elapsed between age x and death. Suppose that T_i's are defined on another probability space $(\Omega_2, \mathcal{F}^2, P_2)$ with the filtration $\mathcal{F}_n^2 = \sigma(T_i \leq k, \ k \leq n, \ i = 1, \ldots, l_x)$.

Denote $p_x(n) = P_2(\{\omega : \ T_i > n\})$, $n = 0, 1, \ldots, N$, the conditional expectation for a policyholder to survive another n years from the age of x (clearly, $p_x(0) = 1$).

From Bayes's formula, we have

$$p_{x+n}(y) = \frac{P_2(\{\omega : \ T_i > y + x\})}{P_2(\{\omega : \ T_i > n\})} = \frac{p_x(y+n)}{p_x(n)},$$

and hence,

$$p_x(y+n) = p_x(n)\, p_{x+n}(y).$$

Denote

$$N_n := \sum_{i=1}^{l_x} I_{\{\omega : \ T_i \leq n\}}$$

the counter of deaths in the given group of policyholders. Then

$$E_2\left(l_x - N_n \big| \mathcal{F}_k^2\right) = (l_x - N_k)\, p_{x+k}(n - k), \quad k \leq n \leq N.$$

Therefore, the discounted value of the total payoff is given by

$$H = \sum_{k=1}^{l_x} Y_k = g(S_N)\, \frac{l_x - N_N}{B_N},$$

where

$$Y_k = g(S_N)\, \frac{I_{\{\omega : \ T_k > N\}}}{B_N}, \quad k \leq l_x.$$

Since we have to price a contingent claim with an insurance component, we introduce the following stochastic basis:

$$(\Omega, \mathcal{F}, \mathbb{F}, P) = (\Omega_1 \times \Omega_2, \ \mathcal{F}^1 \times \mathcal{F}^2, \ \mathbb{F}_1 \times \mathbb{F}_2, \ P_1 \times P_2).$$

Clearly, we have that stochastic sequences (ϵ_n) and (T_i) are independent on this basis.

Now, since probability P_1^* with density

$$Z_N = \exp\left\{ -\left(\frac{\mu - \delta}{\sigma} + \frac{\sigma}{2}\right) \sum_{k=1}^{N} \epsilon_k - \frac{1}{2}\left(\frac{\mu - \delta}{\sigma} + \frac{\sigma}{2}\right)^2 N \right\}$$

is a martingale probability on the (B, S)-market, then the probability $P^* := P_1^* \times P_2$ on $(\Omega, \mathcal{F}, P)$ is such that $(S_n/B_n)_{n \leq N}$ is a martingale under this probability.

Next, using the methodology of hedging in mean square, we obtain

$$
\begin{aligned}
V_n^* &= E^*\left(H|\mathcal{F}_n\right) = E^*\left(g(S_N)B_N^{-1}|\mathcal{F}_n^1\right) E^*\left(l_x - N_N|\mathcal{F}_n^2\right) \\
&= E^*\left(g(S_N)|\mathcal{F}_n^1\right) B_N^{-1}\left[l_x - N_n\right] p_{x+n}(N - n),
\end{aligned}
$$

and

$$
\hat{\gamma}_n = \gamma_n^H = \frac{B_N^{-1} E^*\left(g(S_N) \Delta X_n|\mathcal{F}_{n-1}^1\right) \left[l_x - N_{N-1}\right] p_{x+n-1}(N - n + 1)}{X_{n-1}^2 \left[\exp\{\sigma^2\} - 1\right]}.
$$

Here we also used independence of sequences $(S_n)_{n\leq N}$ and $(T_k)_{k\leq l_x}$, and the equality

$$
E^*\left((\Delta X_n)^2|\mathcal{F}_{n-1}^1\right) = X_{n-1}^2 \left[\exp\{\sigma^2\} - 1\right].
$$

The optimal strategy $\hat{\pi} = (\hat{\gamma}, \hat{\beta})$ and its values have the form

$$
V_n^{\hat{\pi}} = V_n^*, \quad \hat{\beta}_n = V_n^* - \hat{\gamma}_n X_n, \quad n = 1, \ldots, N.
$$

The quantity

$$
V_0^{\hat{\pi}} = p_x(N) \, l_x \, E^*\left(g(S_N)\right) e^{-\delta N}
$$

determines the total premium received by an insurance company.

Now we consider three particular cases of function g and compute premiums there.

Case 1. Let $g(S_N) = S_N$, then

$$
\begin{aligned}
V_n^* &= \left[l_x - N_n\right] X_n \, p_{x+n}(N - n), \\
\hat{\gamma}_n &= \left[l_x - N_{n-1}\right] p_{x+n-1}(N - n + 1), \\
\hat{\beta}_n &= X_n\left\{\left[l_x - N_n\right] p_{x+n}(N - n) - \left[l_x - N_{n-1}\right] p_{x+n-1}(N - n + 1)\right\} \\
&= X_n \, p_{x+n}(N - n) \left\{\left[l_x - N_n\right] - \left[l_x - N_{n-1}\right] p_{x+n-1}(1)\right\} \\
&\quad n = 1, \ldots, N.
\end{aligned}
$$

Premium is therefore determined by

$$
V_0^{\hat{\pi}} = p_x(N) \, l_x \, S_0.
$$

We can compute risk of such strategy:

$$
R_n^{\hat{\pi}} = \left\{\sum_{k=n+1}^{N} e^{k\sigma^2} q_{x+k-1}(1) p_{x+k}(N - k)\right\}
$$
$$
\times p_{x+n}(N - n) \left[l_x - N_n\right] e^{-n\sigma^2} X_n^2,
$$

$$
R_0^{\hat{\pi}} = \left\{\sum_{k=1}^{N} e^{k\sigma^2} q_{x+k-1}(1) p_{x+k}(N - k)\right\} p_x(N) \, l_x \, S_0^2,
$$

where $q_y(1)$ is the probability of death during the year following year y. The latter formula also implies

$$\frac{\left(R_0^{\hat{\pi}}\right)^{1/2}}{l_x} \to 0 \quad \text{as} \quad l_x \to \infty,$$

which means that if there are enough policyholders, then the company's risk associated with this contract is infinitesimal.

Case 2. If $g(S_N) \equiv K = const$, then

$$\begin{aligned}
V_n^{\hat{\pi}} &= V_n^* = K e^{-\delta N} \left[l_x - N_n\right] p_{x+n}(N-n), \\
\hat{\gamma}_n &= 0, \quad \hat{\beta}_n = V_n^{\hat{\pi}} \quad \text{for} \quad n = 1, \dots, N, \\
V_0^{\hat{\pi}} &= K e^{-\delta N} l_x p_x(N),
\end{aligned}$$

which indicates that in this case one has to invest money in a bank account. The risk-sequence here:

$$R_n^{\hat{\pi}} = K^2 e^{-2\delta N} \left[l_x - N_n\right] p_{x+n}(N-n) q_{x+n}(N-n), \quad n \le N.$$

In particular

$$R_0^{\hat{\pi}} = K^2 e^{-2\delta N} l_x p_x(N) q_x(N),$$

and again $\left(R_0^{\hat{\pi}}\right)^{1/2}/l_x \to 0$ as $l_x \to \infty$.

Case 3. Let $g(S_N) = \max\{S_N, K\}$. We can write

$$\max\{S_N, K\} = K + (S_N - K)^+,$$

and therefore we have to compute

$$E^* \left((S_N - K)^+ \Delta X_n \big| \mathcal{F}_{n-1}\right) \quad \text{and} \quad E^* \left((S_N - K)^+ \big| \mathcal{F}_n\right). \qquad (3.6)$$

For the latter, we have

$$\begin{aligned}
&E^* \left((S_N - K)^+ \big| \mathcal{F}_n\right) \\
&= E^* \left(\left(S_n e^{\delta(N-n)} e^{w_{n+1}^* + \cdots + w_N^*} - K\right)^+ \big| \mathcal{F}_n\right) \\
&= E^* \left(\left(S_n e^{\delta(N-n)} e^{\mathcal{N}\left(-\frac{\sigma^2}{2}(N-n), \sigma^2 (N-n)\right)} - K\right)^+ \big| \mathcal{F}_n\right),
\end{aligned}$$

where $w_k^* = \mu - \delta + \sigma \epsilon_k \sim \mathcal{N}\left(-\frac{\sigma^2}{2}, \sigma^2\right)$ with respect to probability P^*. Note that for $\xi \sim \mathcal{N}(0, 1)$ and constants a, b and K one has

$$\begin{aligned}
&E\left(\left(a e^{b\xi - b^2/2} - K\right)^+\right) \\
&= a \, \Phi\left(\frac{\ln(a/K) + b^2/2}{b}\right) - K \, \Phi\left(\frac{\ln(a/K) - b^2/2}{b}\right),
\end{aligned}$$

where

$$\Phi(x) = \frac{1}{\sqrt{2\pi}} \int_{-\infty}^{x} e^{-y^2/2} dy$$

is a standard normal distribution. Hence, we obtain

$$E^* \left((S_N - K)^+ \big| \mathcal{F}_n \right)$$
$$= S_n \, e^{\delta(N-n)} \, \Phi\left(\frac{\ln(S_n/K) + (N-n)\,(\delta + \sigma^2/2)}{\sigma \sqrt{N-n}} \right)$$
$$- K \, \Phi\left(\frac{\ln(S_n/K) + (N-n)\,(\delta - \sigma^2/2)}{\sigma \sqrt{N-n}} \right).$$

Note that for $n = 0$ we have

$$E^* \left(\frac{(S_N - K)^+}{e^{\delta N}} \right) = S_0 \, \Phi\left(\frac{\ln(S_0/K) + N\,(\delta + \sigma^2/2)}{\sigma \sqrt{N}} \right)$$
$$- K \, e^{-\delta N} \, \Phi\left(\frac{\ln(S_0/K) + N\,(\delta - \sigma^2/2)}{\sigma \sqrt{N}} \right),$$

which is the *discrete version of the Black-Scholes formula* for a European call option.

Now we compute the first expectation from (3.6):

$$E^* \left((S_N - K)^+ X_n \big| \mathcal{F}_{n-1} \right) = X_{n-1} \, E^* \left(e^{w_n^*} (S_N - K)^+ \big| \mathcal{F}_{n-1} \right)$$
$$= X_{n-1} \, E^* \left(E^* \left(e^{w_n^*}(S_{n-1} e^{\delta(N-n+1)} e^{w_n^* + \cdots + w_N^*} - K)^+ \big| \mathcal{F}_n \right) \Big| \mathcal{F}_{n-1} \right)$$
$$= X_{n-1} \, E^* \left(e^{w_n^*} \left[S_{n-1} \, e^{\delta(N-n+1)} \, e^{w_n^*} \right. \right.$$
$$\times \Phi\left(\frac{\ln(S_{n-1}/K) + w_n^* + (N-n)\,(\delta + \sigma^2/2) + \sigma^2}{\sigma \sqrt{N-n}} \right)$$
$$\left. \left. - K \, \Phi\left(\frac{\ln(S_{n-1}/K) + w_n^* + (N-n)\,(\delta - \sigma^2/2) + \sigma^2}{\sigma \sqrt{N-n}} \right) \right] \Big| \mathcal{F}_{n-1} \right).$$

Since for $\xi \sim \mathcal{N}\left(-\frac{\sigma^2}{2}, \sigma^2 \right)$ we have

$$E\left(e^\xi \, \Phi(x\,\xi + y) \right) = \Phi\left(\frac{y + \sigma^2 \, x/2}{\sqrt{1 + x^2 \, \sigma^2}} \right),$$

then

$$E^*\left((S_N - K)^+ X_n \big| \mathcal{F}_{n-1}\right)$$

$$= X_{n-1}\left[X_{n-1} e^{\delta N} e^{\sigma^2} \Phi\left(\frac{\ln\left(\frac{S_{n-1}}{K}\right) + (N - n + 1)\left(\delta + \frac{\sigma^2}{2}\right) + \sigma^2}{\sigma\sqrt{N - n + 1}}\right)\right.$$

$$\left. - K\,\Phi\left(\frac{\ln\left(\frac{S_{n-1}}{K}\right) + (N - n + 1)\left(\delta - \frac{\sigma^2}{2}\right) + \sigma^2}{\sigma\sqrt{N - n + 1}}\right)\right].$$

Thus, we obtain the following formulas for the optimal strategy and its capital:

$$V_n^{\hat{\pi}} = V_n^* = e^{-\delta N}\left[l_x - N_n\right] p_{x+n}(N - n)$$

$$\times \left[K + S_n\,e^{\delta(N-n)}\,\Phi\left(\frac{\ln(S_n/K) + (N - n)\,(\delta + \sigma^2/2)}{\sigma\sqrt{N - n}}\right)\right.$$

$$\left. - K\,\Phi\left(\frac{\ln(S_n/K) + (N - n)\,(\delta - \sigma^2/2)}{\sigma\sqrt{N - n}}\right)\right],$$

$$V_0^* = e^{-\delta N}\,l_x\,p_x(N)\left[K + S_0\,e^{\delta N}\,\Phi\left(\frac{\ln(S_0/K) + N\,(\delta + \sigma^2/2)}{\sigma\sqrt{N}}\right)\right.$$

$$\left. - K\,\Phi\left(\frac{\ln(S_0/K) + N\,(\delta - \sigma^2/2)}{\sigma\sqrt{N}}\right)\right],$$

$$\hat{\gamma}_n = \frac{\left[l_x - N_{N-1}\right] p_{x+n-1}(N - n + 1)}{X_{n-1}\left[\exp\{\sigma^2\} - 1\right]}$$

$$\times \left\{ X_{n-1} e^{\delta N}\left[e^{\sigma^2}\,\Phi\left(\frac{\ln\left(\frac{S_{n-1}}{K}\right) + (N - n + 1)\left(\delta + \frac{\sigma^2}{2}\right) + \sigma^2}{\sigma\sqrt{N - n + 1}}\right)\right.\right.$$

$$\left. - \Phi\left(\frac{\ln\left(\frac{S_{n-1}}{K}\right) + (N - n + 1)\left(\delta + \frac{\sigma^2}{2}\right)}{\sigma\sqrt{N - n + 1}}\right)\right]$$

$$+ K\left[\Phi\left(\frac{\ln\left(\frac{S_{n-1}}{K}\right) + (N - n + 1)\left(\delta - \frac{\sigma^2}{2}\right)}{\sigma\sqrt{N - n + 1}}\right)\right.$$

$$\left.\left. - \Phi\left(\frac{\ln\left(\frac{S_{n-1}}{K}\right) + (N - n + 1)\left(\delta - \frac{\sigma^2}{2}\right) + \sigma^2}{\sigma\sqrt{N - n + 1}}\right)\right]\right\},$$

$$\hat{\beta}_n = V_n^* - \hat{\gamma}_n\,X_n, \qquad n = 1, \ldots, N.\;\square$$

Chapter 4

Analysis of Risks: Continuous Time Models

4.1 The Black-Scholes model. "Greek" parameters in risk management, hedging, and optimal investment.

This section is devoted to the rigorous study of the Black-Scholes model of a (B, S)-market with time horizon $T < \infty$.

Let $(\Omega, \mathcal{F}, \mathbb{F}, P)$ be a stochastic basis. Here, filtration $\mathbb{F} = (\mathcal{F}_t)_{t \leq T}$ represents a continuous information flow that is parameterized by a time parameter $t \in [0, T]$ in contrast to the discrete time case of the previous chapter. It is natural to assume that $\mathcal{F}_t$ (being the information up to time t) is a σ-algebra, that is,

1. $\emptyset$, $\Omega \in \mathcal{F}_t$;

2. $A \in \mathcal{F}_t \Rightarrow \Omega \setminus A \in \mathcal{F}_t$ (closed under taking complements);

3. $(A_k)_{k=1}^\infty \subset \mathcal{F}_t \Rightarrow \cup_{k=1}^\infty A_k \in \mathcal{F}_t$ (closed under taking countable unions);

4. $(A_k)_{k=1}^\infty \subset \mathcal{F}_t \Rightarrow \cap_{k=1}^\infty A_k \in \mathcal{F}_t$ (closed under taking countable intersections).

The initial information is usually considered to be trivial: $\mathcal{F}_0 = \{\emptyset, \Omega\}$. It is also customary to assume that the stochastic basis is

- *complete*, that is, $\mathcal{F}$ is P-complete and every $\mathcal{F}_t$ contains all P-null sets of $\mathcal{F}$, and

- *right-continuous*, that is, $\mathcal{F}_t = \mathcal{F}_{t+} = \cap_{s>t} \mathcal{F}_s$.

In this case, we say that the stochastic basis satisfies the *usual conditions*, and we will refer to such stochastic basis as a *standard stochastic basis*. Hereafter, we will always work on a standard stochastic basis. A family of random variables $(X_t)_{t \geq 0}$ is *adapted* to the filtration $\mathbb{F}$ if X_t is $\mathcal{F}_t$-measurable for every $t < T$. Such families are called *stochastic processes*. Recall that stochastic sequences, their discrete counterparts, were studied in the previous chapter.

On a standard stochastic basis, we consider a *Wiener process* (*Brownian motion*) W_t, that is, a process with the following properties:

(W1) $W_0 = 0$;

(W2) $W_t - W_s$ and $W_v - W_u$ are independent for $s < t < v < u$;

(W3) $W_t - W_s \sim \mathcal{N}(0, t - s)$.

It is assumed that all "randomness" of the model is generated by this process, and therefore

$$\mathcal{F}_t = \sigma(W_s, \ s \le t) =: \mathcal{F}_t^W.$$

Note that any stochastic process w is a function of two variables: elementary event $\omega \in \Omega$ and time $t \le T$. For a fixed ω, the function $w(\omega, \cdot)$ is called a *trajectory*. Without loss of generality, one can assume that trajectories of a Wiener process are continuous in t.

Let us divide interval $[0, T]$ into n parts: $0 = t_0 < t_1 < \ldots < t_n = T$, and define

$$\varphi(t, \omega) = \sum_{k=1}^{n} \varphi_{k-1}(\omega) \, I_{(t_{k-1}, t_k]}(t), \qquad (4.1)$$

where φ_{k-1} are square-integrable random variables that are completely determined by σ-algebras $\mathcal{F}_{t_{k-1}}$ (in other words, φ_{k-1} are $\mathcal{F}_{t_{k-1}}$-measurable square-integrable random variables).

Now we define a *stochastic integral* of random function φ with respect to W:

$$(\varphi * W)_t \equiv \int_0^t \varphi(s, \omega) \, dW_s := \sum_{k=1}^{n} \varphi_{k-1}(\omega) \left(W_{t_k \wedge t} - W_{t_{k-1} \wedge t} \right).$$

It has the following properties:

(I1) $\left((\alpha \varphi + \beta \psi) * W\right)_t = \alpha \, (\varphi * W)_t + \beta \, (\psi * W)_t$;

(I2) $E\left((\varphi * W)_T \big| \mathcal{F}_t\right) = (\varphi * W)_t$;

(I3) $E\left((\varphi * W)_t \cdot (\psi * W)_t\right) = E\left(\int_0^t \varphi(s)\psi(s) \, ds \right)$,

for any functions φ and ψ of type (4.1), and any constants α and β.

Next, consider a random function (stochastic process) $(\varphi_t)_{t \le T}$ that is *adapted* to filtration $\mathbb{F} = (\mathcal{F}_t)_{t \le T}$. If

$$E\left(\int_0^t \varphi^2(s, \omega) \, ds < \infty \right),$$

then the stochastic integral is well defined for such function φ as a mean square limit of integrals of functions of type (4.1), and properties (I1)–(I3) hold true.

Thus, one can consider stochastic processes of the following type

$$X_t = X_0 + \int_0^t b(s, \omega) \, ds + \int_0^t a(s, \omega) \, dW_s, \qquad (4.2)$$

where $\int_0^t b(s, \omega) \, ds$ is a usual Lebesgue-type integral for each fixed ω and $\int_0^t a(s, \omega) \, dW_s$ is a stochastic integral. Note that equation (4.2) is often formally written in the equivalent differential form

$$dX_t = b_t \, dt + a_t \, dW_t \, .$$

Let $F(t, x)$ be a real-valued function that is continuously differentiable in t and twice continuously differentiable in x. Then the process $Y_t := F(t, X_t)$ is also of type (4.2), which follows from the celebrated *Kolmogorov-Itô formula*:

$$
\begin{aligned}
F(t, X_t) \;=\;& F(0, X_0) \\
&+ \int_0^t \left[\frac{\partial F}{\partial s}(s, X_s) + b_s \frac{\partial F}{\partial x}(s, X_s) + \frac{1}{2} a_s^2 \frac{\partial^2 F}{\partial x^2}(s, X_s) \right] ds \\
&+ \int_0^t a_s \frac{\partial F}{\partial x}(s, X_s) \, dW_s \, .
\end{aligned}
\tag{4.3}
$$

To sketch the proof of this formula, we note that, by Taylor's formula, increments of a smooth function F can be written in the form

$$\Delta F(t, x) = \frac{\partial F}{\partial x} \Delta x + \frac{\partial F}{\partial t} \Delta t + \frac{1}{2} \frac{\partial^2 F}{\partial x^2} (\Delta x)^2 + \frac{\partial^2 F}{\partial x \partial t} \Delta x \, \Delta t + \frac{1}{2} \frac{\partial^2 F}{\partial t^2} (\Delta t)^2 + \dots \, .$$

Since $\Delta W_t \sim \mathcal{N}(0, \Delta t)$, then increments $\Delta X_t = X_{t+\Delta t} - X_t$ of process X are equivalent (in distribution) to random variable $b \, \Delta t + a \, \varepsilon \sqrt{\Delta t}$, $\varepsilon \sim \mathcal{N}(0, 1)$. Further

$$\left(\Delta X_t \right)^2 = b^2 \, (\Delta t)^2 + a^2 \, \varepsilon^2 \, \Delta t + 2 \, b \, a \, \varepsilon \, (\Delta t)^{3/2}$$

and

$$E\left(\left(\Delta X_t \right)^2 \right) = E\left(a^2 \, \varepsilon^2 \, \Delta t \right) = a^2 \, \Delta t \, .$$

up to terms of higher order in Δt.

Thus, we obtain the following approximation:

$$
\begin{aligned}
\Delta F(t, X_t) \;=\;& F(t + \Delta t, X_{t+\Delta t}) - F(t, X_t) \\[2mm]
\approx\;& \frac{\partial F}{\partial x} \left[b \, \Delta t + a \, \varepsilon \sqrt{\Delta t} \right] + \frac{\partial F}{\partial t} \Delta t + \frac{1}{2} \frac{\partial^2 F}{\partial x^2} a^2 \, \Delta t \\[2mm]
\approx\;& \left(\frac{\partial F}{\partial x} b + \frac{\partial F}{\partial t} + \frac{1}{2} \frac{\partial^2 F}{\partial x^2} a^2 \right) \Delta t + \frac{\partial F}{\partial x} a \, \varepsilon \sqrt{\Delta t} \, ,
\end{aligned}
$$

which implies (4.3).

We say that a process $M = (M_t, \mathcal{F}_t^W)_{t \leq T}$ is a *martingale* on $(\Omega, \mathcal{F}, \mathbb{F}, P)$ if $E(|M_t|) < \infty$, and for all $s \leq t$

$$E\left(M_t | \mathcal{F}_s \right) = M_s \qquad \text{(a.s.).}$$

If filtration $\mathbb{F}$ is generated by a Wiener process W, then any martingale M can be written in the form

$$M_t = M_0 + \int_0^t \varphi_s \, dW_s \,, \tag{4.4}$$

for some random function φ that is adapted to filtration $\mathbb{F}$ and "stochastically" integrable with respect to W according to the definition introduced earlier in this section. This subtle mathematical result is also often referred to as *Clark representation*.

Note that the definition of a stochastic integral with respect to a Wiener process W originates from the martingale property of the constructed process. The martingale representation (4.4) is a subtle result saying that stochastic integrals are the only martingales that exist in this case.

Now we proceed to the Black-Scholes model. Suppose that on a stochastic basis $(\Omega, \mathcal{F}, \mathbb{F}, P)$ processes B_t and S_t are given by

$$B_t = e^{rt} \,, \tag{4.5}$$
$$S_t = S_0 \, e^{(\mu - \sigma^2/2)t + \sigma \, W_t} \,, \quad S_0 > 0 \,,$$

where $r \geq 0$, $\mu \in \mathbb{R}$ and $\sigma > 0$.

Applying the Kolmogorov-Itô formula to the process

$$S_t = S_0 \, e^{X_t} \quad \text{with} \quad X_t = \left(\mu - \frac{\sigma^2}{2} \right) t + \sigma \, W_t \,, \quad X_0 = 0 \,,$$

we obtain

$$
\begin{aligned}
S_t &= S_0 + \int_0^t \left[S_0 \, e^{X_u} \left(\mu - \frac{\sigma^2}{2} \right) + \frac{1}{2} e^{X_u} \sigma^2 \right] du + \int_0^t \sigma \, S_0 \, e^{X_u} \, du \\
&= S_0 + \int_0^t S_u \, \mu \, du + \int_0^t S_u \, \sigma \, dW_u = S_0 + \int_0^t S_u \left(\mu \, du + \sigma \, dW_u \right).
\end{aligned}
$$

Thus, the Black-Scholes model (4.5) can be represented in the following equivalent differential form:

$$dB_t = r \, B_t \, dt \,, \tag{4.6}$$
$$dS_t = S_t \left(\mu \, dt + \sigma \, dW_t \right), \quad S_0 > 0 \,.$$

Parameters r, μ, and σ are referred to as *interest rate*, *profitability (appreciation rate)*, and *volatility* of the (B, S)-market. In practice, parameters μ and σ are unknown and ought to be estimated, say, from the statistics of prices S_t. If time intervals between observations are τ, then we have

$$S_t = S_{t-\tau} \, e^{R_t} \,, \quad \text{where} \quad R_t \sim \mathcal{N}\big((\mu - \sigma^2/2) \, \tau, \, \sigma^2 \, \tau \big) \,,$$

and therefore μ and σ can be estimated using the fact that process R is normally distributed with parameters $(\mu - \sigma^2/2) \, \tau$ and $\sigma^2 \, \tau$.

Remark 4.1

Consider the following linear stochastic differential equation:

$$dX_t = X_t \left(\mu_t \, dt + \sigma_t \, dW_t \right) \equiv X_t \, dY_t \,,$$

where X_0 is a finite (a.s.) random variable,

$$Y_t = \int_0^t \mu_s \, ds + \int_0^t \sigma_s \, dW_s,$$

and (in general, random) functions μ_t and σ_t satisfy some integrability conditions (e.g., μ_t and σ_t are bounded). Now we introduce the *stochastic exponential*

$$\mathcal{E}_t(Y) = \exp\left\{ Y_t - \frac{1}{2} \int_0^t \sigma_s^2 \, ds \right\}.$$

It is not difficult to check that

$$X_t = X_0 \, \mathcal{E}_t(Y), \quad t \geq 0,$$

and that the following properties hold true (compare with the discrete case in Chapter 1):

$(\mathcal{E}1)$ $\mathcal{E}_t(Y) > 0$ (a.s.);

$(\mathcal{E}2)$ $1/\mathcal{E}_t(Y) = \mathcal{E}_t(\widetilde{Y})$, where

$$d\widetilde{Y}_t = -dY_t + \sigma_t^2 \, dt \,;$$

$(\mathcal{E}3)$ If $\mu_t \equiv 0$ (which implies that Y_t is a martingale), then $\mathcal{E}_t(Y)$ is a martingale;

$(\mathcal{E}4)$ The multiplication rule:

$$\mathcal{E}_t(Y^1) \, \mathcal{E}_t(Y^2) = \mathcal{E}_t\left(Y^1 + Y^2 + [Y^1, Y^2]\right),$$

where

$$dY_t^i = \mu_t^i \, dt + \sigma_t^i \, dW_t \,, \quad i = 1, 2, \quad \text{and} \quad d[Y^1, Y^2]_t = \sigma_t^1 \, \sigma_t^2 \, dt \,.$$

As in binomial case, we can write the Black-Scholes model in terms of stochastic exponentials:

$$B_t = B_0 \, \mathcal{E}_t(r \, t) \,,$$
$$S_t = S_0 \, \mathcal{E}_t(\mu \, t + \sigma \, W_t) \,.$$

This representation is useful for studying martingale properties of $(S_t)_{t \geq 0}$ and $(S_t/B_t)_{t \geq 0}$. $\square$

Now we introduce the standard basic notions related to a (B, S)-market. If processes $\beta = (\beta_t)_{t \leq T}$ and $\gamma = (\gamma_t)_{t \leq T}$ are adapted to filtration $\mathbb{F}$, then $\pi = (\pi_t)_{t \leq T} := (\beta_t, \gamma_t)_{t \leq T}$ is called a *portfolio* or *strategy* on a (B, S)-market. The *capital* (or *value*) of strategy π is given by

$$X_t^\pi = \beta_t \, B_t + \gamma_t \, S_t.$$

A *contingent claim* f_T with the repayment date T is defined to be a $\mathcal{F}_T$-measurable non-negative random variable. We say that a strategy π is *self-financing* if

$$dX_t^\pi = \beta_t \, dB_t + \gamma_t \, dS_t.$$

A self-financing strategy π is called a *perfect hedge* for a contingent claim f_T if

$$X_T^\pi \geq f_T \qquad \text{(a.s.)}.$$

We say that a strategy π *replicates* f_T if

$$X_T^\pi = f_T \qquad \text{(a.s.)}.$$

A hedge π^* is called the *minimal hedge* if for any other hedge π and for all $t \leq T$

$$X_t^{\pi^*} \leq X_t^\pi \qquad \text{(a.s.)}.$$

The *price* (*fair price*) of a contingent claim f_T is defined as

$$C_T = X_0^{\pi^*}.$$

A (B, S)-market is *complete* if every contingent claim f_N can be *replicated* by some self-financing strategy. We say that a probability P^* is a *martingale* (*risk neutral*) *probability* if $(S_t/B_t)_{t \geq 0}$ is a martingale with respect to P^*. Similar to the discrete case, P^* is completely determined by its density Z_T^*:

$$Z_T^* = \exp\left\{ -\frac{\mu - r}{\sigma} W_T - \frac{1}{2}\left(\frac{\mu - r}{\sigma}\right)^2 T \right\}. \qquad (4.7)$$

The *Girsanov theorem* states that in this setting the process

$$W_t^* = W_t + \frac{\mu - r}{\sigma} t$$

is a Wiener process with respect to the new probability P^* and the initial filtration $\mathbb{F}$. This implies that probability P^* with density (4.7) is the martingale probability for model (4.5)–(4.6), and hence, this model is complete.

Let F_Y and F_Y^* be distribution functions of a random variable Y with respect to probabilities P and P^*, respectively. Then the equality

$$\begin{aligned}
\mu T + \sigma W_T &= rT + \sigma W_T + (\mu - r)T = rT + \sigma\left(W_T + \frac{\mu - r}{\sigma} T\right) \\
&= rT + \sigma W_T^*
\end{aligned}$$

implies that

$$F^*_{\mu T + \sigma W_T} = F^*_{rT + \sigma W^*_T} = F_{rT + \sigma W_T},$$

and therefore

$$F^*_{S_T} = F^*_{S_0 \exp\left\{\left(\mu - \frac{\sigma^2}{2}\right)T + \sigma W_T\right\}} = F_{S_0 \exp\left\{\left(r - \frac{\sigma^2}{2}\right)T + \sigma W_T\right\}}. \tag{4.8}$$

From the general methodology of pricing contingent claims in complete markets, we have

$$C_T = E^*\left(\frac{f_T}{B_T}\right)$$

for any claim f_T.

For a European call option with $f_T = (S_T - K)^+$, using (4.8) we obtain

$$
\begin{aligned}
C_T = E^*\left(\frac{f_T}{B_T}\right) &= e^{-rT} E^*\left((S_T - K)^+\right) \\[2mm]
&= e^{-rT} E^*\left(\left(S_0 \exp\left\{(\mu - \frac{\sigma^2}{2})T + \sigma W_T\right\} - K\right)^+\right) \\[2mm]
&= e^{-rT} E\left(\left(S_0 \exp\left\{(r - \frac{\sigma^2}{2})T + \sigma W_T\right\} - K\right)^+\right) \\[2mm]
&= e^{-rT} E\left(\left(S_0 \exp\left\{(r - \frac{\sigma^2}{2})T + \sigma \sqrt{T} W_1\right\} - K\right)^+\right) \\[2mm]
&= e^{-rT} E\left(\left(S_0 e^{rT} \exp\left\{-\frac{\sigma^2}{2}T + \sigma \sqrt{T} W_1\right\} - K\right)^+\right) \\[2mm]
&= e^{-rT} E\left(\left(a\, e^{b\xi - b^2/2} - K\right)^+\right),
\end{aligned}
\tag{4.9}
$$

where $a = S_0 e^{rT}$, $b = \sigma \sqrt{T}$, and $\xi \sim \mathcal{N}(0,1)$. Here we also used the following property of a Wiener process:

$$W_T = \sqrt{T}\, W_1.$$

Noting that in (4.9)

$$E\left(\left(a\, e^{b\xi - b^2/2} - K\right)^+\right) = a\, \Phi\left(\frac{\ln(a/K) + \frac{1}{2}b^2}{b}\right) - K\, \Phi\left(\frac{\ln(a/K) - \frac{1}{2}b^2}{b}\right),$$

we arrive at the *Black-Scholes* formula:

$$
\begin{aligned}
C_T &= S_0\, \Phi\left(\frac{\ln(a/K) + \frac{1}{2}b^2}{b}\right) - K\, e^{-rT}\, \Phi\left(\frac{\ln(a/K) - \frac{1}{2}b^2}{b}\right) \\[2mm]
&= S_0\, \Phi(y_+) - K\, e^{-rT}\, \Phi(y_-)
\end{aligned}
\tag{4.10}
$$

with
$$y_\pm = \frac{\ln(S_0/K) + T(r \pm \sigma^2/2)}{\sigma\sqrt{T}}.$$

Thus, we found the "fair" non-arbitrage price of a European call option. As in the case of binomial markets, we have the following *call-put parity* relation:

$$P_T = C_T - S_0 + K e^{-rT}, \qquad (4.11)$$

where P_T is the price of a European put option. Relation (4.11) allows us to compute P_T:

$$
\begin{aligned}
P_T &= -S_0 \left(1 - \Phi(y_+)\right) + K e^{-rT} \left(1 - \Phi(y_-)\right) \\
&= -S_0 \Phi(-y_+) + K e^{-rT} \Phi(-y_-).
\end{aligned}
$$

Note that prices C_T and P_T are functions of K, σ, and S_0. Dividing both sides of the identity

$$(S_T - K)^+ - (K - S_T)^+ = S_T - K$$

by e^{rT} and taking expectations with respect to the risk-neutral probability P^*, we obtain

$$C_T(K, \sigma, S_0) - P_T(K, \sigma, S_0) = E^*(S_T) e^{-rT} - K e^{-rT} = S_0 - K e^{-rT}.$$

Finally, using the Black-Scholes formula (4.10), we write

$$
\begin{aligned}
P_T(K, \sigma, S_0) &= C_T(K, \sigma, S_0) - S_0 + K e^{-rT} \\
&= -S_0 \left(1 - \Phi(y_+)\right) + K e^{-rT} \left(1 - \Phi(y_-)\right) \\
&= (-S_0) \Phi\left(\frac{\ln\left(-S_0/(-K)\right) + T(r + (-\sigma)^2/2)}{-\sigma\sqrt{T}}\right) \\
&\quad -(-K e^{-rT}) \Phi\left(\frac{\ln\left(-S_0/(-K)\right) + T(r - \sigma^2/2)}{-\sigma\sqrt{T}}\right) \\
&= C_T(-K, -\sigma, -S_0),
\end{aligned}
$$

which represents the *duality* of prices of European call and put options.

We also can write the price of a European call option at any time $t \in [0, T]$:

$$C_T(t, S_t) = S_t \Phi(y_+(t)) - K e^{-r(T-t)} \Phi(y_-(t)), \qquad (4.12)$$

where
$$y_\pm(t) = \frac{\ln(S_t/K) + (T-t)(r \pm \sigma^2/2)}{\sigma\sqrt{T-t}}.$$

This suggests the following structure of the minimal hedge π^*:

$$
\begin{aligned}
\gamma_t^* &= \Phi(y_+(t)) = \frac{\partial C_T}{\partial S}(t, S_t), \\
\beta_t^* &= -K e^{-r(T)} \Phi(y_-(t)).
\end{aligned}
$$

Since the option price $C_T(t, S_t)$ in (4.12) is a function of time t, price S_t, rate if interest r and volatility σ, one can consider the following "Greeks" often used by the risk management practitioners:

Theta:

$$\theta = \frac{\partial C_T}{\partial t} = \frac{S_t\, \sigma\, \varphi(y_+(t))}{2\sqrt{T-t}} - K\, r\, e^{-r\,(T-t)}\, \Phi(y_-(t))\,,$$

Delta:

$$\Delta = \frac{\partial C_T}{\partial S} = \Phi(y_+(t))\,,$$

Rho:

$$\rho = \frac{\partial C_T}{\partial r} = K\,(T - t)\, e^{-r\,(T-t)}\, \Phi(y_-(t))\,,$$

Vega:

$$\Upsilon = \frac{\partial C_T}{\partial \sigma} = S_t\, \varphi(y_+(t))\, \sqrt{T - t}\,,$$

where $\varphi(x) = \frac{1}{\sqrt{2\pi}} e^{-x^2/2}$. Note that since there is no Greek letter "vega" we use upsilon instead.

Remark 4.2

In Chapter 2 (Section 2.6), we had a brief discussion of the connection between options pricing and partial differential equations. We now consider the continuous-time model (4.5) and a European call option with the pay-off function $f_T = f(S_T)$, which is appropriately smooth. Denote

$u(t, x)$

$$= E^*\left(e^{-r(T-t)} f\left(S_t \exp\left\{r(T - t) + \sigma(W_T^* - W_t^*) - \frac{\sigma^2}{2}(T - t)\right\}\right)\Big| S_t = x\right)$$

$$= e^{-r(T-t)} \int_{-\infty}^{\infty} f\left(x \exp\left\{(r - \sigma^2/2)(T - t) + \sigma y \sqrt{T - t}\right\}\right) \frac{e^{-y^2/2}}{\sqrt{2\pi}}\, dy\,,$$

where E^* is the expectation with respect to a martingale probability P^* with the density (4.7), $W_t^* = W_t + \frac{\mu-r}{\sigma}t$ is a Wiener process with respect to P^* and $(t, x) \in [0, T] \times \mathbb{R}$. Suppose $u \in C^{1,2}([0, T] \times \mathbb{R})$, which holds true for a call option due to the pricing formula (4.12).

Note that $u(t, S_t)$ is the capital of the unique replicating strategy for the contingent claim $f(S_T)$ and $u(T, x) = f(x)$. Applying Kolmogorov-Itô formula to the process $(u(t, S_t)/B_t)$, we obtain

$$\frac{u(t, S_t)}{B_t} = u(0, S_0) + \int_0^t u_x'(v, S_v)\, d\left(\frac{S}{B}\right)_v$$

$$+ \int_0^t \left[u_t'(v, S_v) + rS_v u_x'(v, S_v) + \frac{1}{2}\sigma^2 S_v^2 u_{xx}''(v, S_v) - ru(v, S_v)\right] B_v^{-1}\, dv.$$

Processes $\left(u(t, S_t)/B_t\right)$ and $\left(S_t/B_t\right)$ are martingales with respect to P^*; hence, the third term in the right-hand side of the equality above must vanish. Thus, we arrive at the following partial differential equation:

$$u_t'(t, x) + rxu_x'(t, x) + \frac{1}{2}\sigma^2 x^2 u_{xx}''(t, x) - ru(t, x) = 0 \qquad (4.13)$$

with the boundary condition $u(T, x) = f(x)$. This equation is usually referred to as the *Black-Scholes differential equation*, and it serves the whole class of European options, including put and call options. $\square$

Next, we consider an optimal investment problem in the framework of the Black-Scholes model, where the optimal strategy is defined by the relation

$$\sup_{\pi \in SF} E\left(\ln X_T^\pi\right) = E\left(\ln X_T^{\pi^*}\right), \qquad X_0^\pi = X_0^{\pi^*} = x.$$

We sketch the solution of this optimal investment problem: find

$$Y_T^*(x) = \sup_Y E\left(\ln Y_T(x)\right),$$

where supremum is taken over the set of all positive martingales with respect to P^*, starting at x.

Let the *optimal* martingale be

$$Y_t^*(x) = E^*\left(\left.\frac{x}{Z_T^*}\right|\mathcal{F}_t\right), \qquad t \in [0, T],$$

where

$$Z_T^* = \exp\left\{-\frac{\mu - r}{\sigma}W_T - \frac{1}{2}\left(\frac{\mu - r}{\sigma}\right)^2 T\right\}$$

is the density of the unique martingale probability P^* with respect to P.

As in the case of the binomial model, it can be shown that

$$E\left(\ln Y_T(x)\right) \le E\left(\ln Y_T^*(x)\right).$$

Then using the martingale characterization of self-financing strategies, we obtain

$$Y_t^* = \frac{X_t^{\pi^*}(x)}{B_t} = X_t^*$$

for some self-financing strategy $\pi^* = \left(\beta_t^*, \gamma_t^*\right)_{t \le T}$. Denote

$$\alpha_t^* = \frac{\gamma_t^* S_t}{X_t^{\pi^*}}$$

the proportion of risky capital in portfolio π^*. By the Kolmogorov-Itô formula, we have

$$dX_t^* = X_t^* \alpha_t^* \sigma\, d\widetilde{W}_t,$$

and therefore

$$X_T^* = x \exp\left\{\sigma \alpha^* W_T + \alpha^* (\mu - r) T - \frac{1}{2}\sigma^2 (\alpha^*)^2 T\right\},$$

where $\alpha_t^* \equiv \alpha^*$.

However,

$$X_T^* = \frac{x}{Z_T^*} = x \exp\left\{\frac{\mu - r}{\sigma} W_T + \frac{1}{2}\left(\frac{\mu - r}{\sigma}\right)^2 T\right\}.$$

Comparing these formulas, we deduce the expression for the optimal proportion:

$$\alpha^* = \frac{\mu - r}{\sigma^2},$$

which is often referred to as *Merton's point*.

Worked Example 4.1 *Find prices of European call and put options on a Black-Scholes market if $r = 0.1$, $T = 215/365$, $S_0 = 100(\$)$, $K = 80(\$)$, $\mu = r$, $\sigma = 0.1$.*

Solution Using Black-Scholes formula, we compute

$$
\begin{aligned}
C_T = C_T(K, S_0, \sigma) \;&=\; S_0\,\Phi(y_+) - K\,e^{-rT}\,\Phi(y_-) \\[2mm]
&=\; 100\,\Phi\left(\frac{\ln(100/80) + \frac{215}{365}(0.1 + (0.1)^2/2)}{0.1\sqrt{215/365}}\right) \\[2mm]
&\quad - 80\,e^{-0.1\frac{215}{365}}\,\Phi\left(\frac{\ln(100/80) + \frac{215}{365}(0.1 - (0.1)^2/2)}{0.1\sqrt{215/365}}\right) \\[2mm]
&=\; 100\,\Phi(3.177) - 80\,e^{-0.1\frac{215}{365}}\,\Phi(3.64) \approx 24.57.
\end{aligned}
$$

The call-put parity can be used now to find the price of a European put option:

$$P_T = P_T(K, S_0, \sigma) = C_T - S_0 + K\,e^{-rT} = 24.57 - 100 + 80\,e^{-0.1\frac{215}{365}} \approx 0.$$

If we increase the rate of interest to $r = 0.2$, then

$$C_T \approx 28.9 \quad \text{and} \quad P_T \approx 0.$$

Increasing volatility to $\sigma = 0.8$ implies higher prices:

$$C_T \approx 35.55 \quad \text{and} \quad P_T \approx 10.97 \quad \text{for} \quad r = 0.1,$$

and

$$C_T \approx 38.05 \quad \text{and} \quad P_T \approx 9.16 \quad \text{for} \quad r = 0.2. \quad \square$$

4.2   Beyond of the Black-Scholes model

Let us consider a model consisting of a bank account B and two risky assets $S^i, i = 1, 2$:

$$
\begin{aligned}
dB_t &= rB_t dt\,, \quad B_0 = 1, \\
dS_t^i &= S_t^i\big(\mu_i dt + \sigma_i dW_t^i\big)\,, \quad S_0^i > 0\,.
\end{aligned}
\tag{4.14}
$$

Here $(W_t^i)_{t \geq 0}$, $i = 1, 2$, are two standard Wiener processes with $Cov(W_t^1, W_t^2) = \rho t$, $-1 < \rho \leq 1$ and μ_i, σ_i are constants.

Noting that every investment strategy in market (4.14) consists of β_t units of B and γ_t^i units of S^i, $i = 1, 2$, we can use definitions of self-financing and hedging strategies, which were introduced in Section 4.1. Let us consider first the case of $|\rho| < 1$. A natural candidate for martingale probability P^* should have a density of the form

$$
Z_T^* = \mathcal{E}_T(N),
$$

where process $N_t \equiv \varphi_1 W_t^1 + \varphi_2 W_t^2$ with some parameters φ_1 and φ_2. To determine these parameters, we use the fact that both processes $(S_t^i/B_t)_{t \geq 0}$, $i = 1, 2$, must be martingales with respect to P^* with density Z_T^*. Thus, we arrive at the following equations:

$$
\begin{aligned}
(\mu_1 - r)t + \sigma_1\varphi_1 t + \sigma_1\varphi_2\rho t &= 0, \\
(\mu_2 - r)t + \sigma_2\varphi_2 t + \sigma_2\varphi_1\rho t &= 0.
\end{aligned}
\tag{4.15}
$$

Solving this system of equations, we obtain the following expressions for φ_1 and φ_2 in terms of the parameters of model (4.14):

$$
\begin{aligned}
\varphi_1 &= \frac{r(\sigma_2 - \sigma_1\rho) + \rho\mu_2\sigma_1 - \mu_1\sigma_2}{\sigma_1\sigma_2(1 - \rho^2)}\,, \\
\varphi_2 &= \frac{r(\sigma_1 - \sigma_2\rho) + \rho\mu_1\sigma_2 - \mu_2\sigma_1}{\sigma_1\sigma_2(1 - \rho^2)}\,.
\end{aligned}
$$

Using these expressions, we arrive at the formula for density

$$
Z_T^* = \exp\left\{\varphi_2 W_T^1 + \varphi_2 W_T^2 - \frac{\sigma_\varphi^2}{2}T\right\},
$$

where

$$
\sigma_\varphi^2 = \varphi_1^2 + \varphi_2^2 + 2\rho\varphi_1\varphi_2.
$$

Note that market (4.14) is not necessarily complete if one of the risky assets is *not tradeable*. In this case, system (4.15) is reduced to only one equation with two unknowns and admits infinitely many solutions, and hence, there are *infinitely many martingale measures*.

Now let $\rho = 1$. This implies that $W_t^1 = W_t^2 = W_t$. In order to prevent the existence of arbitrage opportunities in model (4.14), we assume that

$$\frac{\mu_1 - r}{\sigma_1} = \frac{\mu_2 - r}{\sigma_2}.$$

Then the martingale probability P^* has the following density:

$$Z_T^* = \exp\left\{ -\frac{\mu_1 - r}{\sigma_1} W_T - \frac{1}{2}\left(\frac{\mu_1 - r}{\sigma_1}\right)^2 T \right\}.$$

Let us consider a European option to exchange S^1 with S^2. We can price this option by calculating

$$E^*\left(\frac{(S_T^1 - S_T^2)^+}{B_T}\right). \tag{4.16}$$

We will need the following result.

Lemma 4.1 *Let X and Y be Gaussian random variables with means μ_X and μ_Y, respectively, and the covariance matrix*

$$A = \begin{pmatrix} \sigma_X^2 & \rho_{XY} \\ \rho_{XY} & \sigma_Y^2 \end{pmatrix}.$$

Then

$$E\left[1_{\{X \le x\}} \exp\{-Y\}\right] = \exp\left\{\frac{\sigma_Y^2}{2} - \mu_Y\right\} \Phi(\tilde{x}) \tag{4.17}$$

and

$$E\left[1_{\{X \le x\}} \cdot X \cdot \exp\{-Y\}\right] = \exp\left\{\frac{\sigma_Y^2}{2} - \mu_Y\right\}\left[(\mu_X - \rho_{XY})\Phi(\tilde{x}) - \sigma_x \varphi(\tilde{x})\right], \tag{4.18}$$

where

$$\tilde{x} = \frac{x - (\mu_X - \rho_{XY})}{\sigma_X}$$

and

$$\varphi(x) = \frac{1}{\sqrt{2\pi}} e^{-\frac{x^2}{2}} \quad \text{and} \quad \Phi(x) = \int_{-\infty}^{x} \varphi(y)dy.$$

Proof It is clear that the joint density of vector (X, Y) is

$$f(x, y) = \frac{1}{2\pi\sigma_X\sigma_Y\sqrt{(1 - \rho^2)}} \exp\left\{ -\frac{1}{2} Q(x, y) \right\},$$

where $\rho := \frac{\rho_{XY}}{\sigma_X\sigma_Y}$ and

$$Q(x, y) := \frac{1}{1 - \rho^2}\left[\left(\frac{x - \mu_X}{\sigma_x}\right)^2 - 2\rho\left(\frac{x - \mu_X}{\sigma_X}\right)\left(\frac{y - \mu_Y}{\sigma_Y}\right) + \left(\frac{y - \mu_Y}{\sigma_Y}\right)^2\right].$$

First, we compute

$$E\left[1_{\{X\leq t\}}\exp\{-Y\}\right] = \int\int_{R^2} 1_{\{x\leq t\}}\exp\{-y\}f(x,y)dxdy,$$

where the double integral can be calculated by iterated integration in any order:

$$E\left[1_{\{X\leq t\}}\exp\{-Y\}\right] = \int_{-\infty}^{t}\left(\int_{-\infty}^{+\infty}\exp\{-y\}f(x,y)dy\right)dx. \qquad (4.19)$$

Define

$$J(x) = \int_{-\infty}^{+\infty}\exp\{-y\}f(x,y)dy,$$

so that equation (4.19) takes the form

$$E\left[1_{\{X\leq t\}}\exp\{-Y\}\right] = \int_{-\infty}^{+\infty}J(x)dx. \qquad (4.20)$$

Note that by completing the square we can write

$$Q(x,y) = \frac{1}{1-\rho^2}\left[\left(\frac{y-\mu_Y}{\sigma_Y}\right) - \rho\left(\frac{x-\mu_X}{\sigma_X}\right)\right]^2 + \left(\frac{x-\mu_X}{\sigma_X}\right)^2,$$

which implies

$$J(x) = \frac{1}{2\pi\sigma_X\sigma_Y\sqrt{1-\rho^2}}\int_{-\infty}^{+\infty}\exp\{-y\}\exp\left\{-\frac{1}{2}Q(x,y)\right\}dy$$

$$= \frac{1}{2\pi\sigma_X\sigma_Y\sqrt{1-\rho^2}}\int_{-\infty}^{+\infty}\exp\left\{-y-\frac{1}{2}Q(x,y)\right\}dy$$

$$= \frac{1}{2\pi\sigma_X\sigma_Y\sqrt{1-\rho^2}}\exp\left\{-\frac{1}{2}\left(\frac{x-\mu_X}{\sigma_X}\right)^2\right\}dy$$

$$\times \int_{-\infty}^{+\infty}\exp\left\{-y-\frac{1}{2(1-\rho^2)}\left[\left(\frac{y-\mu_Y}{\sigma_Y}\right)-\rho\left(\frac{x-\mu_X}{\sigma_X}\right)\right]^2\right\}dy.$$

Further, we obtain

$$y + \frac{1}{2(1-\rho^2)}\left[\left(\frac{y-\mu_Y}{\sigma_Y}\right)-\rho\left(\frac{x-\mu_X}{\sigma_X}\right)\right]^2$$

$$= \frac{1}{2\sigma_Y^2(1-\rho^2)}[y-\beta(x)]^2 + \frac{1}{2(1-\rho^2)}\left[\alpha(x)-\frac{\beta(x)^2}{\sigma_Y^2}\right],$$

where

$$\alpha(x) = \rho^2\left(\frac{x-\mu_X}{\sigma_X}\right)^2 + \frac{2\rho\mu_Y}{\sigma_Y}\frac{x-\mu_X}{\sigma_X} + \frac{\mu_Y^2}{\sigma_Y^2}$$

and

$$\beta(x) = \mu_X + \rho\sigma_Y \frac{x - \mu_X}{\sigma_X} - \sigma_Y^2(1 - \rho^2).$$

Defining

$$g(x) := \exp\left\{ -\frac{1}{2}\left(\frac{x - \mu_X}{\sigma_X}\right)^2 - \frac{1}{2(1 - p^2)}\left[\alpha(x) - \frac{\beta(x)^2}{\sigma_Y^2}\right]\right\},$$

we obtain

$$J(x) = \frac{1}{2\pi\sigma_X\sigma_Y\sqrt{1 - \rho^2}} \exp\left\{ -\frac{1}{2}\left(\frac{x - \mu_X}{\sigma_X}\right)^2\right\}$$

$$\times \int_{-\infty}^{+\infty} \exp\left\{ -\frac{1}{2\sigma_Y^2(1 - p^2)}[y - \beta(x)]^2 - \frac{1}{2(1 - \rho^2)}\left[\alpha(x) - \frac{\beta(x)^2}{\sigma_Y^2}\right]\right\} dy$$

$$= \frac{1}{2\pi\sigma_X\sigma_Y\sqrt{1 - \rho^2}} \exp\left\{ -\frac{1}{2}\left(\frac{x - \mu_X}{\sigma_X}\right)^2 - \frac{1}{2(1 - \rho^2)}\left[\alpha(x) - \frac{\beta(x)^2}{\sigma_Y^2}\right]\right\}$$

$$\times \int_{-\infty}^{+\infty} \exp\left\{ -\frac{1}{2\sigma_Y^2(1 - \rho^2)}[y - \beta(x)]^2\right\} dy$$

$$= \frac{1}{2\pi\sigma_X\sigma_Y\sqrt{1 - \rho^2}} g(x) \int_{-\infty}^{+\infty} \exp\left\{ -\frac{1}{2\sigma_Y^2(1 - \rho^2)}[y - \beta(x)]^2\right\} dy$$

$$= \frac{1}{2\pi\sigma_X\sigma_Y\sqrt{1 - \rho^2}} g(x) \int_{-\infty}^{+\infty} \exp\left\{ -\frac{z^2}{2}\right\}\sigma_Y\sqrt{1 - \rho^2} dz$$

$$= \frac{1}{\sqrt{2\pi}\sigma_X} g(x). \tag{4.21}$$

We also have

$$\frac{\beta(x)^2}{\sigma_Y^2} = \alpha(x) - 2\rho\sigma_Y(1 - \rho^2)\frac{x - \mu_X}{\sigma_X} + \sigma_Y^2(1 - \rho^2)^2 - 2\mu_Y(1 - \rho^2)$$

or after rearranging the terms:

$$\alpha(x) - \frac{\beta(x)^2}{\sigma_Y^2} = 2\rho\sigma_Y(1 - \rho^2)\frac{x - \mu_X}{\sigma_X} - \sigma_Y^2(1 - \rho^2)^2 + 2\mu_Y(1 - \rho^2);$$

hence,

$$\frac{1}{(1 - \rho^2)}\left[\alpha(x) - \frac{\beta(x)^2}{\sigma_Y^2}\right] = 2\rho\sigma_Y\frac{x - \mu_X}{\rho_X} - \sigma_Y^2(1 - \rho^2) + 2\mu_Y.$$

Next, we calculate

$$\left(\frac{x-\mu_X}{\sigma_X}\right)^2 + \frac{1}{(1-\rho^2)}\left[\alpha(x) - \frac{\beta(x)^2}{\sigma_Y^2}\right]$$

$$= \left(\frac{x-\mu_X}{\sigma_X}\right)^2 + 2\rho\sigma_Y\frac{x-\mu_X}{\sigma_X} + \rho^2\sigma_Y^2 - \sigma_Y^2 + 2\mu_Y$$

$$= \left[\frac{x-\mu_X}{\sigma_X} + \rho\sigma_Y\right]^2 - \sigma_Y^2 + 2\mu_Y$$

$$= \left[\frac{x-\mu_X + \rho\sigma_Y\sigma_X}{\sigma_X}\right]^2 - \sigma_Y^2 + 2\mu_Y$$

$$= \left[\frac{x-\mu_X + \rho_{XY}}{\sigma_X}\right]^2 - \sigma_Y^2 + 2\mu_Y$$

$$= \left[\frac{x-(\mu_X - \rho_{XY})}{\sigma_X}\right]^2 - \sigma_Y^2 + 2\mu_Y,$$

where we have used the fact that $\rho\sigma_Y\sigma_X = \rho_{XY}$. Thus, we can write $g(x)$ in the following form:

$$g(x) = \exp\left\{-\frac{1}{2}\left[\frac{x-(\mu_X - \rho_{XY})}{\sigma_X}\right]^2\right\}\exp\left\{\frac{\sigma_Y^2}{2} - \mu_Y\right\}.$$

Note that, if we define

$$h(x) = -\frac{1}{2}\left[\frac{x-(\mu_X - \rho_{XY})}{\sigma_X}\right]^2,$$

then (4.21), and the expression for $g(x)$ implies that we can write $J(x)$ as

$$J(x) = -\frac{1}{\sqrt{2\pi}\sigma_X}\exp\{h(x)\}\exp\left\{\frac{\sigma_Y^2}{2} - \mu_Y\right\}.$$

Finally, using equation (4.20), we obtain

$$E\left[1_{\{x\le l\}}\exp\{-Y\}\right] = \int_{-\infty}^{t} J(x)dx$$

$$= \int_{-\infty}^{t}\frac{1}{\sqrt{2\pi}\sigma_X}\exp\{h(x)\}\exp\left\{\frac{\sigma_Y^2}{2} - \mu_Y\right\}dx$$

$$= \exp\left\{\frac{\sigma_Y^2}{2} - \mu_Y\right\}\int_{-\infty}^{t}\frac{1}{\sqrt{2\pi}\sigma_X}\exp\left\{-\frac{1}{2}\left[\frac{x-(\mu_X - \rho_{XY})}{\sigma_X}\right]^2\right\}dx$$

$$= \exp\left\{\frac{\sigma_Y^2}{2} - \mu_Y\right\}\int_{-\infty}^{\tilde{t}}\frac{1}{\sqrt{2\pi}\sigma_X}\exp\left\{-\frac{z^2}{2}\right\}\sigma_X dz$$

$$= \exp\left\{\frac{\sigma_Y^2}{2} - \mu_Y\right\}\Phi(\tilde{t}),$$

where we used the change of variables

$$z = \frac{x - (\mu - \rho_{XY})}{\sigma_X} \quad \text{and} \quad \tilde{t} = \frac{t - (\mu_X - \rho_{XY})}{\sigma_X}.$$

Renaming t as x and $\tilde{t}$ as $\tilde{x}$, we arrive at formula (4.17).

Formula (4.18) will be proved in a similar way. We have

$$E[1_{\{x \leq t\}} \cdot X \cdot \exp\{-Y\}] = \int\int_{\mathbb{R}^2} 1_{\{x \leq t\}} \cdot x \cdot \exp\{-y\} f(x, y) dx dy$$

$$= \int_{-\infty}^{t} x \left(\int_{-\infty}^{+\infty} \exp\{-y\} f(x, y) dy \right) dx = \int_{-\infty}^{t} x J(x) dx$$

$$= \exp\left\{ \frac{\sigma_Y^2}{2} - \mu_Y \right\} \int_{-\infty}^{t} x \frac{1}{\sqrt{2\pi}\sigma_X} \exp\{h(x)\} dx.$$

Note that

$$h'(x) = -\frac{1}{\sigma_X} \left[\frac{x - (\mu_X - \rho_{XY})}{\sigma_X} \right],$$

then

$$\int_{-\infty}^{t} h'(x) e^{h(x)} dx = \int_{-\infty}^{t} \frac{d}{dx} \left[e^{h(x)} \right] dx = e^{h(t)} - \lim_{t \to \infty} e^{h(s)} = e^{h(t)}$$

and

$$\int_{-\infty}^{t} h'(x) e^{h(x)} dx = \int_{-\infty}^{t} -\frac{1}{\sigma_X} \left[\frac{x - (\mu_X - \rho_{XY})}{\sigma_X} \right] e^{h(x)} dx$$

$$= -\frac{1}{\sigma_X^2} \int_{-\infty}^{t} x e^{h(x)} dx + \frac{1}{\sigma_X^2} (\mu_X - \rho_{XY}) \int_{-\infty}^{t} e^{h(x)} dx.$$

Combining the last two expressions, we arrive at

$$\int_{-\infty}^{t} x e^{h(x)} dx = (\mu_X - \rho_{XY}) \int_{-\infty}^{t} e^{h(x)} dx - \sigma_X^2 e^{h(t)}.$$

Using this, we obtain

$$E[1_{\{X \leq t\}} \cdot X \cdot \exp\{-Y\}] = \exp\left\{ \frac{\sigma_Y^2}{2} - \mu_Y \right\} \int_{-\infty}^{t} x \frac{1}{\sqrt{2\pi}\sigma_X} \exp\{h(x)\} dx$$

$$= \exp\left\{ \frac{\sigma_Y^2}{2} - \mu_Y \right\} \int_{-\infty}^{t} x \frac{1}{\sqrt{2\pi}\sigma_X} x e^{h(x)} dx$$

$$= \exp\left\{ \frac{\sigma_Y^2}{2} - \mu_Y \right\} \left\{ (\mu_X - \rho_{XY}) \int_{-\infty}^{t} \frac{1}{\sqrt{2\pi}\sigma_X} e^{h(x)} dx - \sigma_X \frac{1}{\sqrt{2\pi}} e^{h(t)} \right\}$$

$$= \exp\left\{ \frac{\sigma_Y^2}{2} - \mu_Y \right\} \left\{ (\mu_X - \rho_{XY}) \Phi(\tilde{t}) - \sigma_X \varphi(\tilde{t}) \right\},$$

where we used the facts that

$$\int_{-\infty}^{t} \frac{1}{\sqrt{2\pi}\sigma_X} e^{h(x)} dx = \Phi(\tilde{t}) \quad \text{and} \quad \frac{1}{\sqrt{2\pi}} e^{h(t)} = \varphi(\tilde{t}).$$

Renaming t as x and $\tilde{t}$ as $\tilde{x}$, we arrive at formula (4.18). $\square$
   We now calculate price (4.16):

$$
\begin{aligned}
E^* \left(\frac{(S_T^1 - S_T^2)^+}{B_T} \right) &= E^* \left(\left(\frac{S_T^1}{B_T} - \frac{S_T^2}{B_T} \right) \cdot I_{\left\{ \frac{S_T^1}{S_T^2} \geq 1 \right\}} \right) \qquad (4.22) \\
&= E^* \left(\frac{S_T^1}{B_T} \right) - E^* \left(\frac{S_T^2}{B_T} \right) - E^* \left(\frac{S_T^1}{B_T} - \frac{S_T^2}{B_T} \right) I_{\left\{ \frac{S_T^1}{S_T^2} \leq 1 \right\}}.
\end{aligned}
$$

The first two terms in (4.22) are equal to S_0^1 and S_0^2, respectively, since $(S_t^i/B_t)_{t\geq 0}$, $i = 1, 2$, are martingales with respect to P^*. To calculate the third term, we compute

$$
\begin{aligned}
& E^* \left(\frac{S_T^i}{B_T} \cdot I_{\left\{ \frac{S_T^1}{S_T^2} \leq 1 \right\}} \right) \\
&= E^* \left(S_0^i \exp\left\{ \left(\mu_i - r - \frac{\sigma_i^2}{2} \right) T + \sigma_i W_T \right\} \cdot I_{\left\{ \frac{\exp\{(\mu_1 - r - \sigma_1^2/2)T + \sigma_1 W_T\}}{\exp\{(\mu_2 - r - \sigma_2^2/2)T + \sigma_2 W_T\}} \leq \frac{S_0^2}{S_0^1} \right\}} \right) \\
&= E^* \left(S_0^i \exp\left\{ -\frac{\sigma_i^2}{2} T + \sigma_i W_T^* \right\} \cdot I_{\left\{ \exp\{(-\frac{\sigma_1^2}{2} + \frac{\sigma_2^2}{2})T + (\sigma_1 - \sigma_2)W_T^*\} \leq \frac{S_0^2}{S_0^1} \right\}} \right)
\end{aligned}
$$

for $i = 1, 2$. Let us denote

$$Y_i = -\left(-\frac{\sigma_i^2 T}{2} + \sigma_i W_T^* \right) \quad \text{and} \quad X = \left(-\frac{\sigma_1^2}{2} + \frac{\sigma_2^2}{2} \right) T + (\sigma_1 - \sigma_2) W_T^*,$$

and note that they are Gaussian random variables (with respect to P^*). Using (4.22) and Lemma 4.1, we arrive at the following *Margrabe formula* ($\sigma_1 > \sigma_2$):

$$E^* \left(\frac{(S_T^1 - S_T^2)^+}{B_T} \right) = S_0^1 \Phi(b_+(S_0^1, S_0^2, T)) - S_0^2 \Phi(b_-(S_0^1, S_0^2, T)),$$

where

$$b_\pm(S_0^1, S_0^2, T) = \frac{\ln \frac{S_0^1}{S_0^2} \pm (\sigma_1 - \sigma_2)^2 \frac{T}{2}}{(\sigma_1 - \sigma_2)\sqrt{T}}.$$

   Let us discuss another aspect of model (4.14). We can consider a case when the volatility of the first asset S^1 is stochastic and $\sigma_t^2 = S_t^2$. As we mentioned earlier, this market will be incomplete. Hence, using the methodology of option pricing in incomplete markets (see Chapter 3), we can obtain an interval of non-arbitrage prices. Let us consider the Black-Scholes model (4.6):

$$dS_t = S_t(\mu dt + \sigma_t dW_t), \quad S_0 > 0, \tag{4.23}$$

where volatility is a random process such that

$$\sigma_t^2 = \sigma^2 + (-1)^{\Pi_t} \Delta\sigma^2, \quad \Delta\sigma^2 \ll \sigma^2$$

and $(\Pi_t)_{t \geq 0}$ is a Poisson process with unit intensity (formal definition and properties of Poisson process can be found in Section 7.1).

We will now describe the interval of non-arbitrage prices for a European call option $(S_T - K)^+$ in model (4.23). Introducing a new process $(\alpha_t)_{t \geq 0}$ with values $\sigma_{\pm} = \sqrt{\sigma^2 \pm \Delta\sigma^2}$, we can view the price process $(S_t)_{t \geq 0}$ as a *controlled diffusion process* $(S_t^\alpha)_{t \geq 0}$ (see Section 6.1). We assume for simplicity that $r = 0$. Then the capital of minimal hedge for $(S_T - K)^+$ is equal to

$$v(t,x) = \sup_\alpha E^* \left((S_T^\alpha - K)^+ \middle| S_t^\alpha = x \right), \quad x \in \mathbb{R}_+, \ t \geq T.$$

It is shown in Section 6.1 that function $v(t,x)$ satisfies the following Bellmann differential equation:

$$v_t' + \frac{1}{2}\sigma^2 x^2 v_{xx}'' + \frac{1}{2}\Delta\sigma^2 |v_{xx}''| = 0, \tag{4.24}$$

$$v(T,x) = (x - K)^+, \ x \in \mathbb{R}_+.$$

If $\Delta\sigma^2 = 0$, then equation (4.24) becomes the Black-Scholes differential equation. We can construct a solution of (4.24) from a solution of the Black-Scholes equation with some corrections for quantity $\Delta\sigma^2/\sigma^2$ as a "small" parameter. Indeed, changing variables in (4.24),

$$\xi = \ln x - \frac{\sigma^2}{2}(T - t) \quad \text{and} \quad \theta = \sigma^2(T - t),$$

we obtain the following partial derivatives for $v(t,x) = V(\theta, \xi)$:

$$\frac{\partial v}{\partial t} = \sigma^2 \left(\frac{1}{2}\frac{\partial V}{\partial \xi} - \frac{\partial V}{\partial \theta} \right), \qquad \frac{\partial v}{\partial x} = \frac{1}{x}\frac{\partial V}{\partial \xi},$$

$$\frac{\partial^2 v}{\partial x^2} = \frac{1}{x^2} \left(\frac{\partial^2 V}{\partial \xi^2} - \frac{\partial V}{\partial \xi} \right),$$

and thus, we arrive at the following equation for $V(\theta, \xi)$:

$$\frac{\partial V}{\partial \theta} = \frac{1}{2}\frac{\partial^2 V}{\partial \xi^2} + \frac{1}{2}\frac{\Delta\sigma^2}{\sigma^2} \left| \frac{\partial^2 V}{\partial \xi^2} - \frac{\partial V}{\partial \xi} \right|, \tag{4.25}$$

$$V(0, \xi) = (e^\xi - K)^+.$$

It is rather difficult to solve equation (4.25) analytically. We will find an approximate solution

$$V(\theta, \xi) \approx V_0(\theta, \xi) + \Delta\sigma^2 V_1(\theta, \xi) + o(\Delta\sigma^2). \tag{4.26}$$

Substituting (4.26) into (4.25), we obtain

$$\frac{\partial V_0}{\partial \theta} = \frac{1}{2}\frac{\partial^2 V_0}{\partial \xi^2}, \qquad V_0(0, \xi) = (e^\xi - K)^+, \tag{4.27}$$

$$\frac{\partial V_1}{\partial \theta} = \frac{1}{2}\frac{\partial^2 V_1}{\partial \xi^2} + \frac{1}{2\sigma^2}\left|\frac{\partial^2 V_0}{\partial \xi^2} - \frac{\partial V_0}{\partial \xi}\right|, \qquad V_1(0, \xi) = 0. \tag{4.28}$$

Equation (4.27) is the Black-Scholes equation, and its solution has the form

$$V_0(\theta, \xi) = e^{\xi+\frac{\theta}{2}}\Phi\left(\frac{\xi + \theta - \ln K}{\sqrt{\theta}}\right) - K\Phi\left(\frac{\xi - \ln K}{\sqrt{\theta}}\right). \tag{4.29}$$

From (4.29), we calculate the partial derivatives

$$\frac{\partial V_0}{\partial \xi}$$

$$= e^{\xi+\frac{\theta}{2}}\left[\Phi\left(\frac{\xi + \theta - \ln K}{\sqrt{\theta}}\right) + \frac{1}{\sqrt{\theta}}\varphi\left(\frac{\xi + \theta - \ln K}{\sqrt{\theta}}\right)\right] - \frac{K}{\sqrt{\theta}}\varphi\left(\frac{\xi - \ln K}{\sqrt{\theta}}\right),$$

$$\frac{\partial^2 V_0}{\partial \xi^2}$$

$$= e^{\xi+\frac{\theta}{2}}\left[\Phi\left(\frac{\xi + \theta - \ln K}{\sqrt{\theta}}\right) + \left(1 + \frac{2}{\sqrt{\theta}} - \frac{\xi - \ln K}{\sqrt{\theta}}\right)\varphi\left(\frac{\xi + \theta - \ln K}{\sqrt{\theta}}\right)\right]$$

$$+ \frac{K}{\sqrt{\theta}}\left(\frac{\xi - \ln K}{\sqrt{\theta}}\right)\varphi\left(\frac{\xi - \ln K}{\sqrt{\theta}}\right).$$

Then, taking into account the equality

$$\varphi\left(\frac{\xi + \theta - \ln K}{\sqrt{\theta}}\right) = Ke^{-\xi+\frac{\theta}{2}}\varphi\left(\frac{\xi - \ln K}{\sqrt{\theta}}\right),$$

we obtain the following expression for the non-linear term in (4.28):

$$\frac{\partial^2 V_0}{\partial \xi^2} - \frac{\partial V_0}{\partial \xi} = \frac{K}{\sqrt{\theta}}\varphi\left(\frac{\xi - \ln K}{\sqrt{\theta}}\right) \geq 0.$$

Thus,

$$V_1(\theta, \xi) = \frac{K\sqrt{\theta}}{2\sigma^2}\varphi\left(\frac{\xi - \ln K}{\sqrt{\theta}}\right)$$

and therefore

$$v(0, S_0) \approx S_0 \Phi\left(\frac{\ln \frac{S_0}{K} + \frac{\sigma^2}{2}T}{\sigma\sqrt{T}}\right) - K\Phi\left(\frac{\ln \frac{S_0}{K} - \frac{\sigma^2}{2}T}{\sigma\sqrt{T}}\right)$$
$$+ \frac{K\Delta\sigma^2}{2\sigma^2}\sigma\sqrt{T}\,\varphi\left(\frac{\ln \frac{S_0}{K} - \frac{\sigma^2}{2}T}{\sigma\sqrt{T}}\right).$$

This formula determines the upper bound of non-arbitrage prices for a call option.

Another possibility to generalize the Black-Scholes model consists in splitting a bank account with the interest rate r into two accounts: savings account B^1 and credit account B^2 with interest rates r^1 and r^2 ($r^1 \leq r^2$), respectively. As in Section 3.2, we arrive at a continuous time (B^1, B^2, S)-market:

$$\begin{aligned} dB_t^i &= r^i B_t^i dt, \quad B_0^i = 1, \ i = 1, 2, \qquad (4.30)\\ dS_t &= S_t(\mu dt + \sigma dW_t), \quad S_0 \geq 0. \end{aligned}$$

An admissible strategy $\pi_t = (\beta_t^1, \beta_t^2, \gamma_t)$ has the value

$$X_t^\pi = \beta_t^1 B_t^1 + \beta_t^2 B_t^2 + \gamma_t S_t,$$

and strategy π_t is self-financing if

$$dX_t^\pi = \beta_t^1 dB_t^1 + \beta_t^2 dB_t^2 + \gamma_t dS_t.$$

To avoid arbitrage, we suppose that $\beta_t^1 \geq 0$ and $\beta_t^2 \leq 0$. Using the proportion $\alpha_t = \gamma_t S_t / X_t^\pi$, we obtain

$$\begin{aligned} dX_t^\pi(x) &= X_t^\pi(x)\big[(1 - \alpha_t)^+ r^1 - (1 - \alpha_t)^- r^2 + \alpha_t S_t^{-1} dS_t\big]\\ &= X_t^\pi(x)\big[(1 - \alpha_t)^+ r^1 - (1 - \alpha_t)^- r^2 + \alpha_t(\mu dt + \sigma dW_t)\big] \end{aligned}$$

with $X_0^\pi = X_0^\pi(x) = x > 0$. Suppose that $f_T = f(S_T)$ is a contingent claim in market (4.30).

Let us introduce an *auxiliary Black-Scholes market* with the same risky asset S and a bank account B^d with the interest rate $r^d = r_1 + d$, $d \in [0, r^2 - r^1]$, and $B_0^d = 1$. Denote P^d the corresponding martingale probability in this complete market, then the price of f_T is

$$C_T(f_T, r^d) = E^d\big(f_T / B_T^d\big).$$

Proportion α_t defines two self-financing strategies $\pi(\alpha)$ and $\pi(\alpha, d)$ in (B^1, B^2, S)-market and (B^d, S)-market, respectively. We arrive at the following result, which is similar to Lemma 3.1.

Proposition 4.1 *If $X_0^{\pi(\alpha)} = X_0^{\pi(\alpha,d)}$, then the following two conditions are equivalent*

$$X_0^{\pi(\alpha)} = X_0^{\pi(\alpha,d)} \quad \text{for all} \quad t \leq T,$$

and

$$(r^2 - r^1 - d)(1 - \alpha_t)^- + d(1 - \alpha_t)^+ = 0.$$

Using this statement, we can compare capitals of self-financing strategies with given proportions α_t in both market models. Hence, the quantities

$$\inf_d C_T(f_T, r^d) \quad \text{and} \quad \sup_d C_T(f_T, r^d)$$

are the natural bounds for non-arbitrage prices of f_T in the (B^1, B^2, S)-market.

Applying this methodology to a European call option $f_T = (S_T - K)^+$ and using the monotonicity of the Black-Scholes price with respect to the rate of interest, we obtain the quantities

$$C_T(r^i) = S_0 \Phi\left(\frac{\ln\frac{S_0}{K} + (r^i + \frac{\sigma^2}{2})T}{\sigma\sqrt{T}}\right) - Ke^{-r^iT}\Phi\left(\frac{\ln\frac{S_0}{K} + (r^i - \frac{\sigma^2}{2})T}{\sigma\sqrt{T}}\right),$$
$$i = 1, 2,$$

as lower $(i = 1)$ and upper $(i = 2)$ end points of the interval of non-arbitrage prices for this option in model (4.30).

Next, we consider the case when an owner of asset S receives *dividends*. Denote $\widetilde{S}_t$ the process that represents the wealth of the owner of asset S, and let δS_t, $\delta \geq 0$, represent the received dividends. Then the evolution of $\widetilde{S}_t$ is described by the following stochastic equation:

$$d\left(\frac{\widetilde{S}_t}{B_t}\right) = d\left(\frac{S_t}{B_t}\right) + \delta \frac{S_t}{B_t} dt, \quad \delta \geq 0.$$

Using

$$dS_t = S_t\left(\mu\, dt + \sigma\, dW_t\right)$$

and

$$d\left(\frac{S_t}{B_t}\right) = \frac{S_t}{B_t}\left((\mu - r)\, dt + \sigma\, dW_t\right),$$

we obtain

$$d\left(\frac{\widetilde{S}_t}{B_t}\right) = \frac{S_t}{B_t}\left((\mu - r + \delta)\, dt + \sigma\, dW_t\right).$$

Note the analogy of

$$\overline{W}_t := W_t + \frac{\mu - r + \delta}{\sigma} t \quad \text{and} \quad \widetilde{W}_t = W_t + \frac{\mu - r}{\sigma} t,$$

and the analogy of density

$$\overline{Z}_T := \exp\left\{-\frac{\mu - r + \delta}{\sigma} W_T^* + \frac{1}{2}\left(\frac{\mu - r + \delta}{\sigma}\right)^2 T\right\}$$

and density Z_T^*.

We now define a new probability $\overline{P}_T$ with density $\overline{Z}_T$. By Girsanov theorem, $(\overline{W}_t)_{t \leq T}$ is a Wiener process with respect to $\overline{P}_T$. Distribution functions are given by

$$\overline{F}_{\mu T + \sigma W_T} = \overline{F}_{(r-\delta)T + \sigma \overline{W}_T} = F_{(r-\delta)T + \sigma W_T}$$

and

$$\overline{F}_{S_T} = F_{S_0 \exp\{(r-\delta-\sigma^2/2)T + \sigma W_T\}} \,.$$

We compute the price of a European call option

$$
\begin{aligned}
C_T(\delta) &= \overline{E}\left(\frac{(S_T - K)^+}{B_T}\right) = e^{-rT} \overline{E}\left(\left(S_0 e^{(r-\sigma^2/2)T + \sigma W_T} - K\right)^+\right) \\[2mm]
&= e^{-rT} E\left(\left(S_0 e^{(r-\delta-\sigma^2/2)T + \sigma W_T} - K\right)^+\right) \\[2mm]
&= e^{-rT} E\left(\left(S_0 e^{(r-\delta-\sigma^2/2)T + \sigma \sqrt{T} W_1} - K\right)^+\right) \\[2mm]
&= S_0 e^{-\delta T} \Phi\left(\frac{\ln(S_0/K) + T(r - \delta + \sigma^2/2)}{\sigma \sqrt{T}}\right) \\[2mm]
&\quad - K e^{-rT} \Phi\left(\frac{\ln(S_0/K) + T(r - \delta - \sigma^2/2)}{\sigma \sqrt{T}}\right).
\end{aligned}
$$

$$(4.31)$$

In Section 3.2, we studied the binomial model of a market with transaction costs. It was shown in Theorem 3.5 that, if the terminal buy and sell prices of stock S are equal, then there exists a unique strategy that replicates the European call option. This strategy is related to a binomial market without transaction costs where values of profitability (and therefore of volatility) are increased.

We now discuss this problem in the case of the Black-Scholes model. For simplicity, suppose that $B_t \equiv 1$, $t \leq T$, and that capital of portfolio $\pi = (\beta, \gamma)$ is redistributed at discrete times $t_i = iT/N$, $i \leq N$.

Constraints on the redistribution of capital $X_t^\pi = \beta_t + \gamma_t S_t$ of portfolio π can be written in the form of proportional *transaction costs* with parameter $\lambda \geq 0$:

$$\Delta X_t^\pi = \gamma_t \Delta S_t - \lambda S_t |\Delta \gamma_t|.$$

Now consider a European call option that will be hedged in the class of strategies described above. Denote $C^{\mathrm{BS}}(t_i, S_{t_i})$, $i \leq N$, the capital of a Black-Scholes strategy. Then an appropriate hedging strategy π must have capital X_t^π such that

$$X_{t_i}^\pi = C^{\mathrm{BS}}(t_i, S_{t_i}), \quad i \leq N,$$

and approximately (up to infinitesimals of high order of Δt) satisfy equation

$$\frac{\partial X^\pi(t, S_t)}{\partial t} + \frac{\tilde{\sigma}^2}{2} S_t^2 \frac{\partial^2 X^\pi(t, S_t)}{\partial s^2} = 0$$

with parameter

$$\tilde{\sigma}^2 = \sigma^2 \left(1 + \lambda \sqrt{\frac{8}{\sigma^2 \, \pi \, \Delta t}} \right) > \sigma^2 \,.$$

Thus, for pricing European call options in this case, one can use the Black-Scholes formula with the increased volatility.

So far we were studying markets with information flow $\mathbb{F} = \left(\mathcal{F}_t \right)_{t \leq T}$ defined by prices of asset S: $\mathcal{F}_t = \sigma \left(S_0, \ldots, S_t \right)$. If we wish to take into account the *non-homogeneity* of the market, then we assume that some (but not all!) of the market participants have access to a larger information flow. Mathematically, this can mean, for example, that the terminal value S_T is known at time $t < T$ or that S_T belongs to some interval $[S', S'']$ and so forth. Let ξ be a random variable that extends market information $\mathcal{F}_t$ to $\mathcal{F}_t^\xi = \sigma \left(\mathcal{F}_t, \xi \right)$. Then $\mathbb{F}^\xi = \left(\mathcal{F}_t^\xi \right)_{t \leq T}$ is called the *insider* information flow. Now we investigate how this additional information can be utilized by a market participant.

For simplicity, let $r = 0$. Using the formula for Merton's point and the martingale property of stochastic integral, we obtain that the expected utility is given by

$$
\begin{aligned}
v_{\mathbb{F}}(x) &= \sup_{\pi \in SF(\mathbb{F})} E \left(\ln X_T^\pi(x) \right) \\
&= x + E \left(\int_0^T \alpha_s \, \sigma \, dW_s + \int_0^T \mu \, \alpha_s \, ds - \frac{1}{2} \int_0^T \alpha_s^2 \, \sigma^2 \, ds \right) \\
&= x + E \left(\frac{\mu^2}{\sigma^2} T - \frac{\sigma^2}{2} \frac{\mu^2}{\sigma^4} T \right) \\
&= x + \frac{1}{2} \frac{\mu^2}{\sigma^2} T
\end{aligned}
$$

for the information flow $\mathbb{F} = \left(\mathcal{F}_t \right)_{t \leq T}$.

When using the insider information flow $\mathbb{F}^\xi = \left(\mathcal{F}_t^\xi \right)_{t \leq T}$, we cannot assume that process $\left(W_t, \mathcal{F}_t^\xi \right)_{t \leq T}$ is a Wiener process. Nevertheless, it is natural to assume that as in Girsanov theorem, there exists a $\mathbb{F}^\xi$-adapted process $\mu^\xi = \left(\mu_t^\xi \right)_{t \leq T}$ such that

$$\int_0^T |\mu_s^\xi| \, ds < \infty \qquad \text{(a.s.)}$$

and the process

$$\widetilde{W}_t = W_t - \int_0^t \mu_s^\xi \, ds \,, \qquad t \leq T \,,$$

is a Wiener process with respect to $\mathbb{F}^\xi$.

In this case, the additional utility can be expressed in terms of the "information drift" μ^ξ. Indeed, for a self-financing strategy $\pi \in SF(\mathbb{F}^\xi)$, the terminal capital can be written in the form

$$X_T^\pi(x) = x \exp\left\{ \int_0^T \alpha_s \, \sigma \, d\widetilde{W}_s - \frac{1}{2} \int_0^T \alpha_s \, \sigma^2 \, ds + \int_0^T \alpha_s \left(\mu + \sigma \, \mu_s^\xi \right) ds \right\}.$$

Taking into account that

$$E\left(\int_0^T \frac{\mu}{\sigma} \mu_s^\xi \, ds \right) = E\left(\int_0^T \frac{\mu}{\sigma} \left(dW_s - d\widetilde{W}_s \right) \right) = 0,$$

we find the expected utility

$$v_{\mathbb{F}^\xi}(x) \quad = \quad x + \frac{1}{2} E\left(\int_0^T \frac{(\mu + \sigma \, \mu_s^\xi)^2}{\sigma} \, ds \right) = x + \frac{1}{2} E\left(\int_0^T \left[\frac{\mu^2}{\sigma^2} + (\mu_s^\xi)^2 \right] ds \right),$$

given the insider information $\mathbb{F}^\xi$. Thus, the additional utility is given by formula

$$\Delta \, v_{\mathbb{F}^\xi} = v_{\mathbb{F}^\xi}(x) - v_{\mathbb{F}}(x) = \frac{1}{2} E\left(\int_0^T \left(\mu_s^\xi \right)^2 ds \right),$$

which can be written in more detailed form in many particular cases.

Worked Example 4.2 *Find prices of European call and put options in a Black-Scholes market when the owner of a risky asset receives dividends at the rate $\delta = 0.1$ and 0.2. Use the following parameters of the market: $r = 0.1$, $T = 215/365$, $S_0 = 100(\$)$, $K = 80(\$)$, $\mu = r$, $\sigma = 0.1$.*

Solution We can use formula (4.31) directly or, alternatively, we can use the following expressions:

$$C_T(\delta, r) = e^{-\delta T} C_T(0, r - \delta) \qquad \text{and} \qquad P_T(\delta, r) = e^{-\delta T} P_T(0, r - \delta).$$

Let $\delta = 0.1$, then

$$C_T \approx 18.86 \qquad \text{for} \quad r = 0.1,$$

and

$$C_T \approx 23.17 \qquad \text{for} \quad r = 0.2.$$

For $\delta = 0.2$

$$C_T \approx 13.5 \qquad \text{for} \quad r = 0.1,$$

and

$$C_T \approx 17.8 \qquad \text{for} \quad r = 0.2.$$

For given parameters of the model, these results are consistent with our intuitive expectation that bigger rate of dividends implies smaller prices of call options and vice versa. Note that similar calculation for a put option gives us $P_T(\delta, r) \approx 0$. □

4.3 Imperfect hedging and risk measures

Consider the Black-Scholes model (4.6). Let $Y_t \equiv Y_t^\pi := X_t^\pi / B_t \geq 0$ be the discounted value of a self-financing portfolio π. The Kolmogorov-Itô formula implies that

$$dY_t = \phi_t \, dW_t^*, \qquad Y_0 = X_0^\pi,$$

where $\phi_t = \sigma \, \gamma_t \, S_t / B_t$ and $dW_t^* = dW_t + t \, (\mu - r)/\sigma$ is a Wiener process with respect to probability P^*, which is defined by its density (4.7).

The set

$$A = A(x, \pi, f_T) = \{\omega \,:\, X_T^\pi(x) \geq f_T\} = \{\omega \,:\, Y_T^\pi(x) \geq f_T/B_T\}$$

is called the *perfect hedging set* for claim f_T and strategy π with the initial wealth x.

The theory of perfect hedging that was discussed above allows to find a hedge with the initial wealth $X_0 = E^*(f_T/B_T)$ and $P(A) = 1$. However, it is possible that an investor responsible for claim f_T may have initial budget constraints. In particular, an investor's initial capital may be less than amount X_0, which is necessary for successful hedging.

Thus, we arrive at the following problem of *quantile hedging*: among all admissible strategies, find a strategy $\tilde{\pi}$ such that

$$P\big(A(x, \tilde{\pi}, f_T)\big) = \max_\pi P\big(A(x, \pi, f_T)\big)$$

under the budget constraint

$$x \leq x_0 < E^* \left(\frac{f_T}{B_T} \right) = X_0,$$

where x_0 is the initial capital.

The following lemma addresses this problem.

Lemma 4.2 *Suppose perfect hedging set* $\widetilde{A}$ *is such that*

$$P(\widetilde{A}) = \max_\pi P(A), \qquad where \qquad E^* \left(\frac{f_T}{B_T} \, I_A \right) \leq x.$$

Then a perfect hedge $\tilde{\pi}$ *for the claim* $\tilde{f}_T = f_T \, I_{\widetilde{A}}$, *with the initial wealth* x, *yields a solution for the problem of quantile hedging. Furthermore, the perfect hedging set* $A(x, \tilde{\pi}, f_T)$ *coincides with* $\widetilde{A}$.

Proof

Step 1. Let π be an arbitrary admissible strategy with the initial wealth

$$x \leq E^* \left(\frac{f_T}{B_T} \right) = X_0.$$

Its discounted value

$$Y_t = x + \int_0^t \phi_s \, dW_s^*$$

is a non-negative supermartingale with respect to P^*. For a perfect hedging set $A = A(x, \pi, f_T)$, we have

$$Y_t \geq \frac{f_T}{B_T} I_A, \quad (P - \text{a.s.}).$$

Hence,

$$x = E^*(Y_T) \geq E^*\left(\frac{f_T}{B_T} I_A\right),$$

and $P(A) \leq P(\tilde{A})$.

Step 2. Let $\tilde{\pi}$ be a perfect hedge for the claim $\tilde{f}_T = f_T I_{\tilde{A}}$, with the initial wealth x satisfying the inequality

$$E^*\left(\frac{f_T}{B_T} I_{\tilde{A}}\right) \leq x \leq x_0 < E^*\left(\frac{f_T}{B_T}\right) = X_0.$$

We will show that this strategy is optimal for the problem of quantile hedging. Since

$$x + \int_0^t \tilde{\phi}_s \, dW_s^* \geq E^*\left(\frac{f_T}{B_T} I_{\tilde{A}}\right) + \int_0^t \tilde{\phi}_s \, dW_s^* = E^*\left(\frac{f_T}{B_T} I_{\tilde{A}} \Big| \mathcal{F}_t\right) \geq 0,$$

then $\tilde{\pi}$ is an admissible strategy. Denote

$$A' = \left\{\omega : x + \int_0^T \tilde{\phi}_s \, dW_s^* \geq f_T/B_T\right\}$$

the perfect hedging set for $\tilde{\pi}$. Since $\tilde{\pi}$ is a perfect hedge for claim $\tilde{f}_T$, we obtain

$$A' \supseteq \left\{\omega : f_T I_{\tilde{A}} \geq f_T\right\} \supseteq \tilde{A},$$

and hence, $P(A') \geq P(\tilde{A})$.

Step 3. Now we observe that

$$A = \tilde{A}, \quad (P - \text{a.s.}),$$

and taking into account that $\tilde{A}$ is a perfect hedging set for $\tilde{\pi}$, we conclude that $\tilde{\pi}$ is the optimal strategy for the problem of quantile hedging. $\square$

Next, we will use the *fundamental Neumann-Pearson lemma* for construction of a maximal perfect hedging set. Suppose that distributions Q^* and P correspond to hypotheses H_0 and H_1, respectively. Let $\alpha = E_{Q^*}(\phi)$ be the probability of the error of the first kind and $\beta = E_P(\phi)$ be the criterium's power corresponding to a *critical function* ϕ. The Neumann-Pearson criterium has the following structure:

$$\phi = \begin{cases} 1, & dP/dQ^* > c \\ 0, & dP/dQ^* < c \end{cases},$$

and it maximizes β given that the probability of an error of the first kind does not exceed a set level α. Here c is some constant, and values 0 and 1 in the critical function ϕ indicate which of the hypotheses H_0 or H_1 should be preferred.

If we introduce probability Q^* by the relation

$$\frac{dQ^*}{dP^*} = \frac{f_T}{B_T\, E^*\left(f_T/B_T\right)} = \frac{f_T}{E^*\left(f_T\right)},$$

then the constraint in Lemma 4.2 can be written in the form

$$Q^*(A) = \int_A \frac{dQ^*}{dP^*}\, dP^* \le \frac{x}{E^*\left(f_T/B_T\right)} = \alpha.$$

The solution of the corresponding optimization problem is given by

$$\widetilde{A} = \left\{\omega : \frac{dP}{dQ^*} > c\right\} = \left\{\omega : \frac{dP}{dP^*} > c\,\frac{f_T}{E^*\left(f_T\right)}\right\}, \qquad (4.32)$$

where

$$c = \inf\left\{a : Q^*\left(\left\{\omega : \frac{dP}{dQ^*} > a\right\}\right) \le \alpha\right\}.$$

The proof of this claim follows from the fundamental Neumann-Pearson lemma and from the equalities

$$\alpha = E_{Q^*}(\phi) = Q^*(\widetilde{A}) \qquad \text{and} \qquad \beta = E_P(\phi) = P(\widetilde{A}) = \max_\pi P(A).$$

Thus, we arrive at the following theorem.

Theorem 4.1 *An optimal strategy $\widetilde{\pi}$ for the problem of quantile hedging coincides with the perfect hedge for the contingent claim $\widetilde{f}_T = f_T\, I_{\widetilde{A}}$, where the maximal perfect hedging set $\widetilde{A}$ is given by (4.32).*

Next, we consider the problem of quantile hedging for a European call option with $f_T = (S_T - K)^+$. The initial value of a perfect hedge in this case is

$$X_0 = S_0\, \Phi(d_+) - K\, e^{-rT}\, \Phi(d_-).$$

Suppose an investor has an initial capital $x < X_0$. By Theorem 4.1, the optimal strategy for the problem of quantile hedging coincides with the perfect hedge for the contingent claim $f_T I_{\tilde{A}}$, where

$$A = \left\{ \omega : \frac{dP}{dQ^*} > c \right\} = \left\{ \omega : \frac{dP}{dP^*} > c_1 \, f_T \, e^{-rT} \right\}.$$

Since density Z_T^* has the form

$$Z_T^* = \exp\left\{ -\frac{\mu - r}{\sigma} W_T^* + \frac{1}{2}\left(\frac{\mu - r}{\sigma}\right)^2 T \right\},$$

then

$$A = \left\{ \omega : \exp\left\{ \frac{\mu - r}{\sigma} W_T^* - \frac{1}{2}\left(\frac{\mu - r}{\sigma}\right)^2 T \right\} > c_1 \, (S_T - K)^+ \right\}$$

$$= \left\{ \omega : \exp\left\{ \frac{\mu - r}{\sigma^2}\left(\ln S_0 + \left(r - \frac{\sigma^2}{2}\right) T + \sigma W_T^* \right) \right\} \right.$$

$$\times \exp\left\{ -\frac{\mu - r}{\sigma^2}\left(\ln S_0 + \left(r - \frac{\sigma^2}{2}\right) T \right) - \frac{1}{2}\left(\frac{\mu - r}{\sigma}\right)^2 T \right\}$$

$$\left. > c_1 \, (S_T - K)^+ \right\}$$

$$= \left\{ \omega : S_T^{\frac{\mu - r}{\sigma^2}} \exp\left\{ -\frac{\mu - r}{\sigma^2}\left(\ln S_0 + \frac{\mu + r - \sigma^2}{2} T \right) \right\} \right.$$

$$\left. > c_1 \, (S_T - K)^+ \right\}.$$

Now we consider two cases.

Case 1. $\frac{\mu - r}{\sigma^2} \le 1$. Set A can be written in the form

$$A = \{\omega : S_T < d\} = \{\omega : W_T^* < b\}$$

$$= \left\{ \omega : S_T < S_0 \exp\left\{ (r - \sigma^2/2) T + b\sigma \right\} \right\}$$

for some constants b and d under the constraint

$$E^*\left(\frac{f_T}{B_T} I_A \right) = x_0.$$

Taking into account that

$$S_T = S_0 \exp\left\{ \left(r - \frac{\sigma^2}{2}\right) T + \sigma W_T^* \right\},$$

we obtain

$$P(A) = \Phi\left(\frac{b - T(\mu - r)/\sigma}{\sqrt{T}}\right).$$

Constant b can be found from the equality

$$x_0 = E^*\left(e^{-rT} f_T\right) I_A = e^{-rT} F_T^*(S_T),$$

where

$$F_T^* = \frac{1}{\sqrt{2\pi}} \int_{-\infty}^{b/\sqrt{T}} f\left(S_0 \exp\left\{\sigma \sqrt{T} y + \left(r - \frac{\sigma^2}{2}\right) T\right\}\right) e^{-\frac{y^2}{2}} \, dy$$

$$= \frac{1}{\sqrt{2\pi}} \int_{-d_0}^{b/\sqrt{T}} \left(S_0 \exp\left\{\sigma \sqrt{T} y + \left(r - \frac{\sigma^2}{2}\right) T\right\} - K\right)^+ e^{-\frac{y^2}{2}} \, dy;$$

$$d_0 = \frac{\ln(K/S_0) - T(r - \sigma^2)/2}{\sigma \sqrt{T}}.$$

Hence,

$$x_0$$
$$= S_0\left[\Phi(\sigma \sqrt{T} - d_0) - \Phi\left(\sigma \sqrt{T} - \frac{b}{\sqrt{T}}\right)\right] - K e^{-rT}\left[\Phi(d_0) - \Phi\left(-\frac{b}{\sqrt{T}}\right)\right]$$
$$= S_0\left[\Phi(d_+) - \Phi\left(\sigma \sqrt{T} - \frac{b}{\sqrt{T}}\right)\right] - K e^{-rT}\left[\Phi(d_-) - \Phi\left(-\frac{b}{\sqrt{T}}\right)\right].$$

Case 2. $\frac{\mu - r}{\sigma^2} > 1$. Set A can be written in the form

$$A = \{\omega : W_T^* < b_1\} \cup \{\omega : W_T^* > b_2\}$$

for some constants b_1 and b_2. Solving the problem of quantile hedging, we obtain

$$P(A) = \Phi\left(\frac{b_1 - T(\mu - r)/\sigma}{\sqrt{T}}\right) + \Phi\left(\frac{b_2 - T(\mu - r)/\sigma}{\sqrt{T}}\right).$$

Constants b_1 and b_2 can be found from the same equality

$$x_0 = E^*\left(e^{-rT} f_T\right) I_A = e^{-rT} F_T^*(S_T),$$

where now

$$F_T^* = \frac{1}{\sqrt{2\pi}} \int_{-\infty}^{b_1/\sqrt{T}} f\left(S_0 \exp\left\{\sigma \sqrt{T} y + \frac{r - \sigma^2}{2} T\right\}\right) e^{-\frac{y^2}{2}} \, dy$$

$$+ \frac{1}{\sqrt{2\pi}} \int_{b_2/\sqrt{T}}^{\infty} f\left(S_0 \exp\left\{\sigma \sqrt{T} y + \frac{r - \sigma^2}{2} T\right\}\right) e^{-\frac{y^2}{2}} \, dy.$$

Similarly to Case 1,

$$
x_0 = S_0 \left[\Phi(d_+) - \Phi\left(\sigma \sqrt{T} - \frac{b_1}{\sqrt{T}} \right) + \Phi\left(\sigma \sqrt{T} - \frac{b_2}{\sqrt{T}} \right) \right]
$$

$$
- K e^{-rT} \left[\Phi(d_-) - \Phi\left(-\frac{b_1}{\sqrt{T}} \right) + \Phi\left(-\frac{b_2}{\sqrt{T}} \right) \right].
$$

The problem of quantile hedging can be considered from a different perspective: the pay-off function f_T can be interpreted as an investment objective for a given investment period $[0, T]$. Then the terminal value X_T^π of an investment strategy π should be "close enough" to the objective in some probabilistic sense.

If the measure of closedness is chosen to be $E\left(|X_T^\pi - f_T|^2\right)$, then we arrive at the notion of the *mean-variance hedging*. Alternatively, we can use the notion of a loss function $l : \mathbb{R}_+ \to \mathbb{R}_+$, which is commonly used in statistics. In this case, we search for the most appropriate investment strategy by minimizing the *expected losses* $E\left(l((X_T^\pi - f_T)^+)\right)$ over the set of all admissible strategies and under some budget constraints. This type of hedging is usually referred to as *efficient hedging*. Note that quantile hedging is a particular case of efficient hedging with the loss functions $I_{\{\omega : X_T^\pi < f_T\}}$. All these types of hedging are collectively referred to as *imperfect* or *partial hedging*.

The most general and comprehensive description of this area of mathematical finance can be given in terms of quantitative techniques that use the notion of *risk measure*. One of the advantages of this approach is the unified treatment of hedging and investment problems. The industry standard tool for management of financial risks is *Value at Risk (VaR)*. It is based on the statistical notion of a quantile of a random variable X. In our context, we treat X as a return on an investment portfolio and the corresponding risk is identified with X. For simplicity, we assume that the distribution function $F(x)$ of X has density $f(x)$, and then we arrive at the following definition of VaR:

$$
VaR_\lambda(X) = -q_X(\lambda),
$$

where $\lambda \in (0, 1)$ and $q_X(\lambda)$ is the λ-quantile of X; that is, $\lambda = F\left(q_X(\lambda)\right)$.

We now discuss some properties of VaR and some of its financial interpretations. If λ is close to 0, then the corresponding $VaR_\lambda(X)$ defines the lower bound for possible "big positive" returns of the portfolio; if λ is close to 1, then $VaR_\lambda(X)$ defines the upper bound for "big negative" returns. The probabilities of these events are λ and $1 - \lambda$ and they are small.

The following properties of VaR can be easily verified:

1. *Monotonicity*: if X and Y are returns of two portfolios and $X \leq Y$, then $VaR_\lambda(X) \geq VaR_\lambda(Y)$;

2. *Positive homogeneity*: $VaR_\lambda(aX) = aVaR_\lambda(X)$ for any positive constant a; and

3. *Translation invariance*: $VaR_\lambda(X + m) = VaR_\lambda(X) + m$ for any $m \in \mathbb{R}$.

It is natural to expect that risk measures should be useful not just for quantification of risk itself but also for quantification of risk reductions: the *diversification* property. We now explore such property for VaR in the case of Gaussian returns. Suppose $X_i \sim \mathcal{N}(\mu_i, \sigma_i^2)$, $i = 1, 2$, and $\gamma \in (0, 1)$, then their convex linear combination $X = \gamma X_1 + (1 - \gamma)X_2$ is $\mathcal{N}(\mu, \sigma^2)$. Furthermore,

$$\mu = \gamma\mu_1 + (1 - \gamma)\mu_2 \quad \text{and} \quad \sigma = \gamma\sigma_1 + (1 - \gamma)\sigma_2\,,$$

and therefore, for $\lambda \in (0, 1/2]$,

$$
\begin{aligned}
VaR_\lambda(X) &= -\mu - \sigma Z_\lambda = -\gamma\mu_1 - (1 - \gamma)\mu_2 - \sigma Z_\lambda \\
&\leq -\gamma\mu_1 - (1 - \gamma)\mu_2 + \big(\gamma\sigma_1 + (1 - \gamma)\sigma_2\big)\big(-\sigma Z_\lambda\big) \\
&= \gamma(-\mu_1 - \sigma_1 Z_\lambda) + (1 - \gamma)(-\mu_2 - \sigma_2 Z_\lambda) \\
&= \gamma VaR_\lambda(X_1) + (1 - \gamma)VaR_\lambda(X_2)\,, \quad\quad (4.33)
\end{aligned}
$$

where Z_λ is a λ-quantile of $\Phi(x) = \frac{1}{\sqrt{2\pi}}\int_{-\infty}^{x} e^{-y^2/2}\,dy$.

Formula (4.33) proves the *convexity* of VaR for Gaussian random variables when $\lambda \in (0, 1/2]$. Clearly, the convexity of a risk measure allows us to decentralize the problem of risk management by diversifying the risk. Further, applying (4.33) in the case of $\gamma = 1/2$ and using positive homogeneity of VaR, we prove its *sub-additivity*:

$$VaR_\lambda(X_1 + X_2) \leq VaR_\lambda(X_1) + VaR_\lambda(X_2) \quad\quad (4.34)$$

for Gaussian random variables when $\lambda \in (0, 1/2]$. Thus, in this case $VaR_\lambda(X)$ is both convex and sub-additive, which is not necessarily the case in general. Convex risk measures that are also sub-additive are called *coherent* risk measures. We can pose a natural question: are there any risk measures that are coherent for random variables X that are not necessarily Gaussian? The positive answer to this question is intuitively obvious if we look at the definition of VaR via quantiles: VaR does not take into account information on how big "negative"/"positive" returns are. Therefore, any risk measure that takes into account tail distribution of "negative"/"positive" returns returns has a potential to be a measure with better properties than VaR.

The first modification of VaR along this path is the *Average Value at Risk* ($AVaR$):

$$AVaR_\lambda(X) = \frac{1}{\lambda}\int_0^\lambda VaR_\gamma(X)\,d\gamma\,. \quad\quad (4.35)$$

Denoting $x_0 = -VaR_\lambda(X)$ and changing variables: $d\gamma = dF(x) = f(x)dx$ in

(4.35), we find

$$AVaR_\lambda(X) = \frac{1}{\lambda} \int_0^\lambda VaR_\gamma(X)\, d\gamma = -\frac{1}{\lambda} \int_0^\lambda F^{-1}(\gamma)\, d\gamma \tag{4.36}$$

$$= -\frac{1}{\lambda} \int_{-\infty}^{x_0} x f(x)\, dx = -\frac{1}{\lambda} \int_{-\infty}^{x_0} (x_0 - x) f(x)\, dx - \frac{1}{\lambda} \int_{-\infty}^{x_0} x_0 f(x)\, dx$$

$$= \frac{1}{\lambda} \int_{-\infty}^{x_0} (x_0 - x) f(x)\, dx - \frac{x_0}{\lambda} F(x_0) = \frac{1}{\lambda} \int_{-\infty}^{x_0} (x_0 - x) f(x)\, dx - x_0 .$$

We use formula (4.36) to calculate $AVaR_\lambda(X)$ for $X \sim \mathcal{N}(\mu, \sigma^2)$:

$$AVaR_\lambda(X) = \frac{1}{\lambda} \int_{-\infty}^{x_0} \exp\left\{ -\frac{(x-\mu)^2}{2\sigma^2} \right\} (x_0 - x) \frac{dx}{\sigma\sqrt{2\pi}} - x_0$$

$$\tag{4.37}$$

$$= \frac{1}{\lambda} \left[(x_0 - \mu) \Phi\left(\frac{x-\mu}{\sigma} \right) + \frac{\sigma}{\sqrt{2\pi}} \exp\left\{ -\frac{(x-\mu)^2}{2\sigma^2} \right\} \right] - x_0 ,$$

where $x_0 = \mu + \sigma Z_\lambda$.

Worked Example 4.3 *Assume that a portfolio return X is normally distributed. Using the following 20 observations of the return*

$$(x_i)_{i=1,\dots,20}$$
$$= (33, 128, 6, 267, -82, -38, 123, 217, -78, 327, 22, 126, 43, -73,$$
$$-208, 137, -46, -205, 202, 40) ,$$

calculate $VaR_{0.02}$ and $AVaR_{0.02}$.

Solution Calculating the sample mean

$$\bar{\mu} = \frac{1}{20} \sum_{i=1}^{20} x_i = 47.05$$

and the standard deviation

$$\bar{\sigma} = \sqrt{ \sum_{i=1}^{20} \frac{(x_i - \bar{\mu})^2}{19} } = 145.10 ,$$

we find

$$x_0 = \bar{\mu} - \bar{\sigma} \cdot 2.054 = -250.985 \quad \text{and} \quad VaR_{0.02} = 250.985 .$$

Using (4.37), we find

$$AVaR_{0.02} = 50 \left[(-250.985 - 47.05) \Phi\left(\frac{-250.985 - 47.05}{145.10} \right) \right.$$

$$\left. +145.10 \exp\left\{ -\frac{1}{2} \left(\frac{-250.985 - 47.05}{145.10} \right) \right\} / \sqrt{2\pi} \right] + 250.985$$

$$= 335.27 . \ \square$$

We now consider some risk measures in the framework of the Black-Scholes model (4.6):

$$dB_t = r\, B_t\, dt \,,$$
$$dS_t = S_t\left(\mu\, dt + \sigma\, dW_t\right).$$

The capital of the portfolio follows the dynamics:

$$dX_t = X_t\left([(1-\alpha(t))r + \alpha(t)\mu]\,dt + \alpha(t)\sigma\, dW_t\right), \quad X_0 = x\,,$$

where $\alpha(\cdot)$ is the proportion of risky capital in this portfolio (see Section 4.1). If $\alpha(\cdot)$ is constant on $[0,T]$, then

$$X_T = x\,\exp\left\{\left[(1-\alpha)r + \alpha\mu - \frac{\alpha^2\sigma^2}{2}\right]T + \alpha\sigma\, W_t\right\};$$

therefore, the λ quantile of the portfolio at the final time T is

$$q(x,\alpha,T) = x\,\exp\left\{\left[(1-\alpha)r + \alpha\mu - \frac{\alpha^2\sigma^2}{2}\right]T + |\alpha|\sigma\sqrt{T}z_\lambda\right\}.$$

Define another measure of risk, *Capital at Risk (CaR)*, as

$$CaR(x,\alpha,T) = x\,\exp\{r\,T\} - q(x,\alpha,T)\,,$$

then

$$CaR(x,\alpha,T) = x\,\exp\{r\,T\}\left(1 - \exp\left\{\left[\alpha(\mu-r) - \frac{\alpha^2\sigma^2}{2}\right]T + |\alpha|\sigma\sqrt{T}z_\lambda\right\}\right).$$

Generally, we would select λ in the interval $[0, 0.5]$; hence, $z_\lambda < 0$. We also assume that $\mu \geq r$ since, in general, risky assets provide higher returns. Note that CaR attains its maximum at $\alpha = \pm\infty$ and

$$\sup_\alpha\left\{CaR(x,\alpha,T)\right\} = x\,\exp\{r\,T\}\,,$$

which means that the most risky strategy is to borrow or lend money in amounts that are much higher than the portfolio value.

Proposition 4.2   *1. If $(\mu-r)\sqrt{T} > \sigma|z_\lambda|$, then*

$$\min_\alpha\left\{CaR(x,\alpha,T)\right\} = x\,\exp\{r\,T\}\left(1 - \exp\left\{\frac{\left[(\mu-r)T + \sigma\sqrt{T}z_\lambda\right]^2}{2\sigma^2 T}\right\}\right)$$

is attained at

$$\alpha = \frac{(\mu-r)T + \sigma\sqrt{T}z_\lambda}{\sigma^2 T}\,.$$

2. If $\sigma\,|z_\lambda| \geq (\mu - r)\,\sqrt{T} \geq 0$, then $\min_\alpha \{CaR(x, \alpha, T)\} = 0$ is attained at $\alpha = 0$.

Proof Denote

$$f(\alpha) = \left[\alpha(\mu - r) - \frac{\alpha^2 \sigma^2}{2}\right]T + |\alpha|\sigma\,\sqrt{T}z_\lambda\,,$$

so

$$CaR(x, \alpha, T) = x\,\exp\{r\,T\}\Big(1 - \exp\{f(\alpha)\}\Big)$$

and CaR attains its minimum when $f(\alpha)$ attains its maximum. If $\alpha > 0$, then

$$\begin{aligned}
f(\alpha) &= \left[\alpha(\mu - r) - \frac{\alpha^2 \sigma^2}{2}\right]T + \alpha\sigma\,\sqrt{T}z_\lambda \\
&= -\frac{\sigma^2 T}{2}\left[\alpha - \frac{(\mu - r)\,T + \sigma\,\sqrt{T}z_\lambda}{\sigma^2 T}\right]^2 + \frac{[(\mu - r)\,T + \sigma\,\sqrt{T}z_\lambda]^2}{2\sigma^2 T}.
\end{aligned}$$

If $\alpha \leq 0$, then

$$\begin{aligned}
f(\alpha) &= \left[\alpha(\mu - r) - \frac{\alpha^2 \sigma^2}{2}\right]T - \alpha\sigma\,\sqrt{T}z_\lambda \\
&= -\frac{\sigma^2 T}{2}\left[\alpha - \frac{(\mu - r)\,T - \sigma\,\sqrt{T}z_\lambda}{\sigma^2 T}\right]^2 + \frac{[(\mu - r)\,T - \sigma\,\sqrt{T}z_\lambda]^2}{2\sigma^2 T}.
\end{aligned}$$

If $0 \leq (\mu - r)\,\sqrt{T} \leq \sigma|z_\lambda|$, then

$$\frac{(\mu - r)\,T - \sigma\,\sqrt{T}z_\lambda}{\sigma^2 T} \geq 0 \quad \text{and} \quad \frac{(\mu - r)\,T + \sigma\,\sqrt{T}z_\lambda}{\sigma^2 T} \leq 0\,.$$

and $\min_\alpha \{CaR(x, \alpha, T)\} = 0$ is attained when $\alpha = 0$.
   If $(\mu - r)\,\sqrt{T} > \sigma|z_\lambda|$, then

$$\frac{(\mu - r)\,T - \sigma\,\sqrt{T}z_\lambda}{\sigma^2 T} > 0 \quad \text{and} \quad \frac{(\mu - r)\,T + \sigma\,\sqrt{T}z_\lambda}{\sigma^2 T} > 0$$

and

$$\min_\alpha \{CaR(x, \alpha, T)\} = x\,\exp\{r\,T\}\left(1 - \exp\left\{\frac{[(\mu - r)\,T + \sigma\,\sqrt{T}z_\lambda]^2}{2\sigma^2 T}\right\}\right)$$

is attained at

$$\alpha = \frac{(\mu - r)\,T + \sigma\,\sqrt{T}z_\lambda}{\sigma^2 T}\,. \qquad \square$$

Now we consider the optimization problem

$$\max_\alpha E(X_T) \quad \text{subject to} \quad CaR(x, \alpha, T) \leq C, \tag{4.38}$$

where C is a constant.

Since

$$X_T = x \exp\left\{\left[(1-\alpha)r + \alpha\mu - \frac{\alpha^2\sigma^2}{2}\right]T + \alpha\sigma\,W_t\right\},$$

then

$$E(X_T) = x \exp\left\{\left[(1-\alpha)r + \alpha\mu\right]T\right\},$$

and we can prove the following result.

Theorem 4.2 *If*

$$\min_{\alpha}\{CaR(x,\alpha,T)\} < C < x\,\exp\{r\,T\}\,,$$

then the optimization problem (4.38) has a unique solution $\alpha = \varepsilon\,\sigma^{-1}$, *where*

$$\varepsilon = \left(\frac{\mu-r}{\sigma} + \frac{z_\lambda}{\sqrt{T}}\right) + \sqrt{\left(\frac{\mu-r}{\sigma} + \frac{z_\lambda}{\sqrt{T}}\right)^2 - 2\frac{\ln\left(1 - C\exp\{-rT\}/x\right)}{T}}\,,$$

and

$$\max E(X_T) = x\,\exp\left\{\left[r + \varepsilon\frac{\mu-r}{\sigma}\right]T\right\}.$$

Proof The assumption $\min_{\alpha}\{CaR(x,\alpha,T)\} < C < x\,\exp\{rT\}$ implies that the constraint $CaR(x,\alpha,T) \le C$ is not redundant.

Since $E(X_T)$ is an increasing function of α, we just need to find the maximum of α under the constraint $CaR(x,\alpha,T) \le C$. From

$$CaR(x,\alpha,T) = x\,\exp\{r\,T\}\left(1 - \exp\{f(\alpha)\}\right) = C$$

we have

$$f(\alpha) = \ln\left(1 - C\exp\{-rT\}/x\right).$$

Solving

$$\ln\left(1 - C\exp\{-rT\}/x\right) = -\frac{\sigma^2T}{2}\left(\alpha - \frac{(\mu-r)T + \sigma\sqrt{T}z_\lambda}{\sigma^2T}\right)^2$$
$$+ \frac{\left[(\mu-r)T + \sigma\sqrt{T}z_\lambda\right]^2}{2\sigma^2T}\,,$$

we obtain

$$\alpha = \sigma^{-1}\left[\left(\frac{\mu-r}{\sigma} + \frac{z_\lambda}{\sqrt{T}}\right) + \sqrt{\left(\frac{\mu-r}{\sigma} + \frac{z_\lambda}{\sqrt{T}}\right)^2 - 2\frac{\ln\left(1 - C\exp\{-rT\}/x\right)}{T}}\right],$$

which is the value of α where $E(X_T)$ attains its maximum

$$\max E(X_T) = x\,\exp\left\{\left[r + \varepsilon\frac{\mu-r}{\sigma}\right]T\right\}. \quad \square$$

Worked Example 4.4 *Consider an investment of* 1000 *at* $t = 0$. *Suppose that the parameters of the Black-Scholes market are* $r = 0.05$ *per annum,* $\mu = 0.07$, *and* $\sigma = 0.2$. *Suppose that you wish to invest in this market for one year. Find the optimal strategy that minimizes* CaR *at the confidence level* $\lambda = 0.05$ *with* $C = 200$.

Solution Since

$$(\mu - r)\sqrt{T} - \sigma|z_\lambda| = (0.07 - 0.05) - 0.2 \times 1.645 = -0.309 < 0,$$

the maximum of CaR is 0, and it is attained at $\alpha = 0$. The optimal strategy to maximize $E(X_T)$ under the constraint $CaR \leq 200$ is

$$\alpha = 5\big[(0.02/0.2 - 1.645)$$
$$+\sqrt{(0.02/0.2 - 1.645)^2 - 2\ln(1 - 0.2\exp\{-0.05\})}\big]$$
$$= 0.655$$

and

$$\max E(X_T) = x\exp\{\alpha(\mu - r)T + rT\} = 1000\exp\{0.655 \times 0.02 + 0.05\}$$
$$= 1065.14. \ \square$$

Chapter 5

Fixed Income Securities: Modeling and Pricing

5.1 Elements of deterministic theory of fixed income instruments

In this section, we study in continuous time *bonds* or *fixed income instruments* of a financial market. In Chapter 2, we introduced bonds as basic securities with an obligation to make certain payments at certain future times. These payments are called *coupons*. Denote $0 < t_1 < t_2 < \ldots < t_N = T$ the times when coupon payments $c_1, \ldots, c_N$ are made. The last payment c_N is usually denoted A and is referred to as the *principal* (nominal value, par value) of the bond. The bond yields a profit that is fully characterized by these two sequences $(t_i)_{i=1,\ldots,N}$ and $(c_i)_{i=1,\ldots,N}$. Time $t = 0$ is the initial time of the bond and $T = t_N$ is its *maturity time*. The time interval between current time t and the maturity time T is referred to as *time to maturity*.

We now address a very natural question: what should be the price of the bond at the initial time $t = 0$? For simplicity, we make the following assumptions:

- there are no recalls;

- all payments are fixed;

- there is no credit risk (i.e., all payments are guaranteed); and

- there are no transaction costs.

Denote $B(t, T)$ the price of the bond at time t. We define the *yield to maturity y* as a solution of the equation

$$B(0, T) = \sum_{i=1}^{N} \frac{c_i}{(1+y)^{t_i}}. \tag{5.1}$$

Let r be the compounded (annual) rate of interest at the initial time $t = 0$; then we can write

$$B(T, T) = B(0, T)\,(1+r)^T.$$

However,

$$B(T,T) = \sum_{i=1}^{N} \frac{c_i}{(1+y)^{T-t_i}} \, ;$$

thus, using (5.1), we find

$$y = r \, ;$$

that is, the yield to maturity for the period $[0,T]$ coincides with the existing (at time $t = 0$) rate of interest in the market. We illustrate such calculations in the following example.

Worked Example 5.1 *Find yield to maturity for a bond with the following payments:*

Time	0	1	2
Payment	−948	50	1050

Solution In this case, equation (5.1) is reduced to the following:

$$948 = \frac{50}{1+y} + \frac{1050}{(1+y)^2} \, .$$

We introduce function

$$F(y) = 948 - \frac{50}{1+y} - \frac{1050}{(1+y)^2} \, ;$$

therefore, equation (5.1) can be written in the form

$$F(y) = 0 \, .$$

Denote $y_1 = 0.07$ and $y_2 = 0.08$, and note that $F(y_1) = -15.8396$ and $F(y_2) = 1.4979$. Using linear interpolation, we obtain the following approximation:

$$y \approx y_1 - \frac{F(y_1)}{F(y_2) - F(y_1)} (y_2 - y_1) = 0.079 \, . \quad \Box$$

Remark 5.1

A bond is called a *zero-coupon bond* (bond without coupons, pure discounted bond) if there is only one payment at the maturity time T. The yield to maturity of a zero-coupon bond is equal to the risk free rate of interest for the period $[0,T]$ and hence both are denoted $r(T)$. If A is the face value of a zero-coupon bond, then we have

$$B(0,T) = \frac{A}{\left(1 + r(T)\right)^T} \tag{5.2}$$

or equivalently

$$r(T) = \left(\frac{A}{B(0,T)}\right)^{1/T} - 1 \, . \quad \Box$$

This remark gives us a motivation to study certain *collections* of compounded interest rates $r(t_1), \ldots, r(t_N)$ for periods $t_1, \ldots, t_N = T$ (years). Such a collection is referred to as a *term structure of interest rates*. If periods $t_1, \ldots, t_N$ start at some time $t \neq 0$, then the corresponding rates $r(t_1), \ldots, r(t_N)$ are called the *term structure of interest rates with respect to time* t.

Formula (5.2) connects the price of a zero-coupon bond with the interest rate. If the term structure of interest rates $r(t_1), \ldots, r(t_N)$ is known, then it is not difficult to obtain a similar connection for an arbitrary bond B determined by coupons $(c_i)_{i=1,\ldots,N}$ at times $(t_i)_{i=1,\ldots,N}$. We can treat this bond B as a portfolio of zero-coupon bonds $(B_i)_{i=1,\ldots,N}$ with maturities $(t_i)_{i=1,\ldots,N}$ and face values $(A_i)_{i=1,\ldots,N}$. We have

$$B_i(0,T) = \frac{A_i}{\left(1 + r(t_i)\right)^{t_i}}, \quad i = 1, 2, \ldots, N,$$

and

$$B(0,T) = \sum_{i=1}^{N} B_i(0,T) \frac{c_i}{A_i},$$

where each ratio c_i / A_i, $i = 1, 2, \ldots, N$, indicates the number of the i-th zero-coupon bond required in the portfolio. Combining the equalities above, we arrive at the desired formula

$$B(0,T) = \sum_{i=1}^{N} \frac{c_i}{\left(1 + r(t_i)\right)^{t_i}}. \tag{5.3}$$

Note that from a theoretical point of view the term structure of interest rates can be written as a function $r = r(t)$, $t \geq 0$, where $r(t)$ is the risk free rate of interest for the period of t years. The graph of this function is usually referred to as the *yield curve*. Since the real financial market can contain only finite number of zero-coupon bonds, it is impossible to reconstruct the yield curve only from the market prices of these bonds. One of the ways of addressing this issue consists of calculating theoretical values of zero-coupon bonds with different maturities using the available market information about existing zero-coupon bonds. We illustrate this approach in the following example.

Worked Example 5.2 *Suppose that a market contains the following five bonds:*

	Date					
Bond	0.5	1.0	1.5	2.0	2.5	$B(0,T)$
Z1	108					105.27
Z2		121				113.83
Z3	10	11	109			118.71
Z4	11	11	11	120		135.64
Z5	8	8	8	8	108	118.84

Find the term structure of interest rates for the 2.5 (years) period.

Solution For zero-coupon bonds $Z1$ and $Z2$, formula (5.2) implies that their yields to maturity are $r(0.5) = 0.0525$ and $r(1.0) = 0.063$, respectively. Using formula (5.3) and information about bonds $Z1$, $Z2$, and $Z3$, we obtain the following equation for $r(1.5)$, the theoretical value of the risk free rate of interest for 1.5 years:

$$118.71 = \frac{10}{\left(1 + r(0.5)\right)^{0.5}} + \frac{11}{\left(1 + r(1.0)\right)^{1.0}} + \frac{109}{\left(1 + r(1.5)\right)^{1.5}},$$

which implies $r(1.5) = 0.069$. Similarly, using bond $Z4$, we write the equation for $r(2.0)$:

$$135.64 = \frac{11}{\left(1 + 0.0525\right)^{0.5}} + \frac{11}{\left(1 + 0.063\right)^{1.0}} + \frac{11}{\left(1 + 0.069\right)^{1.5}} + \frac{120}{\left(1 + r(2.0)\right)^{2.0}},$$

which gives us $r(2.0) = 0.071$. Finally, using bond $Z5$, we have

$$118.84 = \frac{8}{(1.0525)^{0.5}} + \frac{8}{(1.063)^{1.0}} + \frac{8}{(1.069)^{1.5}} + \frac{8}{(1.071)^{2.0}} + \frac{108}{\left(1 + r(2.5)\right)^{2.5}}.$$

Thus, we arrive at the following term structure of interest rates for the 2.5 (years) period:

$$r(0.5) = 0.0525, \ r(1.0) = 0.063, \ r(1.5) = 0.069, \ r(2.0) = 0.071, \text{ and}$$
$$r(2.5) = 0.079.$$

This method of constructing a term structure of interest rates is referred to as *bootstrapping*. □

Remark 5.2

If the term structure of interest rates $r(t_1), \ldots, r(t_N)$ is known, then using linear interpolation we obtain the following approximation of the theoretical yield curve:

$$r(t) \approx r(t_i) \frac{t_{i+1} - t}{t_{i+1} - t_i} + r(t_{i+1}) \frac{t - t_i}{t_{i+1} - t_i}, \tag{5.4}$$

$$t \in [t_i, t_{i+1}], \ i = 1, \ldots, N - 1. \ \square$$

We now consider the following situation: suppose that the term structure of interest rates is known for periods $t_1, \ldots, t_k$, but the market contains another bond with coupons $c_1, \ldots, c_k, c_{k+1}, \ldots, c_N$ with payment dates $t_1 < \ldots < t_k < t_{k+1} < \ldots < t_N$. Using this information we can construct the term structure of interest rates for periods $t_{k+1}, \ldots, t_N$. Indeed, using (5.4), we obtain the following approximations:

$$r(t_{k+1}) \approx r(t_k) \frac{t_N - t_{k+1}}{t_N - t_k} + r(t_N) \frac{t_{k+1} - t_k}{t_N - t_k}, \ldots,$$

$$r(t_{N-1}) \approx r(t_k) \frac{t_N - t_{N-1}}{t_N - t_k} + r(t_N) \frac{t_{N-1} - t_k}{t_N - t_k}.$$

Substituting $r(t_1), \ldots, r(t_k), r(t_{k+1}), \ldots, r(t_{N-1}), r(t_N)$ in (5.3), we obtain an equation for $r(t_N)$. Solving this equation completes the construction of the term structure of interest rates for periods $t_{k+1}, \ldots, t_N$. The following example illustrates this technique.

Worked Example 5.3 *The term structure of interest rates for the 1.5 (years) period is given by*

$$r(0.5) = 0.06, \quad r(1.0) = 0.07, \quad r(1.5) = 0.08.$$

Find the yield curve for the 2.5 (years) period using a bond with the following payments

Time	0	0.5	1	1.5	2	2.5
Payment	−100	5	5	5	5	105

Solution From formula (5.3) we have

$$100 = \frac{5}{(1.06)^{0.5}} + \frac{5}{(1.07)^{1.0}} + \frac{5}{(1.08)^{1.5}} + \frac{5}{\left(1 + r(2.0)\right)^{2.0}} + \frac{105}{\left(1 + r(2.5)\right)^{2.5}}.$$

Linear interpolation for the interval $(1.5, 2.5)$ gives

$$r(1.5) = 0.08$$
$$r(2.0) = 0.08 \frac{2.5 - 2.0}{2.5 - 1.5} + r(2.5) \frac{2.0 - 1.5}{2.5 - 1.5} = 0.04 + 0.5\, r(2.5).$$

Thus,

$$86.0158 = \frac{5}{\left(1 + 0.04 + 0.5\, r(2.5)\right)^{2.0}} + \frac{105}{\left(1 + r(2.5)\right)^{2.5}},$$

which implies $r(2.5) = 0.105$, and hence, $r(2.0) = 0.092$. $\square$

When we study fixed income instruments without credit risk, it is natural to assume that the only source of risk in this situation is associated with the price changes or changes of interest rates. Consider a bond with coupons $c_1, \ldots, c_N$ that have payment dates $t_1 < \ldots < t_N$. Suppose $t = 0$ is the current time and $t_N = T$ is the maturity time. If the term structure of interest rates is determined by a constant rate r, then the market price of the bond is

$$B(r) = B(r, 0, T) = \sum_{i=1}^{N} \frac{c_i}{(1+r)^{t_i}}. \tag{5.5}$$

In a situation when the interest rate r is perturbed by a small $\triangle r$, the corresponding market price of the bond is

$$B(r + \triangle r) = B(r + \triangle r, 0, T) = \sum_{i=1}^{N} \frac{c_i}{(1 + r + \triangle r)^{t_i}}.$$

Denote $\triangle B(r) = B(r + \triangle r) - B(r)$ and note that the ratio $\triangle B(r)/B(r)$ reflects the sensitivity of bond prices to changes of the interest rate. Using Taylor's formula, we can write

$$\triangle B(r) \approx B'(r) \triangle r \quad \text{and} \quad \triangle B(r) \approx B'(r) \triangle r + \frac{1}{2} B''(r)(\triangle r)^2 .$$

and therefore

$$\triangle B(r)/B(r) \approx \frac{B'(r)}{B(r)} \triangle r, \tag{5.6}$$

$$\triangle B(r)/B(r) \approx \frac{B'(r)}{B(r)} \triangle r + \frac{1}{2} \frac{B''(r)}{B(r)} (\triangle r)^2 .$$

Using (5.5), we obtain

$$B'(r) = -\frac{1}{1+r} \sum_{i=1}^{N} t_i \frac{c_i}{(1+r)^{t_i}} = -\frac{1}{1+r} \sum_{i=1}^{N} t_i c_i^0 ,$$

$$B''(r) = \frac{1}{(1+r)^2} \sum_{i=1}^{N} t_i(t_i + 1) c_i^0 ,$$

and hence,

$$\frac{B'(r)}{B(r)} = -\frac{1}{1+r} \sum_{i=1}^{N} t_i \frac{c_i^0}{B(r)} , \tag{5.7}$$

$$\frac{B''(r)}{B(r)} = \frac{1}{(1+r)^2} \sum_{i=1}^{N} t_i(t_i + 1) \frac{c_i^0}{B(r)} .$$

Denoting

$$D := \sum_{i=1}^{N} t_i \frac{c_i^0}{B(r)} \quad \text{and} \quad C := \sum_{i=1}^{N} t_i(t_i + 1) \frac{c_i^0}{B(r)} ,$$

we rewrite (5.7) in the form

$$\frac{B'(r)}{B(r)} = -D \frac{1}{1+r} , \quad \frac{B''(r)}{B(r)} = C \frac{1}{(1+r)^2} . \tag{5.8}$$

Thus, we arrive at the following approximations of the sensitivity ratio:

$$\triangle B(r)/B(r) \approx -D \frac{\triangle r}{1+r} , \tag{5.9}$$

$$\triangle B(r)/B(r) \approx -D \frac{\triangle r}{1+r} + \frac{1}{2} C \left(\frac{\triangle r}{1+r} \right)^2 .$$

Quantities D and C in expressions (5.8)–(5.9) are naturally associated with the sensitivity of bond prices to changes of the interest rate, and they are commonly referred to as the (Macaulay) *duration* and the *convexity* of the bond, respectively. This is motivated by an obvious interpretation of D as the mean-weighted time to maturity and an observation that the second derivative of B is involved in the definition of C.

We note the following properties of quantities D and C:

1. $D \leq T$ for any bond.

2. $D = T$ for any zero-coupon bond.

3. D and C are non-increasing functions of the yield to maturity r.

4. If all bond payments are shifted by t_0 (years) without changes of r, then values of D and C are adjusted by t_0 and $t_0^2 + 2t_0 D + t_0$, respectively.

Finally, we note that by construction, duration D plays the role of a quantitative measure of risk associated with changes of interest rates. Convexity C reflects the accuracy with which D approximates the ratio $\Delta B(r)/B(r)$. Hence, if we wish to use D as a risk measure, we need to keep C sufficiently small.

Consider again a bond with coupons $c_1, \ldots, c_N$ that have payment dates $t_1 < \ldots < t_N$. Suppose that current time $t \in [t_m, t_{m+1}]$, $1 \leq m \leq N - 1$, then the market price of this bond at time t is equal to the sum of discounted remaining coupons:

$$B(t, T) = \sum_{i=m+1}^{N} \frac{c_i}{\left(1 + r(t_i - t)\right)^{t_i - t}} .$$

If before time t coupons $c_1, \ldots, c_m$ were reinvested, we denote R_t the total cost of such reinvestments. Hence, the sum

$$P(t) = R_t + B(t, T) = \sum_{i=1}^{m} c_i \left(1 + r(t_i - t)\right)^{t - t_i} + B(t, T)$$

can be interpreted as the *total cost of investments* for this bond at time t, and it is convenient to rewrite it in the form

$$P(t) = R_t + P_t , \tag{5.10}$$

where we denoted $P_t := B(t, T)$.

We will now study $P(t)$ comparing the case when the term structure of interest rates is constant and is equal to R with the case when it changes from r to $\tilde{r}$ at the beginning of the term. To emphasize the dependence of $P(t)$ on r or $\tilde{r}$, we will write $P(r, t)$ or $P(\tilde{r}, t)$, respectively.

First, we note that

$$P(r,t) \;=\; R_t(r) + P_t(r) = \sum_{i=1}^{m} c_i \,(1+r)^{t-t_i} + \sum_{i=m+1}^{N} \frac{c_i}{(1+r)^{t_i-t}}$$

$$\;=\; (1+r)^t \sum_{i=1}^{N} \frac{c_i}{(1+r)^{t_i}} = P(r)\,(1+r)^t\,,$$

where $P(r) := \sum_{i=1}^{N} \frac{c_i}{(1+r)^{t_i}}$. Similarly, $P(\tilde{r},t) = P(\tilde{r})\,(1+\tilde{r})^t$.

Now we will use these observations to show that there exists a unique t^* such that

$$P(r,t^*) = P(\tilde{r},t^*)\,. \tag{5.11}$$

Suppose $\tilde{r} > r$, then we have the inequality

$$P(\tilde{r}) = P(\tilde{r},0) < P(r,0) = P(r)\,.$$

However, for $t = T = t_N$, we have

$$P(r,t_N) = \sum_{i=1}^{N} c_i \,(1+r)^{t_N-t_i} < \sum_{i=1}^{N} c_i \,(1+\tilde{r})^{t_N-t_i} = P(\tilde{r},t_N)\,,$$

thus continuity and monotonicity of functions in these inequalities imply the existence of t^*. Solving (5.11) we have

$$\left(\frac{1+\tilde{r}}{1+r}\right)^{t^*} = \frac{P(r)}{P(\tilde{r})}\,,$$

and therefore there is a unique

$$t^* = \frac{\ln\left(P(r)/P(\tilde{r})\right)}{\ln\left((1+r)/(1+\tilde{r})\right)}\,. \quad \square$$

Let $D = D(r)$ be the duration of a bond in the case of a constant term structure of interest rates. We can write

$$P(\tilde{r}, D) = P(\tilde{r})(1+\tilde{r})^D\,.$$

Differentiating with respect to $\tilde{r}$, we obtain

$$P'(\tilde{r}, D) = P'(\tilde{r})(1+\tilde{r})^D + D\,P(\tilde{r})(1+\tilde{r})^{D-1}\,.$$

Taking into account

$$P'(\tilde{r})/P(\tilde{r}) - -D(\tilde{r})\,\frac{1}{1+\tilde{r}}\,,$$

we have

$$P'(\tilde{r}, D) = P(\tilde{r})(1+\tilde{r})^{D-1}\,(D - D(\tilde{r}))\,.$$

If $\tilde{r} > r$, then $D(\tilde{r}) < D(r) = D$ and $P'(\tilde{r}, D) > 0$, that is, $P(\tilde{r}, D)$ is non-decreasing and

$$P(r, D) < P(\tilde{r}, D). \tag{5.12}$$

The case $\tilde{r} < r$ is similar. In particular, if $r_1 < r < r_2$, then

$$t^*(r_1) < D < t^*(r_2),$$

and this inequality can be used for estimating the duration of a bond.

Thus, the inequality (5.12) indicates that the *real cost of investment in a bond* (corresponding to $\tilde{r}$) is bigger than the *expected cost of investment in a bond* (corresponding to r) if the investment period is equal to the duration D. This property is referred to as the *minimization property of duration*, and we will study it in detail for *portfolios of bonds*.

Assume there are m types of bonds in a market and their prices at time $t = 0$ are $P_1, \ldots, P_m$. Denote $\Pi(\Omega_1, \ldots, \Omega_m)$ the *portfolio* of these bonds with the corresponding capital Ω_j, $j = 1, \ldots, m$, invested into j-th bond. Then the sum $\Omega = \sum_{j=1}^m \Omega_j$ is called the *capital* of Π; the quantities $w_j = \Omega_j / \Omega$ and $k_j = \Omega_j / P_j$, $j = 1, \ldots, m$, are called the *proportion* and the *number* of j-th bond in Π.

Denote $t_1 < \ldots < t_N$ the payment times for the portfolio and c_i^j, $i = 1, \ldots, N$, $j = 1, \ldots, m$, the payment amounts for j-th bond at time t_i. Define sequence $(R_i)_{i=1,\ldots,N}$ by

$$R_i = \sum_{j=1}^m \frac{\Omega_j}{P_j} c_i^j,$$

then at time $t = 0$ we can identify $\Pi(\Omega_1, \ldots, \Omega_m)$ with a bond with coupons $R_1, \ldots, R_N$. It is natural then to define r_Π, the *yield to maturity of the portfolio* (with $T = t_N$), as a unique solution to the equation

$$\Omega = \frac{R_1}{(1 + r_\Pi)^{t_1}} + \ldots + \frac{R_N}{(1 + r_\Pi)^{t_N}}.$$

If r is a constant risk-free rate of interest for all future periods, then we can define the *duration* and the *convexity* of portfolio $\Pi(\Omega_1, \ldots, \Omega_m)$ as

$$D_\Pi = \frac{1}{\Omega} \sum_{i=1}^N t_i \frac{R_i}{(1 + r_\Pi)^{t_i}},$$

$$C_\Pi = \frac{1}{\Omega} \sum_{i=1}^N t_i (t_i + 1) \frac{R_i}{(1 + r_\Pi)^{t_i}}.$$

Clearly,

$$D_\Pi = \sum_{j=1}^m w_j D_j, \qquad C_\Pi = \sum_{j=1}^m w_j C_j.$$

and
$$\min_j D_j \le D_\Pi \le \max_j D_j, \qquad \min_j C_j \le C_\Pi \le \max_j C_j,$$

where D_j and C_j are the duration and the convexity of j-th bond, $j = 1, \ldots, m$, in the portfolio.

Furthermore, if we take any $D \in [\min_j D_j, \max_j D_j]$, then, for each $\Omega > 0$, we can construct a portfolio with given duration D. To verify this, we consider the following system

$$\begin{cases} \sum_{j=1}^m \omega_j D_j = D \\ \\ \sum_{j=1}^m \omega_j = 1, \quad \omega_j \ge 0, \ j = 1, \ldots, m. \end{cases}$$

If $D = D_j$ for some $j = 1, \ldots, m$, then the weights

$$\omega_1 = 0, \ \ldots \ \omega_{j-1} = 0, \ \omega_j = 1, \ \omega_{j+1} = 0, \ \ldots \ \omega_m = 0$$

give the desired solution. If $D_j < D < D_{j+1}$ for some $j = 1, \ldots, m-1$, then the desired portfolio is defined by

$$\omega_1 = \ldots = \omega_{j-1} = 0, \ \omega_j = \frac{D_{j+1} - D}{D_{j+1} - D_j}, \ \omega_{j+1} = \frac{D - D_j}{D_{j+1} - D_j},$$

$$\omega_{j+2} = \ldots = \omega_m = 0. \quad \square$$

Finally, we note that if a portfolio was created at time $t = 0$ and the interest rate r was changed by $\triangle r$ after this time, then

$$\triangle \Omega / \Omega \approx -D_\Pi \frac{\triangle r}{1+r}, \tag{5.13}$$

$$\triangle \Omega / \Omega \approx -D_\Pi \frac{\triangle r}{1+r} + \frac{1}{2} C_\Pi \left(\frac{\triangle r}{1+r}\right)^2.$$

Therefore quantities D_Π and C_Π describe the sensitivity of portfolio capital to changes of the interest rate. In particular, D_Π is a quantitative measure of risk associated with changes of interest rates, whereas C_Π reflects the accuracy with which D_Π approximates the ratio $\triangle \Omega / \Omega$.

Since every portfolio can be identified with a bond, we can define the expected and the real costs of investment in the portfolio in the following way:

$$\Omega(r, t) = \sum_{i: t_i \le t} R_i (1+r)^{t-t_i} + \sum_{i: t_i > t} \frac{R_i}{(1+r)^{t_i - t}} = R_t(r) + P_t(r),$$

$$\Omega(\tilde{r}, t) = \sum_{i: t_i \le t} R_i (1+\tilde{r})^{t-t_i} + \sum_{i: t_i > t} \frac{R_i}{(1+\tilde{r})^{t_i - t}} = R_t(\tilde{r}) + P_t(\tilde{r}),$$

where $(R_i)_{i=1,\ldots,N}$ are payments at times $(t_i)_{i=1,\ldots,N}$.

Thus, we arrive at the following portfolio *immunization property*:

$$\Omega(\tilde{r}, D_\Pi) \geq \Omega(r, D_\Pi).\qquad(5.14)$$

The immunization property must take into account the investment horizon T, which can be achieved by using the *Reddington immunization condition*:

$$\sum_{j=1}^{m} w_j D_j = T$$

$$\sum_{j=1}^{m} w_j = 1, \quad w_j \geq 0, \ j = 1,\ldots,m.$$

As we mentioned earlier, such a system admits a solution if $\min_j D_j \leq T \leq \max_j D_j$. In this case, we can construct a portfolio with duration T and the immunization property (5.14) for this portfolio is fulfilled: $\Omega(\tilde{r}, T) \geq \Omega(r, T)$.

Worked Example 5.4 *Consider a portfolio that consists of two types of bonds B_1 and B_2, both with the face value $100 and maturity 2 years. Bonds of type B_1 pay coupons of $2.5 semi-annually and bonds of type B_2 pay coupons of $8.0 annually. The portfolio is created at time $t = 0$ and the risk free rate of interest is assumed to be 9% per annum for all periods. Suppose that $4000 is invested in bonds of type B_1 and $6000 is invested in bonds of type B_2. Given that right after time $t = 0$ the risk free rate of interest is changed to 8% per annum for all periods, find*

1. D_Π *and* C_Π;

2. *Relative change of the portfolio price when the rate of interest reduces from 9% to 8%;*

3. *Expected and real investment costs in this portfolio for $T = 2$ years;*

4. *Expected and real investment costs in this portfolio for $T = D_\Pi$ years.*

Solution

1. Using information about bonds B_1 and B_2, we find

$$P_1 = 93.157, \quad D_1 = 1.925 \text{ (years)}, \quad C_1 = 5.713 \text{ (years)}$$

and

$$P_2 = 98.240, \quad D_2 = 1.925 \text{ (years)}, \quad C_2 = 5.701 \text{ (years)}.$$

Noting that $\Omega(0.9) = 10000$ for our portfolio $\Pi(4000, 6000)$, we calculate $D_\Pi = 1.9252$ and $C_\Pi = 5.706$.

2. Using (5.13) with $r = 0.09$ and $\triangle r = -0.01$, we obtain

$$\frac{\triangle \Omega}{\Omega} = \frac{\Omega(0.8) - \Omega(0.9)}{\Omega(0.9)} \approx 0.0179 \,,$$

which implies $\Omega(0.8) = 10179.02$.

3. Expected and real investment costs for $T = 2$ years have the following values:

$$\Omega(0.09, 2) = \Omega(0.09)(1 + 0.09)^2 \approx 11881.00 \,,$$
$$\Omega(0.08, 2) \approx 11872.85 \,,$$

and we note that the portfolio is not immunized for 2 years against the risk of interest rate changes.

4. Expected and real investment costs for $T = D_\Pi$ years are

$$\Omega(0.09, D_\Pi) = 11804.647 \quad \text{and} \quad \Omega(0.08, D_\Pi) = 11804.683 \,,$$

so this portfolio is immunized against the risk of interest rate changes for the investment horizon of D_Π years. □

In general, portfolio management involves both *active* and *passive* strategies. Active management is based on *dynamic* strategies that reflect investor's response to changes in the market. The immunization procedures are clearly an example of an active strategy. Another example of an active strategy is a *duration strategy*, that is, a strategy that is adapted to the evolution of risk-free rates of interest. An example of such strategy is an investment in coupon-bearing bonds with maturity greater than the investment horizon. The duration changes in this case can be achieved by changing bonds in the portfolio. In financial industry, an agreement between two parties to exchange cash flows in the future in known as a *swap*. Thus, financial contracts that exploit duration control strategies belong to a big class of contracts called *swaps*.

Worked Example 5.5 *Assume that the risk-free interest rate is currently 8% per annum. Consider portfolio $\Pi_0 = \Pi(1000, 1500, 2500, 4000)$ of bonds with durations $D_1 = 1.5$, $D_2 = 2.0$, $D_3 = 3.5$, $D_4 = 5$ (years).*
Find appropriate swaps for the following two scenarios:

(a) *interest rate increases to 9% per annum;*

(b) *interest rate decreases to 7% per annum.*

Solution We can calculate the duration of portfolio Π_0:

$$D_{\Pi_0} = \frac{1.5}{9} + \frac{2}{6} + \frac{5}{18} \times 3.5 + \frac{4}{9} \times 5 = 3.694 \,.$$

(a) The relative change of the portfolio cost is

$$\frac{\Delta\Omega}{\Omega} \approx -D_{\Pi_0}\frac{\Delta r}{1+r} = -3.694 \times \frac{0.01}{1.08} = -0.0342\,,$$

where

$$\Delta\Omega = \Omega_{\Pi_0}(0.09) - \Omega_{\Pi_0}(0.08) = \Omega_{\Pi_0}(0.09) - 9000\,.$$

Hence,

$$\Omega_{\Pi_0}(0.09) \approx \Omega_{\Pi_0}(0.08) - 0.0342 \times \Omega_{\Pi_0}(0.08) = 8692.13\,,$$

and we note that the cost of the portfolio is reduced in this scenario. We now illustrate how investment in more short-term bonds rather than in long-term bonds can mitigate such problem. For example, if we sell all our 5-year bonds and invest this capital in 1.5-year bonds, then we will have portfolio $\Pi_1 = \Pi(5000, 1500, 2500, 0)$, which has the same cost as portfolio Π_0, but its duration is shorter:

$$D_{\Pi_1} = \frac{1.5}{9} \times 5 + \frac{2}{6} + \frac{5}{18} \times 3.5 = 2.139\,.$$

Then

$$\frac{\Delta\Omega}{\Omega} \approx -D_{\Pi_1}\frac{\Delta r}{1+r} = -2.139 \times \frac{0.01}{1.08} = -0.0198\,,$$

and the cost of the new portfolio is

$$\Omega_{\Pi_1}(0.09) \approx \Omega_{\Pi_1}(0.08) - 0.0198 \times \Omega_{\Pi_1}(0.08) = 8821.76\,,$$

which shows smaller reduction in comparison with the original portfolio.

(b) The relative change of the portfolio cost in this case is

$$\frac{\Delta\Omega}{\Omega} \approx -D_{\Pi_0}\frac{\Delta r}{1+r} = -3.694 \times \frac{-0.01}{1.08} = 0.0342\,,$$

therefore

$$\Omega_{\Pi_0}(0.07) \approx \Omega_{\Pi_0}(0.08) + 0.0342 \times \Omega_{\Pi_0}(0.08) = 9307.87\,.$$

Here we can increase the cost our portfolio by selling short-term bonds and buying more long-term bonds. For example, portfolio $\Pi_2 = \Pi(0, 1500, 2500, 5000)$ has the duration

$$D_{\Pi_2} = \frac{2}{6} + \frac{5}{18} \times 3.5 + \frac{5}{9} \times 5 = 4.083$$

and thus

$$\frac{\Delta\Omega}{\Omega} \approx -D_{\Pi_2}\frac{\Delta r}{1+r} = -4.083 \times \frac{-0.01}{1.08} = 0.0378\,,$$

which implies the increased cost $\Omega_{\Pi_2}(0.07) \approx 9340.28$. $\square$

Passive portfolio management is applicable to portfolios that have unchanged structure during the whole investment period. A typical example of such a portfolio is a *dedicated portfolio*. We illustrate the concept of a dedicated portfolio in the following scenario. Suppose that an investor is obliged to make payments $f_1, \ldots, f_n$ at times $t_1, \ldots, t_n$ after time $t = 0$, and suppose that the bond market consists of m bonds with prices $P_1, \ldots, P_m$ at time $t = 0$. Then this investor secures the compulsory payments by investing in a portfolio of bonds such that its structure is defined by the following minimization problem:

$$\text{find } \min f = \min \left(\sum_{j=1}^{m} P_j \, x_j \right)$$

$$\text{subject to } \sum_{j=1}^{m} c_j^i \, x_j \geq f_i \text{ and } x_j \geq 0, \; j = 1, \ldots, m, \; i = 1, \ldots, n,$$

where x_j is the number of j-th bond in the portfolio and c_j^i is the payment of j-th bond at time t_i.

Worked Example 5.6 *Suppose that an investor is obliged to make the following payments*

Time (in years)	1	2	3
Payment	260	660	440

and there are two types of bonds in the market: B_1 and B_2 with payments 10, 10, 110, and 50, 150, 0, and prices 100 and 150, respectively. Find the dedicated portfolio.

Solution We need to solve the following linear programming problem:

$$\min f(x_1, x_2) = \min \left(100 x_1 + 150 x_2 \right)$$

subject to constraints

$$10 x_1 + 50 x_2 \geq 260$$
$$10 x_1 + 150 x_2 \geq 660$$
$$110 x_1 \geq 440, \quad x_1, x_2 \geq 0.$$

It is not difficult to check that the optimal solution is $(x_1, x_2) = (4, 4.4)$ with $f = (4, 4.4) = 1060$; that is, the dedicated portfolio consists of 4 B_1-bonds and 4.4 B_2-bonds. $\square$

Next, we will have a brief discussion of bond markets with *continuous compounding*. For a bond with the payments flow $(c_1, t_1), \ldots, (c_N, t_N)$ and maturity time $T = t_N$, we define the *yield to maturity* y as a unique solution to the equation

$$B(0, T) = \sum_{i=1}^{N} \frac{c_i}{e^{y t_i}}.$$

Usually, payments $c_1, \ldots, c_{N-1}$ are viewed as coupons and the last payment consists of a coupon and the face value A of the bond. The values of a coupon $c_i \equiv c$ are often determined by the *coupon rate* f: $c = f \times A$.

The *duration* of this bond is the weighted average maturity of the bond:

$$D = \sum_{i=1}^{N} t_i\, W_i, \quad \text{where} \quad W_i = \frac{1}{P}\frac{c_i}{e^{yt_i}}, \; P = B(0,T), \; i = 1, \ldots, N.$$

Assuming that the yield curve admits parallel infinitesimal shifts, we also have

$$D = \lim_{\triangle y \to 0} \frac{\triangle P/P}{\triangle y} = -\frac{\partial P/\partial y}{P} \quad \text{and} \quad \frac{\triangle P}{P} \approx -D\,\triangle y.$$

The second formula allows us to interpret the duration D as a measure of risk related to the relative change of bond prices. The duration of a portfolio of bonds is the weighted average of the durations of the bonds in the portfolio, where weights are the proportions of investment in the corresponding bond.

The *convexity* of a bond is the weighted average of its maturity-squares:

$$C = \sum_{i=1}^{N} t_i^2\, W_i,$$

and we have the following expression in terms of the bond price

$$C = \frac{\partial^2 P/\partial y^2}{P},$$

hence

$$\frac{\triangle P}{P} \approx -D\,\triangle y + \frac{1}{2}C(\triangle y)^2.$$

The convexity of a portfolio of bonds is the weighted average of convexities of bonds in the portfolio.

To consider the case of non-parallel shift of a non-flat yield curve, we need to model yields for different maturities. The term structure of interest rates determines the relationship between a yield and the corresponding term to maturity. In the simplest case of a zero-coupon bond, we have

$$P = B(0,T) = \frac{A}{e^{yT}},$$

where the yield to maturity $y = y(T)$ is the continuously compounded rate (zero-coupon rate) and rate $y(T)$ corresponds to the maturity term T. Clearly, the yield to maturity for a coupon bearing bond can be defined as a weighted average of zero-coupon rates.

The notion of a *forward rate* is significant and useful in studies of bond markets. A forward rate $f(t_1, t_2)$, $0 \le t_1 \le t_2$, is the interest rate that is fixed at time $t = 0$ for investments made at time t_1 that mature at time

t_2. The following considerations connect this definition with $y(t_1)$ and $y(t_2)$. By definition of $f(t_1, t_2)$, a one dollar investment at time t_1 will grow to $e^{f(t_1,t_2)(t_2-t_1)}$ during the period $t_2 - t_1$. However, we can consider a zero-coupon paying one dollar at time t_1 that is then reinvested till time t_2 at rate $y(t_2)$. Thus, we arrive at the following equality

$$e^{f(t_1,t_2)(t_2-t_1)} = \frac{e^{y(t_2)t_2}}{e^{y(t_1)t_1}}$$

or equivalently

$$f(t_1, t_2) = y(t_2) + \frac{y(t_2) - y(t_1)}{t_2 - t_1} t_1 .$$

Further, if $t_2 \to t_1$, that is, $t_1 = t$, $t_2 = t + \triangle t$ and $\triangle t \to 0$, then

$$f(t) = \lim_{\triangle t \to 0} f(t, t + \triangle t) = y(t) + \frac{\partial y(t)}{\partial t} t .$$

If this limit exists, then quantity $f(t)$ is called the *instantaneous forward rate*. Note that the zero-coupon rate can be expressed in terms of f:

$$y(t) = \frac{1}{t} \int_0^t f(s) \, ds . \tag{5.15}$$

Both $f(t)$ and $y(t)$ can be used for estimations of the term structure of interest rates. In addition to the bootstrapping method that we mentioned earlier, we now focus on the *Nelson-Siegel approach*, which suggests the following parametrization of the instantaneous forward rate curve:

$$f(t) = \alpha_1 + \alpha_2 e^{-t/\beta} + \alpha_3 e^{-t/\beta} \frac{t}{\beta} . \tag{5.16}$$

Using (5.15), we can write the corresponding formula for y:

$$y(t) = \alpha_1 + (\alpha_2 + \alpha_3) \frac{\beta}{t} \left(1 - e^{-t/\beta}\right) - \alpha_3 e^{-t/\beta} . \tag{5.17}$$

Parameters α_1, α_2, α_3, and $\beta > 0$ in Nelson-Siegel model have some meaningful interpretations. Indeed, for $t = 0$, we have $\alpha_1 + \alpha_2 = f(0)$, so $\alpha_1 + \alpha_2$ is the instantaneous forward rate at time $t = 0$. Taking limits in (5.16)–(5.17) as $t \to \infty$, we obtain $\alpha_1 = y(\infty) = f(\infty)$, which can be interpreted as a *consol rate*. Parameter α_2, being the difference between $\alpha_1 + \alpha_2$ and α_1, can be interpreted as the slope of the term structure of forward rates. Parameter α_3 is responsible for the curvature of the term structure. Finally, parameter β is the speed of convergence of the term structure toward the consol rate.

We now discuss the duration method that captures the interest rate risk when shifts of yield curves are not parallel and not infinitesimal. To describe this approach, we consider at time $t = 0$ a portfolio of bonds with payments at times $t_1 < \ldots < t_N$. Assuming a continuously compounded term structure of instantaneous rates $f(t)$, we allow an instantaneous shift $\triangle f(t)$, so that the perturbed structure is $\hat{f}(t) = f(t) + \triangle f(t)$. In this case, the instantaneous relative change of the current value Ω_0 of the portfolio is

$$
\frac{\triangle \Omega_0}{\Omega_0} = \left. -D_1 \triangle f(0) - D_2 \frac{1}{2} \left(\frac{\partial \triangle f(t)}{\partial t} - (\triangle f(0))^2 \right) \right|_{t=0}
$$

$$
\left. -D_3 \frac{1}{3!} \left(\frac{\partial^2 \triangle f(t)}{\partial t^2} - 3 \triangle f(0) \frac{\partial \triangle f(t)}{\partial t} + (\triangle f(0))^3 \right) \right|_{t=0}
$$

$$
\left. - \ldots - D_k \frac{1}{k!} \left(\frac{\partial^{k-1} \triangle f(t)}{\partial t^{k-1}} + \ldots + (\triangle f(0))^k \right) \right|_{t=0},
$$

where

$$
D_k = \sum_{i=1}^{N} t_i^k W_i, \quad k = 1, \ldots, K; \quad W_i = \frac{1}{V_0} c_i e^{-\int_0^{t_i} f(s) ds}, \quad i = 1, \ldots, N.
$$

To ensure that our portfolio satisfies the immunization property during the investment horizon of T (years), we need to introduce the following system of equalities

$$
D_k = T^k, \quad k = 1, \ldots, K. \tag{5.18}
$$

Denote W_i, $i = 1, \ldots, I$, the proportion of the portfolio capital invested in i-th bond with duration D_k^i, $i = 1, \ldots, I$, $k = 1, \ldots, K$. We clearly have

$$
D_k = \sum_{i=1}^{I} D_k^i \quad \text{and} \quad \sum_{i=1}^{I} W_i = 1, \quad k = 1, \ldots, K. \tag{5.19}
$$

Equalities (5.18)–(5.19) imply that there are three cases:

(a) If $I = K + 1$, then there is a unique solution $W_1, \ldots, W_I$.

(b) If $I < K + 1$, then there are no solutions.

(c) If $I > K + 1$, then there are infinitely many solutions.

It is possible to select a unique immunization solution in the last case by minimizing quadratic function $\sum_{i=1}^{I} W_i^2$ subject to a set of some other constraints.

There are some special forms of the term structure of instantaneous forward rates or zero-coupon yields. One natural choice is the polynomial structure:

$$
y(t) = a_0 + a_1 t + a_2 t^2 + \ldots, \tag{5.20}
$$

where $a_0, a_1, a_2, \ldots$ are the shape parameters of the curve. The corresponding structure of forward rates is

$$
f(t) = a_0 + 2a_1 t + 3a_2 t^2 + 4a_3 t^3 + \ldots. \tag{5.21}
$$

If a shifted curve has a structure that is similar to (5.20):

$$\tilde{y}(t) = \tilde{a}_0 + \tilde{a}_1 t + \tilde{a}_2 t^2 + \ldots,$$

where $\tilde{a}_i = a_i + \triangle a_i$, $i = 0, 1, \ldots$, then the shift of the term structure of zero-coupon yields is

$$\triangle y(t) = \tilde{y}(t) - y(t) = \triangle a_0 + \triangle a_1 t + \triangle a_2 t^2 + \ldots.$$

Similarly, using (5.21), we arrive at the following expression for the shift of the term structure of instantaneous forward rates:

$$\triangle f(t) = \tilde{f}(t) - f(t) = \triangle a_0 + 2\triangle a_1 t + 3\triangle a_2 t^2 + 4\triangle a_3 t^3 + \ldots.$$

Therefore, the relative change of the portfolio value is

$$\begin{aligned}
\frac{\triangle \Omega_0}{\Omega_0} &= -D_1 \triangle a_0 - D_2 \left(\triangle a_1 - \frac{(\triangle a_0)^2}{2} \right) \\
&\quad - D_3 \left(\triangle a_2 - \triangle a_0 \triangle a_1 + \frac{(\triangle a_0)^3}{3!} \right) \\
&\quad - \ldots - D_k \left(\triangle a_{k-1} + \ldots + \frac{(\triangle a_0)^k}{k!} \right).
\end{aligned}$$

Finally, we note that Nelson-Siegel model is another natural choice for the term structure of instantaneous forward rates or zero-coupon yields that can be used in the duration method.

5.2   Stochastic modeling and pricing bonds and their derivatives

Consider a zero-coupon bond maturing at time $T < T^*$, that is, a claim that pays \$1 at time T. Let $B(t, T)$ be its price at time $t \in [0, T]$. Naturally, we have $B(T, T) = 1$ and $B(t, T) < 1$ for all $t \leq T$.

As in deterministic case, price $B(t, T)$ can be written in two equivalent forms:

$$B(t, T) = \exp\left\{ -r(t, T)(T - t) \right\},$$

$$B(t, T) = \exp\left\{ -\int_t^T f(t, s)\, ds \right\}.$$

Function $r(\cdot, T)$ is the *yield to maturity* and function $f(t, s)$ is the *forward rate* for the period $[t, s]$.

Under some reasonable assumptions, we have the following relations:

$$r(t,T) = -\frac{\ln B(t,T)}{T-t},$$

$$f(t,T) = -\frac{\partial}{\partial t} \ln B(t,T) = r(t,T) + (T-t)\frac{\partial}{\partial T}r(t,T).$$

Denote $r_t = f(t,t)$ the *instantaneous short rate* at t. This rate of interest can be a stochastic process; therefore, bonds must be studied as risky assets since their prices depend on interest rates.

Let $r = (r_t)_{t\geq 0}$ be a stochastic process on some stochastic basis $(\Omega, \mathcal{F}, \mathbb{F}, P)$. Defining a bank account by

$$B_t = \exp\left\{ \int_0^t r_s\,ds \right\},$$

we arrive at the notion of a *bonds market* as a family $\left(B_t, B(t,T) \right)_{t\leq T\leq T^*}$. As in the case of the studied above (B,S)-market ("shares" market), we can consider discounted bond prices:

$$\overline{B}(t,T) = \frac{B(t,T)}{B_t}$$

and construct a probability P^* that is equivalent to the initial probability P and such that the process $\overline{B} = (\overline{B}(t,T))_{t\geq 0}$ is a martingale with respect to P^*. If such probability exists, then we say that the bonds market is *arbitrage free*. We can interpret the absence of arbitrage as the impossibility of making profit without risk.

Taking into account that $B(T,T) = 1$, we obtain

$$E^*\left(B_T^{-1}\big|\mathcal{F}_t\right) = \frac{B(t,T)}{B_t},$$

and therefore we have the representation

$$B(t,T) = E^*\left(\exp\left\{ -\int_0^t r_s\,ds \right\}\big|\mathcal{F}_t \right),$$

which allows one to study the structure of prices $B(t,T)$ by specifying process $r = (r_t)_{t\geq 0}$.

Here we list some of the frequently used models:

Merton

$$dr_t = \alpha\,dt + \gamma\,dW_t, \qquad \alpha, \gamma \in \mathbb{R};$$

Vasiček

$$dr_t = (\alpha - \beta r_t)\,dt + \gamma\,dW_t, \qquad \alpha, \beta, \gamma \in \mathbb{R};$$

Ho-Lee
$$dr_t = \alpha(t)\, dt + \gamma\, dW_t\,, \qquad \gamma \in \mathbb{R};$$

Black-Derman-Toy
$$dr_t = \alpha(t)\, dt + \gamma(t)\, dW_t\,,$$

Hull-White
$$dr_t = (\alpha(t) - \beta\, r_t)\, dt + \gamma\, dW_t\,, \qquad \beta,\, \gamma \in \mathbb{R};$$

Another way of specifying process r is given by the *Schmidt model*: let functions f and g be continuous and functions T and F be continuous and strictly increasing. Then define

$$r_t = F\big(f(t) + g(t)\, W_{T(t)}\big)\,.$$

All the models listed above can be obtained from the Schmidt model by choosing appropriate functions F, f, g, and T.

An equivalent alternative way of describing the structure of bond prices is based on specifying the evolution of forward rate:

$$df(t,T) = \sigma^2\, (T - t)\, dt + \sigma\, dW_t\,,$$

or

$$f(t,T) = f(0,T) + \sigma^2\, t\, (T - t/2) + \sigma\, W_t\,,$$

where $f(0,T)$ is the present forward rate. This implies

$$dr_t = \left(\frac{\partial}{\partial t} f(0,t) + \sigma^2\, t\right) dt + \sigma\, dW_t\,,$$

or

$$r_t = f(0,t) + \frac{\sigma^2}{2}\, t^2 + \sigma\, W_t\,.$$

Substituting the expression for $f(t,s)$ into formula

$$B(t,T) = \exp\left\{ -\int_t^T f(t,s)\, ds \right\}, \quad t \le T\,,$$

we obtain

$$
\begin{aligned}
\int_t^T f(t,s)\, ds &= \int_t^T \Big[f(0,s) + \sigma^2\, t\, (s - t/2) \Big]\, ds + \sigma\, (T - t)\, W_t \\
&= \int_t^T f(0,s)\, ds + \frac{\sigma^2}{2}\, t\, T\, (T - t) + \sigma\, (T - t)\, W_t\,,
\end{aligned}
$$

and hence,

$$
\begin{aligned}
B(t,T) &= \exp\left\{ - \int_t^T f(0,s)\,ds - \frac{\sigma^2}{2}\, t\,T\,(T-t) + \sigma\,(T-t)\,W_t \right\} \\
&= \frac{B(0,T)}{B(0,t)} \exp\left\{ - \frac{\sigma^2}{2}\, t\,T\,(T-t) + \sigma\,(T-t)\,W_t \right\}.
\end{aligned}
$$

We can also rewrite it in the form

$$
B(t,T) = \frac{B(0,T)}{B(0,t)} \exp\left\{ (T-t)\,f(0,T) - \frac{\sigma^2}{2}\, t\,(T-t)^2 - (T-t)\,r_t \right\}.
$$

Note that this model is a particular case of the *Heath-Jarrow-Morton model*, and it is not difficult to check that the initial probability is a martingale probability.

Now we proceed to a detailed study of Vasiček model. According to this model, the interest rate oscillates around α/β: r_t has positive drift if $r_t < \alpha/\beta$, and negative if $r_t > \alpha/\beta$. If $\alpha/\beta = 0$, then r_t is a stationary (Gaussian) Ornstein-Uhlenbeck process.

Applying the Kolmogorov-Itô formula, we obtain

$$
r_t = e^{-\beta t} \left[r_0 + \int_0^t \alpha\, e^{\beta s}\,ds + \int_0^t \gamma\, e^{\beta t}\,dW_s \right].
$$

Using the Markov property of r_t, we can write

$$
\begin{aligned}
B(t,T) &= E\left(\exp\left\{ - \int_t^T r_s\,ds \right\} \Big| \mathcal{F}_t \right) = E\left(\exp\left\{ - \int_t^T r_s\,ds \right\} \Big| r_t \right) \\
&= \exp\left\{ \frac{\gamma^2}{2} \int_t^T \left(\int_s^T e^{-\beta(u-s)}\,du \right)^2 ds \right. \\
&\qquad \left. -\alpha \int_t^T \int_t^u e^{-\beta(u-s)}\,ds\,du - r_t \int_t^T e^{-\beta(u-t)}\,du \right\} \\
&\equiv \exp\left\{ a(t,T) - r_t\, b(t,T) \right\},
\end{aligned}
$$

where

$$
a(t,T) := \frac{\gamma^2}{2} \int_t^T \left(\int_s^T e^{-\beta(u-s)}\,du \right)^2 - \alpha \int_t^T \int_t^u e^{-\beta(u-s)}\,ds\,du
$$

$$
b(t,T) := \int_t^T e^{-\beta(u-t)}\,du.
$$

This gives us a general structure of bond prices. Now, in the framework of

Vasiček model, we consider a European call option with the exercise time $T' \le T \le T^*$ and payoff function

$$f = \left(B(T',T) - K\right)^+,$$

where K is the strike price.

The price of this option is given by

$$C(T',T) = B(0,T)\,\Phi(d_+) - K\,B(0,T')\,\Phi(d_-),$$

where

$$d_\pm = \frac{\ln\frac{B(0,T)}{K\,B(0,T')} \pm \frac{1}{2}\sigma^2(T',T)\left(\int_{T'}^T e^{-\beta(u-T')}\,du\right)^2}{\sigma(T',T)\int_{T'}^T e^{-\beta(u-T')}\,du},$$

$$\sigma(T',T) = \left(\int_{T'}^T \left(\int_s^T \gamma\, e^{-\beta(u-s)}\,du\right)^2 ds\right)^{1/2}.$$

We need to compute

$$\begin{aligned}
C(T',T) &= E\left(e^{-\int_0^{T'} r_u\,du}\left(B(T',T) - K\right)^+\right)\\[2mm]
&= E\left(I_{\{\omega:\,B(T',T)>K\}}\; e^{-\int_0^{T'} r_u\,du}\,B(T',T)\right)\\[2mm]
&\quad - K\,E\left(I_{\{\omega:\,B(T',T)>K\}}\; e^{-\int_0^{T'} r_u\,du}\right).
\end{aligned}$$

Note that

$$\begin{aligned}
\{\omega:\,B(T',T) > K\} &= \{\omega:\,a(T',T) - r_{T'}\,b(T',T)) > \ln K\}\\[2mm]
&= \{\omega:\,r_{T'} \le r'\},
\end{aligned}$$

where

$$r' = \frac{\ln K - a(T',T)}{-b(T',T)}.$$

Letting

$$\xi = r_{T'} \quad \eta = \int_0^T r_u\,du \quad \zeta = \int_0^{T'} r_u\,du,$$

we obtain

$$C(T',T) = E\left(I_{\{\omega:\,\xi \le r'\}}\,e^{-\eta}\right) - K\,E\left(I_{\{\omega:\,\xi \le r'\}}\,e^{-\zeta}\right).$$

Note that the quantitative characteristics of ξ, η, and ζ are given by

$$\mu_\xi = E(r_{T'}) = e^{-\beta T'} \left(r_0 + \alpha \int_0^{T'} e^{-\beta s} \, ds \right),$$

$$\mu_\eta = E\left(\int_0^T r_u \, du \right) = r_0 \int_0^T e^{-\beta u} \, du + \alpha \int_0^T \int_0^u e^{-\beta(u-s)} \, ds \, du,$$

$$\mu_\zeta = E\left(\int_0^{T'} r_u \, du \right) = r_0 \int_0^{T'} e^{-\beta u} \, du + \alpha \int_0^{T'} \int_0^u e^{-\beta(u-s)} \, ds \, du,$$

$$\sigma_\xi^2 = V(r_{T'}) = \gamma^2 \int_0^{T'} e^{-2\beta(T'-s)} \, ds,$$

$$\sigma_\eta^2 = V\left(\int_0^T r_u \, du \right) = \gamma^2 \int_0^T \left(\int_s^T e^{-\beta(u-s)} \, du \right)^2 ds,$$

$$\sigma_\zeta^2 = V\left(\int_0^{T'} r_u \, du \right) = \gamma^2 \int_0^{T'} \left(\int_s^{T'} e^{-\beta(u-s)} \, du \right)^2 ds,$$

$$\rho_{\xi\zeta} = Cov\left(r_{T'}, \int_0^{T'} r_u \, du \right) = \gamma^2 \int_0^{T'} e^{-\beta(T'-s)} \int_s^{T'} e^{-\beta(u-s)} \, du \, ds,$$

$$\rho_{\xi\eta} = Cov\left(r_{T'}, \int_0^T r_u \, du \right) = \rho_{\xi\zeta} + \sigma_\xi^2 \int_{T'}^T e^{-\beta(u-T')} \, du.$$

Applying Lemma 4.1, we obtain

$$
\begin{aligned}
C(T',T) &= E\left(I_{\{w:\,\xi \le r'\}} \, e^{-\eta} \right) - K\, E\left(I_{\{w:\,\xi \le r'\}} \, e^{-\zeta} \right) \\
&= \exp\left\{ \frac{\sigma_\eta^2}{2} - \mu_\eta \right\} \Phi\left(\frac{r' - (\mu_\xi - \rho_{\xi\eta})}{\sigma_\xi} \right) \\
&\quad - K \exp\left\{ \frac{\sigma_\zeta^2}{2} - \mu_\zeta \right\} \Phi\left(\frac{r' - (\mu_\xi - \rho_{\xi\zeta})}{\sigma_\xi} \right).
\end{aligned}
$$

Substituting expressions for μ_ξ, μ_η, μ_ζ, σ_ξ^2, σ_η^2, σ_ζ^2, $\rho_{\xi\zeta}$, and $\rho_{\xi\eta}$ into the latter formula, gives us the final expression for the price $C(T',T)$.

Using the observation

$$\left(K - B(T',T) \right)^+ = \left(B(T',T) - K \right)^+ - B(T',T) + K,$$

we compute the price of a European put option in a $(B_t, B(t,T))$-market:

$$P(T',T) = K\, B(0,T')\, \Phi(-d_-) - B(0,T)\, \Phi(-d_+). \tag{5.22}$$

Note that formula (5.22) is a natural analogue of the Black-Scholes formula (4.12).

Now we discuss one of the approximation methods for pricing such assets with fixed income. Consider a zero-coupon bond with face value 1 and terminal date $T = 1$ (year). For simplicity, suppose that $P^* = P$ (i.e., the initial probability is a martingale probability; see, for example, Vasiček model). The bond price is given by

$$B(t,T) = E\left(\exp\left\{ -\int_t^T r_s \, ds \right\} \Big| \mathcal{F}_t \right).$$

In our case, $t = 0$ and $T = 1$; hence,

$$B(0,1) = E\left(r_0 \exp\left\{ -\int_0^1 r_s \, ds \right\} \Big| \mathcal{F}_0 \right) = E\left(\exp\left\{ \ln r_0 - \int_0^1 r_s \, ds \right\} \right).$$

Suppose that the evolution of the interest rate is described by

$$r_t = r_0 \, e^{at + \sigma W_t} = e^{\ln b + Y_t} = b \, e^{Y_t},$$

where

$$r_0 = b \quad \text{and} \quad Y_t = a \, t + \sigma \, W_t.$$

Our further discussion is based on the following methodology. Let $f = f(x)$, $x \in \mathbb{R}$ be a convex function, $(\xi_s)_{0 \le s \le 1}$ be a Gaussian process, $X := \int_0^1 e^{\xi_s} \, ds$, and $\xi \sim \mathcal{N}(0,1)$.

From Jensen's inequality, we have

$$E\big(f(X)\big) = E\Big(E\big(f(X)\big)|\xi\Big) \ge E\Big(f\big(E(X|\xi)\big)\Big).$$

Now choose

$$f(x) = e^{-bx}$$

and

$$\xi = \int_0^1 W_s \, ds \Big/ \sqrt{V\left(\int_0^1 W_s \, ds\right)},$$

which is clearly a Gaussian random variable. Then

$$E(\xi) = E\left(\int_0^1 W_s \, ds \Big/ \sqrt{V\left(\int_0^1 W_s \, ds\right)} \right)$$

$$= \int_0^1 E(W_s) \, ds \Big/ \sqrt{V\left(\int_0^1 W_s \, ds\right)} = 0.$$

Using the Kolmogorov-Itô formula, we write

$$\left(\int_0^1 W_s \, ds\right)^2 = 2 \int_0^1 \int_0^t W_s \, ds \, d\left(\int_0^t W_u \, du\right) = 2 \int_0^1 \int_0^t W_s \, ds \, W_t \, dt$$

$$= 2 \int_0^1 \int_0^t W_t \, W_s \, ds \, dt$$

$$= 2 \int_0^1 \int_0^t \big[W_s + (W_t - W_s)\big] W_s \, ds \, dt.$$

Since increments of W are independent and $V(W_t) = t$, then

$$E\left(\int_0^1 W_s \, ds\right)^2 = 2 \int_0^1 \int_0^t E(W_s^2) \, ds \, dt = 2 \int_0^1 \int_0^t s \, ds \, dt$$

$$= \int_0^1 t^2 \, dt = \frac{1}{3}.$$

Using the theorem on normal correlation, we can write

$$E(Y_t | \xi) = a t + k_t \xi,$$

where

$$k_t = \text{Cov}(Y_t, \xi) = \sqrt{3} \, \sigma \, \text{Cov}\left(W_t, \int_0^1 W_s \, ds\right) = \sqrt{3} \, \sigma \int_0^1 (1 - s) \, ds$$

$$= \sqrt{3} \, \sigma \left(t - \frac{t^2}{2}\right).$$

Also

$$V(Y_t | \xi) = V(Y_t) - k_t^2 = \sigma^2 \left(t - 3 t^2 + 3 t^3 - 3 t^4 / 4\right) = \nu_t,$$

$$\text{Cov}(Y_t \, Y_s | \xi) = \sigma^2 \, \min\{t, s\} - k_t \, k_s = \nu_{ts}.$$

Now consider

$$h(\xi) = E\left(\int_0^1 e^{Y_s} \, ds \,\Big|\, \xi\right) = \int_0^1 e^{as + k_s \xi + \nu_s / 2} \, ds.$$

Computing

$$LB_1 = \int_{-\infty}^\infty h(z) \frac{e^{-z^2/2}}{\sqrt{2\pi}} \, dz,$$

gives us the lower estimate for the bond price.

To find the upper estimate UB_1, we note that there exists a random variable η such that

$$E(f(X)) = E\left(f(E(X|\xi))\right) + E\left([X - E(X|\xi)] \, f'(E(X|\xi))\right)$$

$$+ \frac{1}{2} E\left([X - E(X|\xi)]^2 \, f''(\eta)\right).$$

This implies the estimates

$$E(f(X)) \leq f(E(X|\xi)) + \frac{1}{2} E\left([X - E(X|\xi)]^2 \, f''(\eta)\right)$$

and

$$E(f(X)) \leq E\left(f(E(X|\xi))\right) + \frac{1}{2} E\left([X - E(X|\xi)]^2 \, \sup_x f''(x)\right).$$

Thus,

$$UB_1 = LB_1 + \frac{1}{2} c^2 E\left(V(X|\xi)\right),$$

where

$$c^2 = \sup_x f''(x).$$

One can compute LB_1 using standard approximation methods for computing integrals. Thus, this methodology allows one to approximate bond prices and to compute the corresponding error estimates.

This methodology can also be used for computing prices of options. For example, for a European call option, we have

$$f(x) = \left(e^{-bx} - K\right)^+,$$

and one has to approximate

$$LB_2 = \int_{-\infty}^{\infty} f\left(h(z)\right) \frac{e^{-z^2/2}}{\sqrt{2\pi}} \, dz.$$

Chapter 6

Implementations of Risk Analysis in Various Areas of Financial Industry

6.1 Real options: pricing long-term investment projects

Long-term investment projects play a significant role in modern economy. Development of a new enterprise is a typical example of such a project. A company that plans an investment of this type is often *not obliged* to realize the project. In this sense, such investment activities are similar to a call option on a financial asset. In both cases an investor has the *right* to gain some outcomes of a project in return for invested capital (e.g., buy shares at a strike price). Such investment programs in "real economy" are referred to as *real options*.

This similarity suggests that methods of managing risk related to contingent claims may be helpful in managing risk related to long-term investment projects.

Let us consider a project with a fixed implementation date T. As before, we will use the notion of a *basic asset*, which represents the expected result of the project. Let S_t be its price, then it is natural to expect that the price of the project is given by some function $F(S_T)$. Clearly, this quantity must reflect the discounted yield generated by the basic asset S.

Studying the *profitability* of an investment project is essential for making a decision about its realization. If I is a fixed capital of the proposed investment, then it must be compared with some *level of profitability* R, which depends on $F(S_T)$:

> If $I \leq R$, then the project is accepted for realization;
>
> If $I > R$, then the project is rejected.

How to find sensible values of R? If evolution of the basic asset is deterministic, then its price can be written in the form

$$S_t^{\text{det}} = \exp\{s_t\},$$

where s_t is a deterministic function of $t \in [0, T]$. If r is the rate of interest, then the level of profitability can be defined as

$$R = R_0^{\text{det}} = e^{-rT} F(S_T^{\text{det}}).$$

If evolution of price of S is not deterministic, then we model it in terms of S_t^{det} perturbed by a Gaussian white noise with mean zero and variance σ^2 (see Figure 6.1) . Then the expectation of price S_t^{noise} will coincide with price's deterministic component:

$$E\left(S_t^{\text{noise}}\right) = S_t^{\text{det}} = \exp\{s_t\}, \quad S_0 = 1.$$

The evolution of prices is given by

$$S_t^{\text{noise}} = S_t^{\text{det}} \exp\left\{\sigma W_t - \frac{\sigma^2}{2}t\right\} = \exp\left\{s_t + \sigma W_t - \frac{\sigma^2}{2}t\right\},$$

where W is a Wiener process.

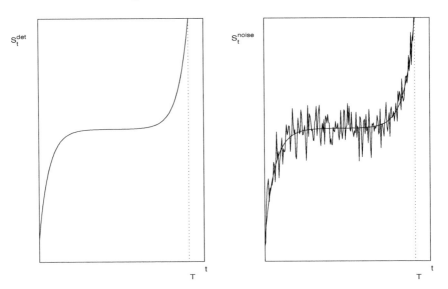

FIGURE 6.1: Dynamics of S.

Then it is natural to define

$$
\begin{aligned}
R_\sigma^{\text{noise}} &= E\left(e^{-rT} F\left(S_T^{\text{det}}\right)\right) \\
&= e^{-rT} \int_{-\infty}^{\infty} F\left(S_T^{\text{det}} \exp\left\{\sqrt{T}\sigma x - \frac{\sigma^2}{2}T\right\}\right) \frac{e^{-x^2/2}}{\sqrt{2\pi}}\, dx \\
&= \frac{e^{-rT}}{\sqrt{2\pi}} \int_{-\infty}^{\infty} F\left(\exp\left\{s_T + \sqrt{T}\sigma x - \frac{\sigma^2}{2}T\right\}\right) e^{-x^2/2}\, dx \\
&\to R_0^{\text{det}} \quad \text{as} \quad \sigma \to 0.
\end{aligned}
$$

Alternatively, one can define R as the expectation of $F(S_T)$ with respect to a risk-neutral probability P^* (see Section 4.1) such that its density with respect to P is

$$Z_T^* = \exp\left\{\frac{r}{\sigma}W_T - \frac{1}{2}\left(\frac{r}{\sigma}\right)^2 T\right\}.$$

By Girsanov theorem, the process

$$W_t^* = W_t - \frac{r}{\sigma}t$$

is a Wiener process with respect to P^*. Thus, we obtain another value of R:

$$
\begin{aligned}
R = R^* &= E^*\left(e^{-rT}F(S_T)\right) = E^*\left(e^{-rT}F\left(\exp\left\{s_T + \sigma W_T - \frac{\sigma^2}{2}T\right\}\right)\right) \\
&= e^{-rT}E^*\left(F\left(e^{s_T+rT}\exp\left\{\sigma W_T - rT - \frac{\sigma^2}{2}T\right\}\right)\right) \\
&= e^{-rT}E^*\left(F\left(e^{s_T+rT}\exp\left\{\sigma\left(W_T - \frac{r}{\sigma}T\right) - \frac{\sigma^2}{2}T\right\}\right)\right) \\
&= e^{-rT}E^*\left(F\left(e^{s_T+rT}\exp\left\{\sigma W_T^* - \frac{\sigma^2}{2}T\right\}\right)\right) \\
&= \frac{e^{-rT}}{\sqrt{2\pi}}\int_{-\infty}^{\infty}F\left(\exp\left\{s_T + rT + \sqrt{T}\,\sigma x - \frac{\sigma^2}{2}T\right\}\right)e^{-x^2/2}\,dx.
\end{aligned}
$$

In some types of long-term investment projects, it is natural to assume that the total value of an investment and the implementation date are not known in advance. For example, investments in scientific research or in energy production are projects of this type.

Suppose we know the final cost of the basic asset, and let X_t, $t \geq 0$ be the amount of capital necessary for completion of the project. As a particular example, consider the *Pindyck model*, where random process X satisfies the following stochastic differential equation:

$$dX_t = -\alpha_t\,dt + \beta\sqrt{\alpha_t X_t}\,dW_t, \quad X_0 = x,$$

where $\beta > 0$ and α_t is the intensity of the investment flow.

Since investment potential is often limited, and it is not possible to reverse the investment flow, then it is natural to assume that $\alpha = (\alpha_t)$ is a bounded random variable. For simplicity, say, $\alpha_t \in [0,1]$. Process α plays the role of control in the process of spending the investment capital $X_t = X_t^\alpha$. Choosing α from the set of all admissible processes

$$\beth = \{\alpha : \alpha_t \in [0,1]\}$$

implies defining a natural implementation time

$$\tau = \tau^\alpha = \inf\left\{t : X_t = X_t^\alpha = 0\right\}$$

for a project.

If V is the final cost of the project and r is the rate of interest, then the quantity

$$V e^{-r\tau} - \int_0^\tau \alpha_t\, e^{-rt}\, dt$$

represents the profit gained by choosing the investment strategy α. The average profit is given by

$$v^\alpha(x) = E_x\left(V e^{-r\tau} - \int_0^\tau \alpha_t\, e^{-rt}\, dt\right),$$

where notation E_x for mathematical expectation indicates that the initial investment x was necessary for completion of this project.

Since all control strategies belong to class J, it is natural to define the *optimal strategy* α^* from

$$v(x) \equiv \sup_{\alpha \in \mathrm{J}} v^\alpha(x) = v^{\alpha^*}(x). \tag{6.1}$$

Problems of this type are usually solved by the method of *dynamic programming*, where one of the main tools is the *Bellman principle*. We will briefly sketch this method. Suppose that the controlled process $X_t = X_t^\alpha$ satisfies the stochastic differential equation

$$dX_t \equiv dX_t^\alpha = b_\alpha(X_t^\alpha)\, dt + \sigma_\alpha(X_t^\alpha)\, dW_t, \quad X_0 = x,$$

where b_α and σ_α are some reasonable functions (for example, satisfying Lipschitz condition), and α is a control process that is adapted to a σ-algebra generated by X_t.

For estimating the quality of control α, we introduce function $f^\alpha(x)$, $\alpha \in [0,1]$, $x \in \mathbb{R}$, which is interpreted as the intensity of the profit flow. Then the total profit on interval $[0,t]$ is equal to

$$\int_0^t f^\alpha(X_s^\alpha)\, ds.$$

Denoting

$$v^\alpha(x) = E_x\left(\int_0^\infty f^\alpha(X_s)\, ds\right)$$

its expectation on $[0, \infty)$, we will find the optimal control α^* from condition (6.1):

$$v(x) \equiv \sup_{\alpha \in \mathrm{J}} v^\alpha(x) = v^{\alpha^*}(x).$$

We use the *Bellman principle*:

$$v(x) = \sup_{\alpha \in \mathrm{J}} E_x\left(\int_0^t f^\alpha(X_s^\alpha)\, ds + v(X_t^\alpha)\right), \quad t > 0, \tag{6.2}$$

for determining the *price* $v(x)$, and now we will briefly explain the motivation for using it. Let us write the total profit of using strategy α in the form

$$\int_0^\infty f^\alpha(X_s)\,ds = \int_0^t f^\alpha(X_s)\,ds + \int_t^\infty f^\alpha(X_s)\,ds\,.$$

If this strategy was used only up to time t, then the first term in the right-hand side represents the profit on interval $[0, t]$. Suppose the controlled process has value $y = X_t$ at time t. If we wish to alter the control process after time t with the aim of maximizing the profit over the whole of $[0, \infty)$, then we have to maximize the expectation

$$E_y\left(\int_t^\infty f^\alpha(X_s)\,ds\right),$$

where α also denotes the continuation of the control process to $[t, \infty)$. Changing variable $s = t + u$, $u \geq 0$, and using independence and stationarity of increments of the Wiener process, we obtain

$$E_{X_t}\left(\int_t^\infty f^\alpha(X_s)\,ds\right) = v^\alpha(X_t) \leq v(X_t)\,.$$

Thus, a strategy that is optimal after time t, gives the average profit such that

$$E_x\left(\int_0^t f^\alpha(X_s^\alpha)\,ds + v(X_t^\alpha)\right) \geq v^\alpha(x)\,.$$

One can choose α_s, $s \geq t$, so that the corresponding profit is close enough to the average profit. Hence, taking supremum of both sides of the latter inequality yields the Bellman principle (6.2). If we *a priori* assume that the *Bellman function* is smooth enough, then the Bellman principle can be written in the following differential form:

$$v(X_t^\alpha) = v(x) + \int_0^t \left[\frac{\partial v}{\partial x} b_\alpha(X_s^\alpha) + \frac{1}{2}\frac{\partial^2 v}{\partial x^2}\sigma_\alpha^2(X_s^\alpha)\right]ds + \int_0^t \frac{\partial v}{\partial x}\sigma_\alpha(X_s^\alpha)\,dW_s,$$

where we also used the Kolmogorov-Itô formula. Since the last term in the right-hand side is a martingale, then we obtain

$$v(x) = \sup_{\alpha \in \exists} E_x\left(\int_0^t f^\alpha(X_s^\alpha)\,ds + v(X_t^\alpha)\right)$$

$$= \sup_{\alpha \in \exists} E_x\left(\int_0^t \left[\frac{\partial v}{\partial x} b_\alpha(X_s^\alpha) + \frac{1}{2}\frac{\partial^2 v}{\partial x^2}\sigma_\alpha^2(X_s^\alpha) + f^\alpha(X_s^\alpha)\right]ds + v(x)\right).$$

Hence

$$\sup_{\alpha \in \exists}\left[L_\alpha v(x) + f^\alpha(x)\right] = 0\,,$$

where

$$L_\alpha v = \frac{\partial v}{\partial x} b_\alpha + \frac{1}{2} \frac{\partial^2 v}{\partial x^2} \sigma_\alpha^2.$$

The latter relation is usually referred to as *Bellman differential equation*.

Note that the considered investment problem controls process X_t^α only up to time

$$\tau = \tau_D^\alpha$$

of its exit from region D. Thus, this problem can be written in the following general form

$$v(x) = \sup_{\alpha \in \mathbb{J}} E_x \left(\int_0^{\tau_D^\alpha} f^\alpha(X_s^\alpha) \, e^{-rs} \, ds + g(X_{\tau_D^\alpha}) \, e^{-r\tau_D^\alpha} \right),$$

where $g = g(x)$ is some function defined on the boundary ∂D of set D. In this case, we again arrive at the following Bellman differential equation

$$\sup_{\alpha \in \mathbb{J}} \left[\frac{1}{2} \frac{\partial^2 v}{\partial x^2} \sigma_\alpha^2(x) + \frac{\partial v}{\partial x} b_\alpha(x) - r \, v(x) + f^\alpha(x) \right] = 0,$$

which is satisfied by Bellman function v for sufficiently wide class of coefficients $b_\alpha(x)$ and $\sigma_\alpha(x)$, $\alpha \in [0,1]$, $x \in \mathbb{R}$.

Consider again an investment problem in the Pindyck model. We note that

$$f^\alpha(x) = \alpha, \quad g(x) = V, \quad D = \{x : x > 0\}.$$

Then Bellman differential equation has the form

$$r \, v(x) = \sup_{\alpha \in \mathbb{J}} \left[-\alpha - \alpha \frac{\partial v}{\partial x} + \frac{\beta^2}{2} \alpha x \frac{\partial^2 v}{\partial x^2} \right],$$

or, taking into account linearity in α,

$$r \, v(x) = \begin{cases} -1 - \frac{\partial v}{\partial x} + \frac{\beta^2}{2} x \frac{\partial^2 v}{\partial x^2} & \text{if } -1 - \frac{\partial v}{\partial x} + \frac{\beta^2}{2} x \frac{\partial^2 v}{\partial x^2} > 0 \\ 0 & \text{otherwise} \end{cases}.$$

It is clear that the investment strategy

$$\alpha_t^* = \begin{cases} 1 & \text{if } X_t < x^* \\ 0 & \text{if } X_t \geq x^*, \end{cases}$$

where x^* is a solution of $v(x^*) = 0$, is a candidate for being optimal.

Consider differential equation

$$r \, v(x) = -1 - \frac{\partial v}{\partial x} + \frac{\beta^2}{2} x \frac{\partial^2 v}{\partial x^2}.$$

Its general solution has the form

$$v(x) = c_1 \, x^{\nu/2} \, J_\nu \left(2 - \sqrt{b \, x} \right) + c_2 \, x^{\nu/2} \, H_\nu^1 \left(2 \sqrt{-b \, x} \right) + (1 - \nu) \, b,$$

where $\nu = 1 + 2/\beta^2$, $b = 2\,r/\beta^2$,

$$J_\nu(x) = \sum_{k=0}^{\infty} \frac{(-1)^k\,(x/2)^{\nu+2k}}{k!\,\Gamma(\nu+k+1)}$$

is the Bessel function of the first kind, Γ is the gamma-function, and $H_\nu^{(1)}$ is the Hankel function of the first kind.

This solution can be also written in terms of modified Bessel functions:

$$I_\nu(x) \;=\; \sum_{k=0}^{\infty} \frac{(x/2)^{\nu+2k}}{k!\,\Gamma(\nu+k+1)}\,,$$

$$K_\nu(x) \;=\; \frac{\pi}{2}\,\frac{I_{-\nu}(x) - I_\nu(x)}{\sin(\pi\nu)}\,, \qquad \nu \notin \mathbb{Z}\,,$$

$$K_n(x) \;=\; (-1)^{n+1}\,I_n(x)\,\ln(x/2) + \frac{1}{2}\sum_{k=0}^{n-1} \frac{(-1)^k\,(n-k-1)!}{k!}\,(x/2)^{2k-n}$$

$$+ \frac{(-1)^n}{2}\sum_{k=0}^{n} \frac{(x/2)^{2k+n}}{k!\,(n+k)!}\,\Big[\Psi(n+k+1) + \Psi(k+1)\Big]\,, \qquad n \in \mathbb{Z}\,,$$

where Ψ is the logarithmic derivative of Γ. We have

$$v(x) = c_1\,(-1)^{\nu/2}\,x^{\nu/2}\,I_\nu\big(2\sqrt{b\,x}\big) + c_2\,\frac{2}{\pi}\,(-1)^{(\nu+1)/2}\,x^{\nu/2}\,K_\nu\big(2\sqrt{b\,x}\big) + \frac{1-\nu}{b}\,.$$

Since

$$x^{\nu/2}\,I_\nu\big(2\sqrt{b\,x}\big) = x^{\nu/2}\sum_{k=0}^{\infty} \frac{(b\,x)^{\nu/2+k}}{k!\,\Gamma(\nu+k+1)} \to 0 \quad \text{as } x \to 0\,,$$

then the initial condition $v(0) = V$ allows to compute

$$c_2 = \begin{cases} \sin(\pi\nu)\,\Gamma(1-\nu)\,b^{\nu/2}\,(-1)^{-(\nu+1)/2}\left(V + \frac{1}{r}\right) & \text{if } \nu \notin \mathbb{Z} \\[2mm] \frac{\pi}{(n-1)!}\,b^{\nu/2}\,(-1)^{-(\nu+1)/2}\left(V + \frac{1}{r}\right) & \text{if } \nu \in \mathbb{Z}\,. \end{cases}$$

Note that we are solving a problem with an unknown boundary. In the theory of differential equations, such problems are referred to as *Stefan problems*. The methodology of dealing with such problems involves the ideas of continuity and smooth gluing on the boundary $x = x^*$:

$$v(x^*) = 0 \qquad \text{and} \qquad v'(x^*) = 0\,.$$

This implies

$$c_1 \;=\; (-1)^{-\nu/2}\,\frac{K_{\nu-1}\big(2\sqrt{b\,x^*}\big)}{I_{\nu-1}\big(2\sqrt{b\,x^*}\big)}\,,$$

$$v(x) \;=\; \frac{K_{\nu-1}\big(2\sqrt{b\,x^*}\big)}{I_{\nu-1}\big(2\sqrt{b\,x^*}\big)}\,x^{\nu/2}\,I_\nu\big(2\sqrt{b\,x}\big) + c\,K_\nu\big(2\sqrt{b\,x}\big) + (1-\nu)/b\,,$$

where $c = \frac{2}{\pi} c_2$.

Now we have to check that the constructed function v and control α^* indeed solve the initial investment problem. The *verification conditions* in this case are

1) $v^\alpha(x) \le v(x)$ for any α and x;

2) $v^{\alpha^*}(x) = v(x)$ for $x \ge 0$.

Here is the sketch of this verification. From the properties of Bessel functions, we have that the solution to

$$\frac{\beta^2}{2} x \frac{\partial^2 v}{\partial x^2} - \frac{\partial v}{\partial x} - r v - 1 = 0$$

is a smooth function. Also

$$\frac{\beta^2}{2} \alpha x \frac{\partial^2 v}{\partial x^2} - \alpha \frac{\partial v}{\partial x} - r v - \alpha \le 0$$

for $\alpha \in [0, 1]$. Further, using the Kolmogorov-Itô formula, we have

$$e^{-r(t \wedge \tau)} v(X_{t \wedge \tau}) = v(x) + \int_0^{t \wedge \tau} e^{-rs} \beta \sqrt{\alpha_s X_s}\, v(X_s)\, dW_s$$

$$+ \int_0^{t \wedge \tau} e^{-rs} \left[\frac{\beta^2}{2} \alpha_s X_s \frac{\partial^2 v}{\partial x^2}(X_s) - \alpha_s \frac{\partial v}{\partial x}(X_s) - r\, v(X_s) \right] ds$$

$$\le v(x) + \int_0^{t \wedge \tau} e^{-rs} \beta \sqrt{\alpha_s X_s}\, v(X_s)\, dW_s + \int_0^{t \wedge \tau} e^{-rs} \alpha_s\, ds .$$

Taking expectations and using the martingale property of stochastic integrals, we obtain

$$v(x) \ge E_x \left(e^{-r(t \wedge \tau)} v(X_{t \wedge \tau}) \right) - E_x \left(\int_0^{t \wedge \tau} e^{-rs} \alpha_s\, ds \right)$$

and hence

$$v(x) \ge v^\alpha(x)$$

due to convergence

$$E_x \left(e^{-r(t \wedge \tau)} v(X_{t \wedge \tau}) \right) \to V\, e^{-r\tau} \qquad \text{as} \quad t \to \infty .$$

Establishing second verification property, we note that it clearly holds true for $X_t \geq x^*$. For $X_t < x^*$, we use the Kolmogorov-Itô formula:

$$v(x) = E_x\left(e^{-r(t \wedge \tau)}\, v\left(X_{t \wedge \tau}\right)\right)$$

$$+E_x\left(\int_0^{t \wedge \tau} e^{-rs}\,\beta\,\sqrt{\alpha_s\,X_s}\,v(X_s)\,dW_s\right)$$

$$+E_x\left(\int_0^{t \wedge \tau} e^{-rs}\left[\frac{\beta^2}{2}\,\alpha_s\,X_s\,\frac{\partial^2 v}{\partial x^2}(X_s) - \alpha_s\,\frac{\partial v}{\partial x}(X_s) - r\,v(X_s)\right]ds\right)$$

$$= E_x\left(e^{-r(t \wedge \tau)}\,v\left(X_{t \wedge \tau}\right)\right)$$

$$+E_x\left(\int_0^{t \wedge \tau} e^{-rs}\left[\frac{\beta^2}{2}\,\alpha_s\,X_s\,\frac{\partial^2 v}{\partial x^2}(X_s) - \alpha_s\,\frac{\partial v}{\partial x}(X_s) - r\,v(X_s)\right]ds\right).$$

Passing to the limit as $t \to \infty$ and choosing $\alpha = \alpha^*$ completes the verification.

Finally, we note that the existence of x^* as a solution to $v(x^*) = 0$, follows from analyzing this equation with the help of the following asymptotic representations of the modified Bessel functions:

$$I_\nu(x) = \frac{e^x}{\sqrt{2\pi x}}\left(1 + \mathcal{O}(1/x)\right) \quad \text{and} \quad K_\nu(x) = \frac{\pi}{\sqrt{2x}}\,e^{-x}\left(1 + \mathcal{O}(1/x)\right)$$

as $x \to \infty$.

6.2   Technical analysis in risk management

The study of *market activities* is an essential part of analysis of financial markets. The collection of methods and tools of qualitative analysis of market prices forms an important part of modern *financial engineering* and is usually referred to as *technical analysis*. Recent developments in financial mathematics provide significant theoretical support to empirical methodologies of technical analysis and hopefully will encourage development of new trends in this area.

Technical analysts believe that market prices depend on psychology of market participants, and therefore various types of financial information are often used in technical analysis. Forecasting future price *trends* is the major goal of technical analysis. All relevant current information is represented in the form of *indicators* and expressed in graphs, mnemonic rules, and mathematical functions.

To make an informed investment decision, one has to identify the most probable *trends* in the market, estimate effectiveness of operations and risk of having losses, determine volumes of transactions given information on the liquidity of stocks taking into account transaction costs, and other factors.

Charts are traditional forms of visualization of dynamics of prices and indices. The most widely used forms of charts are *bar charts* and (Japanese) *candlestick charts*. For example, a candlestick consists of a line that represents the price range from the lowest to the highest, and of a rectangle that measures the difference between opening and closing prices: it is white if the closing price is higher than the opening price, and it is black otherwise.

The most important elements of a chart are *trend lines, support lines,* and *resistance lines*. Uptrends have ascending sequences of local maximums and minimums, downtrends correspond to descending sequences, and sideways trends correspond to constant sequences. *Support* is represented by a horizontal line that indicates the level from which prices start growing. It "supports" the graph of the price trend from below. *Resistance* is represented by a horizontal line that bounds the graph of the price trend from above. It indicates the price level when selling pressure overcomes buying pressure and prices start going down.

Support and resistance lines can move up and down, which corresponds to increasing or decreasing price trends. It is extremely important to identify the moments when a trend line breaks, that is, becomes decreasing after being increasing or vice versa, since most financial gains and losses happen at such moments.

More complex patterns on charts are usually described in terms of *figures*. The most popular are *head and shoulders*, various types of *triangles* and *flags*.

One of most essential axioms of technical analysis is that prices "remember" their past. This makes the concept of trend the key element of technical analysis: one has to identify trends in an appropriately chosen past and use them for forecasting future prices.

Quantitative realization of these ideas is given by indicators. One of the most popular indicators is the *moving average*, whose simplest and most commonly used version is defined by

$$\frac{S_{t-n} + \ldots + S_{t+n}}{2n + 1},$$

where t is the current time, n defines time horizon, S_t is the price of stock S at time t.

Moving average is widely used in identifying trends, in making decisions about buying or selling stock, and in constructing other indicators. If the stock's price moves above moving average, then it is recommended to buy this stock, and to sell otherwise. Thus, moving average is designed to keep one's position in the boundaries of the main trend, and parameter n must correspond to the length of the market cycle.

Another important indicator is *divergence*. Fluctuation of prices reflects market's instability and is represented by a sequence of rises and drops. It is essential to determine as quickly as possible which of the rises or drops indicate changes in the main trend. If the price line reaches its new peak but the indicator does not, this indicates that the market activity is becoming slow and is called *bearish divergence* (or negative divergence). The symmetric *bullish divergence* (or positive divergence) corresponds to a situation when prices continue to drop but the indicator does not.

Technical analysis of various averaging indicators and individual stock prices is often complemented by the study of trading *volumes*. Volume-based indicators are based on the hypothesis that changes in trading volumes precede changes of prices. Thus, observation of a *change point* in the dynamics of a volume indicator can be naturally interpreted as a change in the price trend. One of the key indicators here is called the *accumulation-distribution* indicator, which is defined by the formula

$$\frac{S_2 - S_1}{\max S - \min S} V + I,$$

where S_1 and S_2 are opening and closing prices, $\max S$ and $\min S$ are price's maximum and minimum taken over a specified period of time, V is trading volume and I is the previous value of the indicator.

We can summarize that one of the key problems of technical analysis consists in detection of change points in price trends. We will use quantitative methods for dealing with this problem, and we need to introduce some notions and assumptions.

Let a stochastic process $X = (X_t)_{t\in[0,T]}$ represent the evolution of prices. We wish to identify a moment of time θ when process X changes its probabilistic characteristics. This can be a point in $[0, T]$ where X attains its maximum; that is, prices change the ascending tendency to descending (see Figure 6.2).

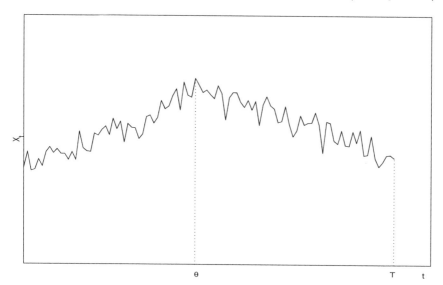

FIGURE 6.2: Prices change the ascending tendency to descending.

Then we need to choose a stopping time τ^* adapted to the observed information, such that τ^* is sufficiently close to θ and the values of X at these points are also close in some sense: for example, the variance of the difference $X_\theta - X_{\tau^*}$ is minimal. In 1900 Bachelier suggested that the evolution of prices can be modeled with the help of a standard Wiener process (Brownian motion): $X_t = W_t$, $t \in [0, T]$.

For simplicity, let $T = 1$. We construct an approximation of quantity W_θ using W_τ, where stopping time τ is adapted to filtration $\mathbb{F} = (\mathcal{F}_t)_{t \geq 0}$ generated by Wiener process W:

$$\mathcal{F}_t = \sigma(W_t), \qquad t \geq 0.$$

Let us introduce the following notation:

$$S_t := \max_{0 \leq s \leq t} W_s,$$

and

$$V^* := \inf_\tau E\Big(G\big(S_1 - W_\tau\big)\Big),$$

where G is an observations cost function, and the infimum is taken over all stopping times τ, that is, over all random variables such that

$$\{\omega : \tau \leq t\} \in \mathcal{F}_t \quad \text{for all} \quad t \geq 0.$$

Our aim is to find an optimal stopping time τ^*, so that

$$V^* = E\Big(G\big(S_1 - W_\tau^*\big)\Big).$$

The existence and structure of quantities τ^* and V^* are given by the following theorem.

Theorem 6.1 *For the cost function $G(x) = x^2$, the optimal stopping time τ^* is defined by the formula*

$$\tau^* = \inf \big\{t \leq 1 : \ S_t - W_t \geq z^* \sqrt{1 - t} \big\}.$$

Here $z^* \approx 1.12$ is a solution to equation

$$4\,\Phi(z) - 2\,z\,\phi(z) - 3 = 0,$$

where

$$\Phi(z) = \int_{-\infty}^z \phi(x)\,dx \qquad \text{and} \qquad \phi(z) = \frac{1}{\sqrt{2\pi}}\,e^{-z^2/2}.$$

In this case,

$$V^* = 2\,\Phi(z^*) - 1 \approx 0.73.$$

Remark 6.1

1. The quantitative characteristics of the optimal stopping time are

$$E(\tau^*) = \frac{z^{*2}}{1 + z^{*2}} \approx 0.55 \,,$$

$$V(\tau^*) = \frac{2\,z^{*4}}{(1 + z^{*2})\,(3 + 6\,z^{*2} + z^{*4})} \approx 0.05 \,.$$

2. For an arbitrary time interval $[0, T]$, the optimal stopping time and price are given by

$$\tau^*(T) = \inf \left\{ t \le T : \; S_t - W_t \ge z^*\,\sqrt{T - t} \,\right\} \quad \text{and} \quad V^*(T) = V^*\,T \,,$$

respectively.

Proof of Theorem 6.1 From the strong Markov property of Wiener process, for any stopping time τ and for any cost function $G = G(x)$ with

$$E(|G(S_1 - W_t)|) < \infty,$$

we have

$$E\Big(G(S_1 - W_\tau)\big|\mathcal{F}_\tau\Big) = E\Big(G\big(\max\big\{\max_{u \le \tau} W_u, \max_{\tau < u \le 1} W_u\big\} - W_\tau\big)\big|\mathcal{F}_\tau\Big)$$

$$= E\Big(G\big(\max\big\{s, \max_{0 \le \tau \le 1-t} W_\tau + x\big\} - x\big)\Big)\Big|_{x = W_\tau, s = S_\tau, t = \tau}$$

$$= E\Big(G\big(\max\big\{s, \eta + x\big\} - x\big)\Big)\Big|_{x = W_\tau, s = S_\tau, t = \tau},$$

where random variable

$$\eta = \max_{0 \le \tau \le 1-t} W_\tau$$

has the following distribution

$$dF_\eta(t, y) = 2\,\phi\Big(\frac{y}{\sqrt{1 - t}}\Big)\frac{dy}{\sqrt{1 - t}} \,.$$

Thus,

$$E\Big(G(S_1 - W_\tau)\big|\mathcal{F}_\tau\Big) = G(s - x)\,F_\eta(t, s - x) + \int_{s-x}^{\infty} G(y)\,dF_\eta(t, y)$$

$$= G(s - x) + \int_{s-x}^{\infty} \big[G(y) - G(s - x)\big]\,dF_\eta(t, y)\,,$$

where $x = W_\tau$, $s = S_\tau$, $t = \tau$. $\square$

Using this representation, we compute

$$V^* = \inf_{\tau} E\Big(G(S_1 - W_{\tau})\Big)$$

$$= \inf_{\tau} E\left(G(S_{\tau} - W_{\tau})\left[2\Phi\left(\frac{S_{\tau} - W_{\tau}}{\sqrt{1-\tau}}\right) - 1\right]\right.$$

$$\left. +2\int_{\frac{S_{\tau}-W_{\tau}}{\sqrt{1-\tau}}}^{\infty} G(z\sqrt{1-\tau})\phi(z)dz\right).$$

In particular, for $G(x) = x^2$, we have

$$V^* = \inf_{\tau} E\left((1-\tau)\left(\frac{S_{\tau} - W_{\tau}}{\sqrt{1-\tau}}\right)^2 \left[2\,\Phi\left(\frac{S_{\tau} - W_{\tau}}{\sqrt{1-\tau}}\right) - 1\right] + 2\int_{\frac{S_{\tau}-W_{\tau}}{\sqrt{1-\tau}}}^{\infty} z^2\,\phi(z)\,dz\right).$$

Taking into account that distributions of processes $S - W$ and $|W|$ coincide, we obtain

$$V^*$$

$$= \inf_{\tau} E\left((1-\tau)\left(\frac{|W_{\tau}|}{\sqrt{1-\tau}}\right)^2 \left[2\,\Phi\left(\frac{|W_{\tau}|}{\sqrt{1-\tau}}\right) - 1\right] + 2\int_{\frac{S_{\tau}-W_{\tau}}{\sqrt{1-\tau}}}^{\infty} z^2\,\phi(z)\,dz\right)$$

$$= \inf_{\tau} E\left((1-\tau)\,H_2\left(\frac{|W_{\tau}|}{\sqrt{1-\tau}}\right)\right),$$

where $H_2(z) = z^2 + 4\int_z^{\infty} u\left(1 - \Phi(u)\right) du$.
   Let us introduce new time $s \geq 0$ by

$$1 - t = e^{-2s}, \qquad t \in [0,1],$$

then

$$\frac{|W_{\tau}|}{\sqrt{1-\tau}} = e^s\, W_{1-e^{-2s}} =: Z_s.$$

Using the Kolmogorov-Itô formula, we represent Z_s in the following differential form:

$$dZ_s = Z_s\, ds + \sqrt{2}\, d\beta_s,$$

where

$$\beta_s = \frac{1}{\sqrt{2}}\int_0^{1-e^{-2s}} \frac{dW_{\tau}}{\sqrt{1-\tau}}, \qquad s \geq 0,$$

is a new Brownian motion.
   Let $Z_0 = Z$ and

$$V^*(z) = \inf_{s} E\Big(e^{-2s}\, H_2\big(|Z_s|\big)\Big),$$

then $V^*(0)$ is a solution to the initial problem.

The optimization problem for diffusion process Z_s is reduced to the following Stefan problem:

$$L_Z V(z) = 2 V(z), \qquad z \in (-z^*, z^*), \tag{6.3}$$

$$V(\pm z^*) = H_2(z^*), \quad V'(\pm z^*) = \pm H_2'(z^*),$$

where $L_Z = z \frac{d}{dz} + \frac{d^2}{dz^2}$ is the generating operator of the diffusion process Z_s, $s \geq 0$.

General solution of equation (6.3) is given by

$$V(z) = c_1 \, e^{-z^2/2} \, M\left(3/2, 1/2; z^2/2\right) + c_2 \, z \, e^{-z^2/2} \, M\left(2, 3/2; z^2/2\right),$$

where

$$M(a, b; x) = 1 + \frac{a}{b} x + \frac{a(a+1)}{b(b+1)} \frac{x^2}{2!} + \dots$$

is the hypergeometric Kummer function.

Since $V^*(z)$ is an even function, we have $c_2 = 0$. From boundary conditions, we also have

$$c_1 = e^{z^{*2}/2} \, \frac{H_2(z^*)}{M\left(3/2, 1/2; z^{*2}/2\right)},$$

where z^* is a strictly positive solution of equation

$$\frac{H_2'(z)}{H_2(z)} + z = 3 z \, \frac{M\left(5/2, 1/2; z^2/2\right)}{M\left(3/2, 1/2; z^2/2\right)}.$$

Thus, we obtain the following expression for the price $V^*(z)$:

$$V^*(z) = \begin{cases} H_2(z) \exp\left\{\frac{z^{*2} - z^2}{2}\right\} \frac{M\left(3/2, 1/2; z^2/2\right)}{M\left(3/2, 1/2; z^{*2}/2\right)}, & \text{if } |z| \leq z^* \\[2mm] H_2(|z|), & \text{if } |z| \geq z^* \end{cases}.$$

The corresponding stopping time is then defined as the first exit time from the ball

$$\sigma^* = \inf\left\{s \geq 0 : |Z_s| \geq z^*\right\}$$

with radius z^*.

The standard verification procedure can be used now to prove that these quantities are optimal. Then we can write solutions to the initial problem. Indeed, changing back to time $t = 1 - e^{-2s}$ and using the facts that $Z_t = W_t/\sqrt{1-t}$ and that the distributions of $S - W$ and $|W|$ coincide, we obtain the following expression for the optimal stopping time:

$$\tau^* = \inf\left\{t \leq 1 : S_t - W_t \geq z^* \sqrt{1-t}\right\}.$$

Since evolution of prices in the problem of *quickest detection of tenden-cies* is described in terms of stochastic processes, one can consider a slightly different setting of that problem. Namely, one solves a problem of quickest detection of time when the probabilistic characteristics of the stochastic process change. This problem is referred to as a *change point* problem, and it was introduced by Kolmogorov and Shiryaev.

It is convenient to specify process X in the following way:

$$dX_t = \begin{cases} \sigma \, dW_t, & \text{if } t < \theta \\ r \, dt + \sigma \, dW_t, & \text{if } t \geq \theta \end{cases}.$$

The stopping time τ adapted to observations of process X can be interpreted as an *alarm time* by considering the following events:

$$\{\omega : \tau < \theta\} \quad \text{and} \quad \{\omega : \tau \geq \theta\},$$

where the first event corresponds to a false alarm, and the second indicates that the change point has been passed and one has to make a decision as promptly as possible. One of the natural criteria for making such a decision can be formulated in the following form: for a fixed $c > 0$,

(a) find

$$V(c) = \inf_\tau \left\{ P(\{\omega : \tau < \theta\}) + c E((\tau - \theta)^+) \right\},$$

where τ is a stopping time adapted to filtration $(\mathcal{F}_t^X)$ generated by the observed price process X_t;

(b) find a stopping time τ^* such that

$$V(c) = P(\{\omega : \tau^* < \theta\}) + c E((\tau^* - \theta)^+).$$

This criterium has a clear and natural meaning: the decision to stop is made at a time when the probability of a false alarm and the average delay after the change point θ are minimal.

Suppose that random variable θ has an exponential *a priori* distribution with parameter $\lambda > 0$:

$$P(\{\omega : \theta = 0\}) = \pi \in [0, 1]$$

$$P(\{\omega : \theta \geq t \mid \theta > 0\}) = e^{-\lambda t}.$$

Posterior distribution of θ is denoted

$$\pi_t = P(\{\omega : \theta \leq t\} \mid \mathcal{F}_t^X).$$

It gives rise to a new statistic

$$\varphi_t = \frac{\pi_t}{1 - \pi_t},$$

and now we will study its structure.

Denote P_θ the conditional distribution of X with respect to θ. Note that P_0 corresponds to the case when $dX_t = r\,dt + \sigma\,dW_t$, and P_∞ corresponds to the case when $dX_t = \sigma\,dW_t$. Introducing statistics

$$L_t = \frac{dP_0}{dP_\infty}(t, X),$$

we can write

$$\frac{dP_\theta}{dP_\infty} = \frac{L_t}{L_\theta}, \qquad \theta \le t.$$

By Bayes's formula we obtain

$$
\begin{aligned}
\varphi_t(\lambda) \;=\; \varphi_t &= \frac{P\big(\{\omega:\ \theta \le t\}\,|\,\mathcal{F}_t^X\big)}{P\big(\{\omega:\ \theta > t\}\,|\,\mathcal{F}_t^X\big)} \\[2mm]
&= \frac{\pi}{1-\pi}\, e^{\lambda t}\, \frac{dP_0}{dP_\infty}(t, X) + e^{\lambda t} \int_0^t \frac{dP_\theta}{dP_\infty}(t, X)\, \lambda e^{-\lambda \theta}\, d\theta \\[2mm]
&= \frac{\pi}{1-\pi}\, e^{\lambda t}\, L_t + \lambda e^{\lambda t} \int_0^t \frac{L_t}{L_\theta}\, e^{-\lambda \theta}\, d\theta .
\end{aligned}
$$

Now, taking into account

$$dL_t = \frac{r}{\sigma^2}\, L_t\, dX_t,$$

and using the Kolmogorov-Itô formula, we obtain

$$d\varphi_t = \lambda\,(1 + \varphi_t)\, dt + \frac{r}{\sigma^2}\, \varphi_t\, dX_t, \qquad \varphi_0 = \frac{\pi}{1-\pi}.$$

Taking into account the relationship between φ_t and π_t, and using the Kolmogorov-Itô formula, we arrive at the following stochastic differential equation for the posterior probability π_t:

$$d\pi_t = \left(\lambda - \frac{r}{\sigma^2}\,\pi_t^2\right)(1 - \pi_t)\, dt + \frac{r}{\sigma^2}\,\pi_t\,(1 - \pi_t)\, dX_t, \qquad \pi_0 = \pi.$$

Now we solve the problem (a)–(b) in this Bayes's setting with the *a priori* probability π. Rewrite $V(c) = V(c, \pi)$ in the form

$$V(c, \pi) = \inf_\tau E\left((1 - \pi_\tau) + c \int_0^\tau \pi_s\, ds\right) = \rho^*(\pi).$$

Consider the following *innovation* representation of process X:

$$dX_t = r\,\pi\,dt + \sigma\,d\overline{W}_t,$$

where $\overline{W}$ is some new Brownian motion with respect to filtration $\big(\mathcal{F}_t^X\big)$.

Using this representation, we can rewrite stochastic differential equation for π_t in the form

$$d\pi_t = \lambda(1 - \pi_t)\,dt + \frac{r}{\sigma^2}\,\pi_t(1 - \pi_t)\,d\overline{W}_t\,.$$

Noting that

$$\int_0^t \pi_s\,ds = \frac{\pi_0 - \pi_t}{\lambda} + \frac{1}{\lambda}\frac{r}{\sigma}\int_0^t \pi_s(1 - \pi_s)\,d\overline{W}_t + t\,,$$

we arrive at the following expression for the price function

$$V(c, \pi) = \rho^*(\pi) = \inf_\tau E\left(\left(1 + \frac{c}{\lambda}\pi\right) - \left(1 + \frac{c}{\lambda}\right)\pi_\tau + c\tau\right).$$

Thus, π_t is a diffusion process generated by operator

$$L = a(\pi)\frac{d}{d\pi} + \frac{1}{2}b^2(\pi)\frac{d^2}{d\pi^2}\,,$$

where $a(\pi) = \lambda(1 - \pi)$ and $b(\pi) = \pi(1 - \pi)r/\sigma$.

Now we can apply the standard method of solving the change point problem, which reduces to the Stefan problem:

$$\begin{aligned}
L\,\rho(\pi) &= -c\,\pi\,, & \pi \in [0, B),\\
\rho(B) &= 1 - B\,, & \pi \in [B, 1],\\
\rho'(B) &= -1\,, \ \rho'(0) = 0\,.
\end{aligned}$$

A general solution of this problem depends on two unknown constants. Another unknown parameter is constant B, which defines the *a priori* unknown boundary of the region in this free-boundary problem. Having one boundary condition for ρ at $\pi = B$ and two conditions for derivatives $\rho'(B)$ and $\rho'(0)$ (conditions of smooth sewing of a solution), we can write solution in the explicit form:

$$\rho(\pi) = \begin{cases}
(1 - B^*) - \int_\pi^{B^*} y^*(x)\,dx\,, & \pi \in [0, B^*)\\[2mm]
1 - \pi\,, & \pi \in [B^*, 1]\,,
\end{cases}$$

where

$$y^*(x) = -C\int_0^x e^{-\Lambda\,[G(x) - G(y)]}\,\frac{dy}{y(1 - y)^2}$$

with

$$G(y) = \log\frac{y}{1 - y} - \frac{1}{y}\,, \quad \Lambda = \frac{\lambda}{r^2/2\sigma}\,, \quad C = \frac{c}{r^2/2\sigma}\,,$$

and B^* is a solution to

$$C\int_0^{B^*} e^{-\Lambda\,[G(B^*) - G(y)]}\,\frac{dy}{y(1 - y)^2} = 1\,.$$

The standard verification technique can be used to show that this function $\rho(\pi)$ coincides with $\rho^*(\pi)$, and

$$\tau^* = \tau^*(B) = \inf\{t : \pi_t \geq B^*\}$$

is an optimal stopping time, such that

$$\rho^*(\pi) = E\left((1 - \pi_{\tau^*}) + c\int_0^{\tau^*} \pi_s\, ds\right),$$

$$V(c, \pi) = P_\pi(\{w : \tau^* < 0\}) + c\, E_\pi((\tau^* - \theta)^+),$$

where notation P_π and E_π reflects the presence of an *a priori* distribution with $P(\{w : \theta = 0\}) = \pi$.

Finally, we note that the same methodology can be applied when the evolution of prices is represented by process $X_t = \mu t + W_t$. Using Girsanov theorem, we can construct a new probability P^* such that process $\mu t + W_t$ is a Brownian motion with respect to it.

6.3 Performance measures and their applications

So far we used risk measures for quantification of risk associated with certain financial positions. In this section, we use the concept of a risk measure to estimate the performance of investment managers, which, in turn, would facilitate ranking the managers. We use the term "manager" in a rather broad sense: we include all types of investment managers from individual investors to investment firms, mutual funds, and so forth. Risk measures that are used in this area of financial analysis are usually referred to as *performance measures* and the area itself is referred to as *performance analysis*. Clearly, performance measures must take into account the balance between profitability of investment decisions and the corresponding levels of risk, thus performance measures are usually called *performance ratios*.

Suppose that both profitability of investment in some company and the corresponding risks are adequately represented by company's shares and their returns. In order to compare performance of different investment managers, we will fix a time interval $[0, T]$ and we will consider values of the corresponding investment portfolios at times $t = 0, \triangle, 2\triangle, \ldots, [T/\triangle]\triangle$ (in a trivial case when all capital is invested in one company, we will consider prices of shares of this company). Recall that the notion of *return* was introduced for binomial markets in Chapter 2:

$$\rho_t^\triangle = \frac{S_{t+\triangle} - S_t}{S_t}, \quad t \leq T.$$

Thus, the collection of random variables $\left(\rho_t^\triangle\right)_{t=0,\triangle,2\triangle,\ldots,[T/\triangle]\triangle}$ is the only information that is required for our calculations and assessments. Also note that the stability of the market can be indicated by small values of differences $S_{t+\triangle} - S_t$ compare to values of S_t. Therefore, it is reasonable to use the following logarithmic approximation:

$$\frac{S_{t+\triangle} - S_t}{S_t} \approx \ln\left(1 + \frac{S_{t+\triangle} - S_t}{S_t}\right) = \ln\left(\frac{S_{t+\triangle}}{S_t}\right),$$

which is usually referred to as *logarithmic return* for the period $[t, t+\triangle]$. The sequence of random variables

$$R_t^{\triangle} := \frac{S_{t+\triangle} - S_t}{S_t} \times \frac{1}{\triangle} \approx \frac{\ln\left(\frac{S_{t+\triangle}}{S_t}\right)}{\triangle}, \qquad t = 0, \triangle, 2\triangle, \ldots, [T/\triangle]\triangle,$$

is then naturally called the *rate of return* and is often used in constructions of performance measures.

We now give an example of such construction using the Black-Scholes model (4.5) with discrete times $t = 0, \triangle, 2\triangle, \ldots, [T/\triangle]\triangle$. For simplicity, let $\triangle = 1$, so that the rate of return $(R_n)_{n=0,1,\ldots,N}$, $N = [T]$, coincides with the return in this case. Note that in the context the Black-Scholes model (4.5), it is a sequence of independent identically distributed random variables $R_n \sim \mathcal{N}(m, \sigma^2)$, $m = \mu - \sigma^2/2$, with density

$$f_R(x) = \varphi\left(\frac{x - m}{\sigma}\right), \qquad \text{where} \quad \varphi(y) = \frac{1}{\sqrt{2\pi}} e^{-y^2/2}.$$

One of the most common ratios that measures returns against risks is the *Sharp ratio (ShR)*. In our setting, it is defined as

$$ShR = \frac{E(R) - r}{\sqrt{Var(R)}};$$

that is, it is the expected return in excess of interest per unit of time measured in terms of standard deviations of the rate of return. Using our notation, we write

$$ShR = \frac{m - r}{\sigma} = z.$$

Define the *downside risk (DR)* as

$$DR = \int_{-\infty}^{r} (r - x)^2 f_R(x)\, dx.$$

Then using DR instead of Var, we arrive at the notion of the *Sortino ratio*:

$$SoR = \frac{E(R) - r}{\sqrt{DR}}.$$

Calculating

$$
\begin{aligned}
DR &= \int_{-\infty}^{r} (r-x)^2 \, f_R(x) \, dx = \int_{-\infty}^{(r-m)/\sigma} (r - \sigma y - m)^2 \, \varphi(y) \, dy \\
&= \sigma^2 \int_{-\infty}^{-z} (-z-y)^2 \, \varphi(y) \, dy = \sigma^2 \left((z^2+1) \, \Phi(-z) - z \, \varphi(z) \right),
\end{aligned}
$$

where $\Phi(x) = \frac{1}{\sqrt{2\pi}} \int_{-\infty}^{x} e^{-y^2/2} \, dy$, we obtain the following expression for SoR is terms of z:

$$
SoR = \frac{m-r}{\sigma \sqrt{(z^2+1) \, \Phi(-z) - z \, \varphi(z)}} = \frac{z}{\sqrt{(z^2+1) \, \Phi(-z) - z \, \varphi(z)}}.
$$

Introducing the *Upside Potential (UP)* as

$$
UP = \int_{r}^{\infty} (x-r) \, f_R(x) \, dx,
$$

we now define the third performance ratio

$$
UPR = \frac{UP}{\sqrt{DR}},
$$

which is called the *Upside Potential ratio*. The following calculations connect it with the Sharpe ratio. We have

$$
\begin{aligned}
UP &= \int_{r}^{\infty} (x-r) \, f_R(x) \, dx = \int_{-z}^{\infty} (\sigma y + m - r)\varphi(y) \, dy \\
&= \sigma \int_{-z}^{\infty} (y+z)\varphi(y) \, dy = \sigma \left(\varphi(z) + z\Phi(z) \right),
\end{aligned}
$$

and hence,

$$
UPR = \frac{\varphi(z) + z\Phi(z)}{\sqrt{(z^2+1) \, \Phi(-z) - z \, \varphi(z)}}.
$$

Using the expressions for SoR and UPR, we find

$$
(SoR)'_z = \frac{\Phi(-z)}{\left[(z^2+1) \, \Phi(-z) - z \, \varphi(z) \right]^{3/2}} > 0
$$

and

$$
(UPR)'_z = \frac{\Phi(-z)\left(\Phi(z) - z \, \varphi(z) \right) + \varphi^2(z)}{\left[(z^2+1) \, \Phi(-z) - z \, \varphi(z) \right]^{3/2}} > 0
$$

as $\Phi(z) - z \, \varphi(z) > 0$ for all z.

Thus, we observe that a stock with higher Sharpe ratio will also have higher Sortino and Upside Potential ratios.

Remark 6.2

Real financial market contains various risky stocks that can be used for investments and therefore be included in an investment portfolio. The structure of an investment portfolio reflects the proportion of each risky stock in the portfolio's cost. Since the number of risky stocks is sufficiently large, we can compare the riskiness of stocks and some other characteristics in order to assess the transparency of the market. Risky stocks are included in various market indexes, such as S&P500, which is based on 500 most common stocks traded in the USA or the family of RUSSELL indexes that represent various distinct segments of the market. Since these indexes are objectively constructed on the basis of a set of transparent rules, they are commonly used as *investment benchmarks*, which are useful for assessment of investment managers' performance. However, these indexes can be viewed as investment portfolios that are sufficiently safe since their rate of return is very close to the risk-free rate of interest in the market. Thus, it is natural to replace the Sharpe ratio with another ratio that involves both the portfolio's rate of return R^Π and the benchmark rate R^B:

$$IR = \frac{E(R^\Pi) - E(R^B)}{\sqrt{Var(R^\Pi)}},$$

which is called the *information ratio*. We also mention here that there are some other performance measures, such as *Jensen's alpha* and *Treynor ratio*, which are based on quantities α and β from the capital asset pricing model; *Calmar ratio* and *Sterling ratio*, which use the maximum drawdown instead of the standard deviation, and so forth.

When ranking investment managers, a manager with the higher Sharpe ratio (or other appropriate performance measure) will be ranked higher. The following example illustrates that, if the distribution of investment returns is not normal, then different performance measures can produce opposite results.

Worked Example 6.1 *Consider two managers who have investment portfolios of 20 stocks each and their returns are given in the following table.*

MN1	MN2
0.07	0
0.06	0.06
−0.03	0.01
0.01	−0.01
0.05	0.05
−0.01	0.04
0.03	0.09
0.04	0.06
0.08	0.08
0.04	0.09
0.03	0.07
0.05	0
0.08	0.01
0.02	0
0.06	0.06
0.05	0.05
0.04	0.04
0.05	0.05
0.08	−0.01
0.07	0.06

Assuming that risk-free interest rate is 4% per annum, rank these managers using Sharpe and Sortino ratios.

Solution We note that skewness and kurtosis of returns for $MN1$ are -0.61304 and -0.13255, respectively; for $MN2$: -0.97499 and 0.48808, respectively. So both returns have distributions that are negatively skewed, but second distribution has bigger kurtosis. Calculating performance ratios, we find that

$$ShR_{MN1} = 0.11893 \quad \text{and} \quad SoR_{MN1} = 2.43266$$

for $MN1$ and

$$ShR_{MN2} = 0.12851 \quad \text{and} \quad SoR_{MN2} = 2.41921$$

for $MN2$, thus $MN2$ is ranked higher by the Sharpe ratio and $MN1$ is ranked higher by the Sortino ratio. □

We conclude this section with a brief discussion of possible applications of performance measures in assessment of companies' *merger or acquisition* proposals. This will illustrate their usefulness for *strategic management* in corporate finance. First, we note that *merger* of two companies assumes establishment of one company whose assets consist of combined assets of two merging companies, whereas *acquisition* of one company by another means that assets of the acquired company are controlled by the acquiring company. Both types of scenarios can give an opportunity to use managers that are the most effective.

In order to assess the quality of management and the effectiveness of a merger or acquisition, we need to take into account both profitability and riskiness of the corresponding companies. Suppose company $C1$ considers a proposition to acquire either company $C2$ or company $C3$. We identify each company with the portfolio of its assets: $PC1$, $PC2$, and $PC3$, respectively. As a result of acquisition, the new company will be associated with either portfolio $(PC1, PC2)$ or portfolio $(PC1, PC3)$. Thus, in order to assess the effectiveness of these acquisitions, we need to choose an appropriate performance measure μ and then calculate

$$\max \{\mu(PC1, PC2), \mu(PC1, PC3)\} - \mu(PC1),$$

which, clearly, must be positive for an effective acquisition.

Chapter 7

Insurance and Reinsurance Risks

7.1 Modeling risk in insurance and methodologies of premium calculations

Insurance is a contract (*policy*) according to which one party (a *policy-holder*) pays an amount of money (*premium*) to another party (*insurer*) in return for an obligation to compensate some possible losses of the policy-holder. The aim of such a contract is to provide the policyholder with some protection against certain *risks*. Death, sickness, disability, motor vehicle accident, loss of property, and so forth are some typical examples of such risks. Each policy contract specifies the policy term and the method of compensation. Usually compensation is provided in the form of payment of an amount of money. Any event specified in the policy contract that takes place during its term can result in such an *insurance claim*. If none of the events specified in the policy contract happen during the policy term, then the policyholder has no monetary compensation for paid premiums.

The problem of premium calculation is one of the key issues in the insurance business. If the premium rate is too high, an insurance company will not have enough clients for successful operation. If the premium rate is too low, the company also may not have sufficient funds to pay all the claims.

To study this problem, we need the following basic notions:

- x, the initial capital of an insurance company;

- non-negative sequence of random variables $\sigma_0 = 0 \leq \sigma_1 \leq \ldots$, time moments of receiving claims. Sequence $T_n = \sigma_n - \sigma_{n-1}$, $n \geq 1$, represents time intervals between claims arrivals;

- $N(t) = \sup\{n : \sigma_n \leq t\}$ is the total number of claims up to time t. It is obviously connected with sequence (σ_n):

$$\{\omega : N(t) = n\} = \{\omega : \sigma_n \leq t < \sigma_{n-1}\};$$

- sequence of independent identically distributed random variables (X_n), where each X_n represents the size of claim at time σ_n;

- $X(t) = \sum_{i=1}^{N(t)} X_i$ is the aggregate claim amount up to time t. Usually X is referred to as the *risk process*, and $X(t) = 0$ if $N(t) = 0$;

- denote $\Pi(t)$ the total premium income up to time $t \geq 0$;

- the capital of an insurance company at time $t \geq 0$ is given by

$$R(t) = x + \Pi(t) - X(t).$$

Naturally, we want to measure and to compare risks. The most common measure of risk in insurance is the *probability of ruin*:

$$1 - P\big(\{\omega : \ R(t) \geq 0, \ t \in [0,T]\}\big),$$

where T is some time horizon.

Next, we introduce some natural assumptions regarding process $N(\cdot)$:

1. $N(0) = 0$;

2. $N(t) \in \{0, 1, 2, \dots\}$;

3. $N(t) \leq N(t+h)$.

Thus, the quantity $N(t+h) - N(t)$ describes the number of claims received during the time interval $(t, t+h)$.

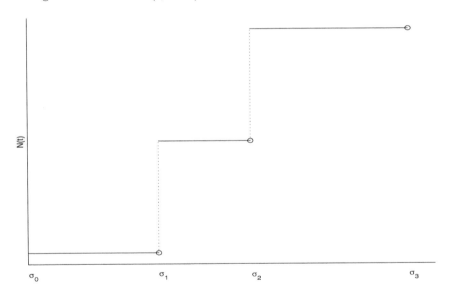

FIGURE 7.1: Process N.

Usually it is assumed that process $N(\cdot)$ can have only unit jumps; that is, it is not possible to receive two or more claims simultaneously (see Figure 7.1). Consider the distribution of $N(\cdot)$:

$$p_k(t) = P\big(\{\omega : \ N(t) \leq k\}\big) = P\left(\left\{\omega : \ \sum_{i=1}^{k} T_i \leq t < \sum_{i=1}^{k+1} T_i\right\}\right).$$

Probabilities $p_k(t)$ can be explicitly computed under some additional assumption on sequence (T_n). If (T_n) is a sequence of independent identically distributed random variables with the distribution function

$$F_T(x) = P(\{\omega : T_n \le x\}),$$

then sequence (σ_n) is called a *renewal process*. A typical example of such a process is a Poisson renewal process, when (T_n) has the exponential distribution with a parameter $\lambda > 0$, and therefore the distribution of $N(t)$ has the form

$$p_k(t) = e^{-\lambda t} \frac{(\lambda t)^k}{k!}, \qquad k = 0, 1, \dots .$$

In this case, $E(N(t)) = \lambda t$, $V(N(t)) = \lambda t$.

For example, if $\lambda = 2$, then we have the following values of $p_k(t)$ (see Figure 7.2):

k	$t = 0.1$	$t = 0.2$	$t = 1$	$t = 2$
0	0.8187	0.6703	0.1353	0.0183
1	0.1637	0.2681	0.2707	0.0733
2	0.0164	0.0536	0.2707	0.1465
3	0.0011	0.0072	0.1804	0.1954
4	0.0001	0.0007	0.0902	0.1954
5	0	0.0001	0.0361	0.1563
6	0	0	0.0120	0.1042
7	0	0	0.0034	0.0595
8	0	0	0.0009	0.0298
9	0	0	0.0002	0.0132
10	0	0	0	0.0053

We will assume that claims are paid instantaneously at the time of arrival, although in reality, there may be a time delay related to estimation of the amount of a claim. Sometimes these delays can be rather significant, for example, in insurance against catastrophic events.

The exact distribution of claims is often unknown. It is assumed that it can be described by some parametric family. Hence, one of the primary tasks in modeling insurance risks is estimating these parameters.

Here are examples of some widely used distributions:

Poisson

$$P(\{\omega : X = x\}) = e^{-\lambda} \frac{\lambda^x}{x!}, \qquad x = 0, 1, 2, \dots, \ \lambda > 0,$$

is often used for modeling the number of claims;

Binomial

$$P(\{\omega : X = x\}) = \binom{m}{x} q^x (1 - q)^{m-x}, \qquad x = 0, 1, 2, \dots, m,$$

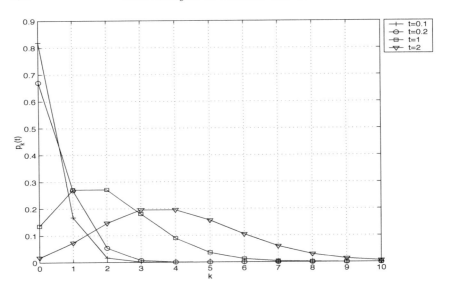

FIGURE 7.2: Values of $p_k(t)$ for $\lambda = 2$.

represents the number of claims for a portfolio of m independent policies, where q is the *probability of receiving a claim* (if $m = 1$, then it is called the **Bernoulli** distribution);

Normal

$$P\big(\{\omega:\ X \le x\}\big) = \int_{-\infty}^{x} \frac{1}{\sqrt{2\pi}\,\sigma}\, e^{-(x-\mu)^2/2\sigma^2}\, dx\,, \qquad \mu \in \mathbb{R}\,,\ \sigma > 0;$$

Exponential

$$P\big(\{\omega:\ X \le x\}\big) = 1 - e^{-\lambda x}\,, \qquad x \ge 0,\ \lambda > 0\,,$$

has various applications, for example, models the distribution of jumps of a Poisson process with intensity λ;

Gamma

$$P\big(\{\omega:\ X \le x\}\big) = \int_{0}^{x} \frac{\beta^\alpha}{\Gamma(\alpha)}\, x^{\alpha-1}\, e^{-\beta x}\, dx\,, \qquad \beta > 0;$$

Pareto

$$P\big(\{\omega:\ X \le x\}\big) = 1 - \left(\frac{\lambda}{\lambda + x}\right)^{\alpha}\,, \qquad x \ge 0,\ \alpha > 0,\ \lambda > 0\,,$$

has a "heavy" tail, and hence is often used in modeling large claims;

Lognormal

$$P(\{\omega: \ X \le x\}) = \int_0^x \frac{1}{\sqrt{2\pi}\,\sigma}\, e^{-(\log y - \mu)^2/2\sigma^2}\, dy\,.$$

Denote

$$F_{X(t)} = P(\{\omega: \ X \le x\}) = P\left(\left\{\omega: \ \sum_{i=1}^{N(t)} X_i \le x\right\}\right),$$

the distribution of the risk process.

To compute $F_{X(t)}(x)$, one needs some additional assumptions. Usually processes (X_n) and $N(\cdot)$ are assumed to be independent. Then we can write

$$F_{X(t)}(x) = P(\{\omega: \ X \le x\}) = \sum_{k=0}^{\infty} p_k(t)\, F_X^{*k}(x),$$

where

$$F_X^{*k}(x) = P(\{\omega: \ X_1 + \ldots + X_k \le x\})\,.$$

Premium calculation or determination of process $\Pi(t)$ is one of the most essential and complex tasks of an insurer. Premium flow must guarantee payments of claims; however, premiums must be competitive. One of the most widely used ways of computing Π on interval $[0, t]$ is given by

$$\Pi(t) = (1 + \theta)\, E\big(N(t)\big)\, E(X),$$

where X is a random variable with the same distribution as X_i, and θ is the *security loading coefficient*. This formula says that the average premium income should be greater than the average aggregate claims payment. If they are equal, then such premium is called *net-premium*, and the method of its computing is referred to as the *equivalence principle*.

The *bonus-malus* system is an example of a different approach to premium calculations. In this case, all policyholders are assigned certain ratings according to their claims history, and they can be transferred from one group to another. This system is typically used by the motor vehicle insurance companies.

A calculation of adequate premium involves the construction of process $\Pi(t)$ given $F_{X(t)}$, the distribution function of the risk process. In this case, we will write $\Pi(F_X)$ or simply $\Pi(X)$.

Process Π has the following properties:

- $\Pi(a) = a$ for any constant a if $\theta = 0$;

- $\Pi(a\,X) = a\,\Pi(X)$ for any constant a;

- $\Pi(X + Y) \le \Pi(X) + \Pi(Y)$;

- $\Pi(X + a) = \Pi(X) + a$ for any constant a;

- if $X \leq Y$, then $\Pi(X) \leq \Pi(Y)$;

- for any $p \in [0, 1]$ and any random variable Z

$$\Pi(X) = \Pi(Y)$$

implies that

$$\Pi\big(p\,F_X + (1 - p)\,F_Z\big) = \Pi\big(p\,F_Y + (1 - p)\,F_Z\big).$$

We list some widely used actuarial *principles of premium calculations*:

Expectation principle

$$\Pi(X) = (1 + a)\,E(X),\ a > 0;$$

Variance principle

$$\Pi(X) = E(X) + a\,Var(X);$$

Standard deviation principle

$$\Pi(X) = E(X) + a\,\sqrt{Var(X)};$$

Modified variance principle

$$\Pi(X) = \begin{cases} E(X) + a\,Var(X)/E(X), & E(X) > 0 \\ 0, & E(X) = 0; \end{cases}$$

Exponential utility principle

$$\Pi(X) = \frac{\log E(e^{aX})}{a};$$

Quantile principle

$$\Pi(X) = F_X^{-1}(1 - \varepsilon);$$

Absolute deviation principle

$$\Pi(X) = E(X) + a\,\kappa_X, \qquad \text{where}\quad \kappa_X = E\big(\big|X - F_X^{-1}(1/2)\big|\big);$$

Zero utility principle

$$E\big(\upsilon(\Pi(X) - X)\big) = \upsilon(0),$$

where υ is a given utility function.

Note that the exponential principle is a particular case of the zero utility principle with

$$v(x) = \frac{1 - e^{-ax}}{a}.$$

The notion of *risk* is the key ingredient of insurance theory and practice. Risk exposure gives rise to insurance companies that manage risks and provide some protection against these risks to their clients. Reinsurance companies provide similar services to insurance companies. There are several approaches to modeling the risk process.

Consider a *portfolio* that consists of n policy contracts with claim payments ("risks") $X_1, \ldots, X_n$ being independent non-negative random variables. Then the risk process

$$X^{\text{ind}} = \sum_{i=1}^{n} X_i,$$

has distribution $F_{X_1} * \ldots * F_{X_n}$. This model of risk is referred to as *individual*.

Suppose that an insurance company issues n insurance contracts that terminate at some time t, for example, in one year time. Each contract allows no more than one claim. Claim payments $X_1, \ldots, X_n$ are non-negative random variables. The total amount of claims incurred over this period is represented by the risk process $X^{\text{ind}} = \sum_{i=1}^{n} X_i$. It is also assumed that all claims are payable at the termination time. Therefore, the *probability of ruin* is given by

$$P(\{\omega: \ X^{\text{ind}} > x + \Pi\}),$$

where x is the initial capital of the company and Π is the premium income.

Thus, the model of individual risk is based on the following assumptions:

- time horizon is relatively short;

- number of insurance contracts is deterministic and fixed;

- premiums are payable at the time of contracts issue;

- the distribution of claim payments is known.

Example 7.1

Consider a model of individual risk with a sufficiently large number of insurance contracts. Since exact calculation of the probability of ruin is technically complicated, we use the central limit theorem for its approximation.

Using the net-premium principle, we have that $\Pi = X^{\text{ind}}$. Then we compute the probability of ruin:

$$
\begin{aligned}
P(\{\omega: \ X^{\text{ind}} > \Pi\}) \ &= \ P(\{\omega: \ X^{\text{ind}} - E(X^{\text{ind}}) > 0\}) \\
&= \ P\left(\left\{\omega: \ \frac{X^{\text{ind}} - E(X^{\text{ind}})}{\sqrt{Var(X^{\text{ind}})}} > 0\right\}\right) \\
&\approx \ 1 - \Phi(0) = 0.5,
\end{aligned}
$$

where

$$\Phi(x) = \frac{1}{\sqrt{2\pi}} \int_{-\infty}^{x} e^{-y^2/2} \, dy \,.$$

This means that the net-premium principle cannot be used in this situation. The standard deviation principle gives

$$\Pi = E(X^{\text{ind}}) + a \sqrt{Var(X^{\text{ind}})}$$

and

$$P(\{\omega : X^{\text{ind}} > \Pi\}) = P\left(\left\{\omega : \frac{X^{\text{ind}} - E(X^{\text{ind}})}{\sqrt{Var(X^{\text{ind}})}} > a\right\}\right) \approx 1 - \Phi(a)\,.$$

In this case, for any fixed level of risk ε, we can find a parameter a such that $\Phi(a) = 1 - \varepsilon$, so that the probability of ruin is

$$P(\{\omega : X^{\text{ind}} > \Pi\}) \approx \varepsilon \,. \ \square$$

Now we consider a situation when N, the number of possible claims, is unknown. We can single out two types of insurance contracts: static and dynamic. In the static case, claims are payable at the terminal time, and therefore N is an integer-valued random variable. In the dynamic model, $N = N(t)$ is a stochastic process that counts the number of claims incurred during the time interval $[0, t]$. Both these models of risk are referred to as *collective*. The risk process has the form

$$X^{\text{col}} = \sum_{i=1}^{N} X_i \,,$$

where claims amounts X_i are positive and independent of N. Clearly, the collective model of risk is more realistic than the individual model, and it gives more flexibility in managing risk for an insurance company.

Some essential differences between the two models are summarized in the following table.

individual model	collective model
n, the number of insurance contracts is known *a priori* and all claims are payable at the same time	the process of receiving claims is represented by a stochastic process
each contract admits no more than one claim	there is no restriction on the number of claims per contract
all claims are assumed to be independent	it is assumed that the amounts of incurred claims are independent

Worked Example 7.1 *Suppose an insurance company issues 1-year contracts. All policyholders are divided into four groups:*

k	q_k	b_k	n_k
1	0.02	1	500
2	0.02	2	500
3	0.1	1	300
4	0.1	2	500

Here n_k is the number of policyholders in group k, q_k is the probability of making a claim by a member of this group and b_k is the amount of the corresponding claim. Using normal approximation, find the value of the security loading coefficient that will reduce the probability of ruin to 0.05.

Solution The total number of policyholders is 1800, so the total amount of claims is

$$S = X_1 + \ldots + X_{1800}.$$

We will find parameter θ from the equation

$$P(\{\omega : S \le (1+\theta) E(S)\}) = 0.95,$$

which can be written in the form

$$P\left(\left\{\omega : \frac{S - E(S)}{\sqrt{Var(S)}} \le \frac{\theta E(S)}{\sqrt{Var(S)}}\right\}\right) = 0.95.$$

Since the total number of policyholders is reasonably large, then the quantity $(S-E(S))/\sqrt{Var(S)}$ can be accurately approximated by the standard normal distribution. Hence, we obtain the equation

$$\frac{\theta E(S)}{\sqrt{Var(S)}} \approx 1.645.$$

The following table contains expectations $\mu_k = b_k q_k$ and variances $\sigma_k^2 = b_k^2 q_k (1 - q_k)$ for each policy.

k	q_k	b_k	μ_k	σ_k^2	n_k
1	0.02	1	0.02	0.0196	500
2	0.02	2	0.02	0.0784	500
3	0.1	1	0.1	0.09	300
4	0.1	2	0.2	0.36	500

Thus,

$$E(S) = \sum_{i=1}^{1800} E(X_i) = \sum_{k=1}^{4} n_k \mu_k = 160,$$

$$Var(S) = \sum_{i=1}^{1800} Var(X_i) = \sum_{k=1}^{4} n_k \sigma_k^2 = 256,$$

and

$$\theta \approx 1.645 \frac{\sqrt{V(S)}}{E(S)} = 1.645 \frac{16}{160} = 0.1645 . \ \square$$

Note that situations where the number of claims is a random variable are typical for life insurance, and they are studied later.

Another example of a collective risk model is the *Cramér-Lundberg model.* The claims flow is modeled here as a Poisson process $N(t)$ with intensity λ, and claims amounts are independent random variables that are also independent of $N(t)$. The premium income is a linear function of time t: $\Pi(t) = ct$. The risk process

$$X(t) = \sum_{i=1}^{N(t)} X_i$$

is a compound Poisson process. It turns out that if the initial probability P that describes the distribution of claims is replaced by an equivalent probability Q under which X is also a compound Poisson process, then applying the equivalence principle with respect to this new probability Q, we obtain all the above-mentioned traditional principles of premium calculations.

Indeed, define a positive process

$$M_t^\beta = \exp \left\{ X^\beta(t) - \lambda t \, E_P \big(\exp\{\beta(X_1)\} - 1 \big) \right\}, \quad M_0^\beta = 1, \ t \in [0, T],$$

where $X^\beta(t) = \sum_{k=1}^{N(t)} \beta(X_k)$, function $\beta : \mathbb{R}_+ \to \mathbb{R}$ is such that $E_P\big(\exp\{\beta(X_1)\}\big) < \infty$, and the expectation E_P is taken with respect to probability P.

Process M^β is a martingale with respect to the natural filtration $(\mathcal{F}_t)_{t\geq 0}$, since for $s \leq t$ we have

$$E_P\big(M_t^\beta | \mathcal{F}_s\big)$$
$$= M_s^\beta \, E_P\big(\exp\{X^\beta(t) - X^\beta(s)\}\big) \, e^{-\lambda(t-s)} \, E_P\big(\exp\{\beta(X_1)\} - 1 \big) = M_s^\beta.$$

Hence, $E_P\big(M_t^\beta\big) = 1$, and for each such function β one can define new probability Q with density M_t^β. Note that any other probability Q under which the risk process $X(t)$ is a compound Poisson process must have exactly the same structure with some appropriate function β.

Thus, we can use this function β and the corresponding probability Q for calculating premium. This calculation is based on the condition that the difference

$$X(t) - ct$$

between the total amount of claims and the total premium income is a martingale with respect to Q. This agrees with the equivalence principle in insurance and with the no-arbitrage principle in finance.

So we obtain that

$$c = E_Q\big(X(1)\big),$$

and since $X(t)$ is a compound Poisson process, then

$$E_Q(X(1)) = E_Q(N(1)) E_Q(X_1),$$

where

$$E_Q(N(1)) = \lambda E_P(\exp\{\beta(X_1)\}),$$

$$E_Q(X_1) = E_P(X_1 \exp\{\beta(X_1)\})/E_P(\exp\{\beta(X_1)\}).$$

Finally, we deduce

$$c = \lambda E_P(X_1 \exp\{\beta(X_1)\}).$$

Choosing appropriately β, we then obtain all the traditional actuarial principles of premium calculations. For example, the expectation principle corresponds to $\beta(x) = \ln(1 + a)$, and we have

$$c = \lambda E_P(X_1 \exp\{\beta(X_1)\}) = \lambda E_P(X_1 (1 + a)) = (1 + a) E_P(X_1) = \Pi(X).$$

7.2   Risks transfers via reinsurance

Reinsurance is a mechanism that insurance companies use to transfer some or all of their risks to reinsurance companies. The primary aim of reinsurance is to protect the solvency of the insurance company by minimizing the probability of ruin. Some typical examples of situations when such solvency protection is required include receiving very large claims (e.g., in the cases of big man-made disasters such as an airplane crash); receiving a large number of claims from policies affected by the same event (e.g., in the case of natural disasters such as earthquakes, hurricanes, floods); sudden changes in the premiums flow (say, due to inflation) or in the number of policyholders; the need to access some additional capital so that the insurance company can take on larger risks and therefore attract more clients.

As in Section 7.1, we consider a risk process

$$X(t) = \sum_{i=1}^{N(t)} X_i,$$

which represents the aggregate amount of claims up to time t.

One of the main characteristics of a reinsurance contract is quantity $h(X)$ that determines the amount of claims payment made by the insurance company. The remainder $X - h(X)$ is paid by the reinsurance company and is naturally referred to as an *insured proportion* of risk X. The insurance company pays premium to the reinsurance company in order to transfer this part of its risk.

Function $h(X)$ is called the *retention function*. It has the following properties:

(a) $h(X)$ and $X - h(X)$ are non-decreasing functions;

(b) $0 \le h(X) \le X, \quad h(0) = 0.$

There are two basic forms of reinsurance: *proportional reinsurance* and *non-proportional reinsurance*. The main types of proportional reinsurance are *quota share* and *surplus*. A quota share reinsurance transfers all risks in the same proportion, whereas in a surplus reinsurance proportions of transfer may vary. The typical examples of non-proportional reinsurance are *stop-loss* reinsurance and *excess of loss* reinsurance. They provide protection when claims exceed a certain agreed level.

The following retention functions

1. $h(x) = a\,x \quad 0 < a \le 1,$

2. $h(x) = \min\{a,\, x\} \quad a > 0,$

correspond to the quota share reinsurance and to the stop-loss reinsurance, respectively.

From the reinsurance company's point of view, a reinsurance contract is just a usual insurance against risk $X - h(X)$. Hence, one can calculate the corresponding premium level using the methodology described earlier in this chapter:

$$\widetilde{\Pi} = \Pi\big(X - h(X)\big).$$

Let us consider a quota share reinsurance contract in the framework of the individual risk model. If X_i is the amount of an individual claim, then $a X_i$ is paid by the insurance company and $(1 - a)\,X_i$ is paid by the reinsurance company. Thus, the total amount of claim

$$S = X_1 + \ldots + X_n$$

received by the insurance company is reduced to

$$a\,S = a\,(X_1 + \ldots + X_n).$$

Suppose both insurance and reinsurance companies use the expectation principle in the premium calculation (see Section 7.1) and security loading coefficients are θ and θ^*, respectively.

Prior to entering the reinsurance contract the capital of the insurance company is

$$x + (1 + \theta)\,E(S).$$

After paying the premium $(1 + \theta^*)(1 - a)E(S)$, the capital becomes

$$x + \left[\theta - \theta^* + a(1 + \theta^*)\right]E(S).$$

Now we compare the probabilities of ruin as measures of risk to which the insurance company is exposed when it purchases the reinsurance contract and when it does not. In the first case, it is

$$P\left(\{\omega: \ aS < x + \left[\theta - \theta^* + a(1 + \theta^*)\right]E(S)\}\right)$$

$$= P\left(\left\{\omega: \ S < \frac{x + \left[\theta - \theta^* + a(1 + \theta^*)\right]E(S)}{a}\right\}\right),$$

and in the second:

$$P\left(\{\omega: \ S < x + (1 + \theta)E(S)\}\right).$$

This allows us to manage the risk of the insurance company, since if $(\theta - \theta^*)E(S) < x$, then the probability of ruin can be reduced by purchasing the reinsurance contract.

Next, we consider a stop-loss reinsurance contract with the retention level a. According to this contract, if the amount of an individual claim $X_i \le a$, then it is paid by the insurance company; otherwise, the insurance company pays a and the reinsurance company pays the remainder $X_i - a$. So by purchasing such a reinsurance contract the insurance company protects itself from paying more than a per individual claim.

Suppose that the insurance company issues N identical insurance contracts, so that the independent identically distributed random variables $X_1, \ldots, X_N$ represent the amounts of corresponding claims. Under the stop-loss reinsurance contract, the total amount of claim

$$S = X_1 + \ldots + X_N$$

received by the insurance company is reduced to

$$S^{(a)} = X_1^{(a)} + \ldots + X_n^{(a)}, \qquad \text{where} \qquad X^{(a)} = \min\{X, a\}.$$

For example, the sequence of payments made by the insurance company may look like

$$X_1, X_2, a, a, X_5, \ldots,$$

and the corresponding sequence of payments made by the reinsurance company is

$$0, 0, X_3 - a, X_4 - a, 0, \ldots.$$

Note that the number of claims paid by the reinsurance company may be less than N. Nevertheless, we can represent the risk process of the reinsurance company in the form $\sum_{i=1}^{N} Z_i$, where some Z_i may be equal to zero.

Again we assume that both insurance and reinsurance companies use the expectation principle in the premium calculation, and that security loading coefficients are θ and θ^*, respectively.

The capital of the insurance company prior to entering the reinsurance contract is given by

$$N\,p = N\,(1+\theta)\,p_0 \equiv N\,(1+\theta)\,E(X)\,.$$

After paying the premium,

$$N\,(1+\theta^*)\,(E(X)-E(X^{(a)}))\,,$$

it becomes

$$N\,(1+\theta)\,E(X) \quad - \quad N\,(1+\theta^*)\,(E(X)-E(X^{(a)}))$$
$$= N\,(\theta-\theta^*)\,E(X) + N\,(1+\theta^*)\,E(X^{(a)})\,.$$

Hence, the probability of ruin is

$$P\Big(\{\omega:\ S^{(a)} > N\,(\theta-\theta^*)\,E(X) + N\,(1+\theta^*)\,E(X^{(a)})\}\Big)\,.$$

It is rather difficult to compute this probability explicitly. We use the central limit theorem for computing its approximation:

$$P\left(\left\{\omega:\ \frac{S^{(a)}-E(S^{(a)})}{\sqrt{V(S^{(a)})}} > \frac{N\,(\theta-\theta^*)\,E(X) + N\,\theta^*\,E(X^{(a)})}{\sqrt{N\,V(X^{(a)})}}\right\}\right)$$

$$\approx 1 - \Phi\left(\sqrt{N}\,\frac{(\theta-\theta^*)\,E(X) + \theta^*\,E(X^{(a)})}{\sqrt{V(X^{(a)})}}\right),$$

where

$$\Phi(x) = \frac{1}{\sqrt{2\pi}} \int_{-\infty}^{x} e^{-y^2/2} dy$$

is the standard normal distribution.

Suppose that the insurance company can vary the retention level a (variation of a, of course, changes the premium payable to the reinsurance company). Suppose that $\tilde{a}$ is the maximum of function

$$\varphi(a) = \frac{\left[(\theta-\theta^*)\,E(X) + \theta^*\,E(\min\{X,a\})\right]^2}{Var(\min\{X,a\})}\,,$$

then the stop-loss reinsurance contract with the retention level $\tilde{a}$ minimizes the probability of ruin.

Worked Example 7.2 *Suppose that a random variable representing the amount of an individual claim is uniformly distributed in $[0, x]$. Consider a stop-loss reinsurance contract and find the value of $\tilde{a}$ that minimizes the probability of ruin.*

Solution Clearly, we can assume that the retention level $a < x$. Then we have

$$E(X) = \frac{x}{2}, \quad E(X^2) = \frac{x^2}{3}, \quad Var(X) = \frac{x^2}{12},$$

and

$$E(X^{(a)}) = a - \frac{a^2}{2x}, \quad E((X^{(a)})^2) = a^2 - \frac{2a^3}{3x}, \quad Var(X^{(a)}) = \frac{a^3}{3x} - \frac{a^4}{4x^2},$$

which implies

$$\varphi(a) = \frac{\left[(\theta - \theta^*)\frac{x}{2} + \theta^*\left(a - \frac{a^2}{2x}\right)\right]^2}{\frac{a^3}{3x} - \frac{a^4}{4x^2}}.$$

Because of no-arbitrage considerations, we have that $\theta \leq \theta^*$, since otherwise the insurance company can transfer all the risk to the reinsurance company and make a non-zero profit with zero initial capital. If we additionally assume that $\theta^* < 3\theta$, then it is not difficult to see that function $\varphi(a)$ has a unique maximum on $[0, x]$:

$$\tilde{a} = \frac{3x\left[\theta^* - \theta\right]}{2\theta^*}, \quad \text{so that} \quad \varphi(\tilde{a}) = \frac{(\theta^*)^2}{9}\frac{9\theta - \theta^*}{\theta^* - \theta}. \quad \square$$

There are large risks (e.g., a jumbo jet or an oil drilling platform accidents) of a magnitude that makes it impossible for most single insurance companies to insure the whole risk without sharing this risk exposure. In this case, an insurer transfers some risk to a reinsurer. The reinsurer itself may also reinsure this risk, which is often referred to as *retrocession*. Thus, one can summarize that the insurance market has at least three levels:

1. Primary market (insurance companies)

2. Reinsurance market (reinsurance companies)

3. Retrocession market (reinsurance companies that provide insurance to other reinsurance companies)

Clearly, the retrocession market can consist of more than one level. For each nth-level reinsurance company, the *risk transfer time*, that is, time between receiving a risk from a $(n-1)$st-level company and passing it to a $(n+1)$st-level company, is a random variable with some distribution F, and it is independent of risk transfer times of another companies. Denote $\Sigma(X_n)$ the total number of nth-level companies, and $R_{n,i}$ the number nth-level companies that insured the ith company from $(n-1)$st level. We assume that $R_{n,i}$ are independent random variables with distribution $(p_k)_{k=0,1,\dots}$. Note that

$$X_n = 1, \qquad X_{n+1} = \sum_{i=1}^{X_n} R_{n,i}.$$

Denote $\hat{g}_{X_n}(s)$ and $\hat{g}_R(s)$ the generating functions of X_n and R, respectively. For $|s| < 1$ and all $n \in \mathbb{N}$, we have the following relation:

$$\hat{g}_{X_{n+1}}(s) = \hat{g}_{X_n}\big(\hat{g}_R(s)\big) = \hat{g}_R\big(\hat{g}_{X_n}(s)\big) = \underbrace{\hat{g}_R\Big(\hat{g}_R\big(\dots\hat{g}_R(s)\big)\Big)}_{n \text{ times}},$$

which follows from the equality

$$\hat{g}_{X_{n+1}}(s) \equiv E\big(s^{X_{n+1}}\big) = \sum_{j=0}^{\infty} E\left(s^{\sum_{i=1}^{X_n} R_{n,i}} \Big| X_n = j\right) P(\{\omega : X_n = j\})$$

$$= \hat{g}_{X_n}\big(\hat{g}_R(s)\big),$$

that is applied n times.

Using the latter formula and taking into account that

$$E(X_{n+1}) = E(X_n)\, E(R)$$

and

$$Var(X_{n+1}) = Var(X_n)\,\big(E(R)\big)^2 + E(X_n)\, Var(R),$$

we obtain

$$E(X_n) = \big(E(R)\big)^2 \equiv \mu_R^n,$$

and

$$Var(X_n) = \begin{cases} \dfrac{\mu_R^{n-1}\,(\mu_R^n - 1)}{\mu_R - 1}\, Var(R), & \mu_R \neq 1 \\[2mm] n\, Var(R), & \mu_R = 1 \end{cases}.$$

Let $\Sigma\big(X(t)\big)$ be the number of companies involved in the contract up to time t, and $X_{n,i}(t)$ the number auxiliary companies that insured the ith company from nth level after time t. It is clear that

$$X(t) = \begin{cases} 1, & t < T \\[2mm] \sum_{i=1}^{R} X_{2,i}(t - T), & t \geq T \end{cases}.$$

Let $\mu(t) = E\big(X(t)\big)$, the average number of companies involved in a reinsurance project at time t. The following result holds true.

Proposition 7.1 *We have*

$$\mu(t) = \overline{F}(t) + \mu_R \int_0^t \mu(t - \nu) \, dF(\nu) \,, \tag{7.1}$$

where $F(x)$ is the distribution function of the risk transfer time, and $\overline{F}(x) = 1 - F(x)$.

Proof Suppose that random variable T represents the time between issuing the primary insurance contract by a primary insurance company and the time when this risk is reinsured by a next-level company. Then, using properties of conditional expectations, we obtain

$$
\begin{aligned}
\mu(t) &= E\Big(E\big(X(t)\,|\,T\big)\Big) = \int_0^\infty E\big(X(t)\,|\,T = \nu\big) \, dF(\nu) \\
&= \int_0^t E\big(X(t)\,|\,T = \nu\big) \, dF(\nu) + \int_t^\infty E\big(X(t)\,|\,T = \nu\big) \, dF(\nu) \,.
\end{aligned}
$$

Note that the conditional expectation in the second integral in the right-hand side is equal to 1. To compute the first term, we write

$$E\big(X(t)\,|\,T = \nu\big) = \sum_{j=0}^\infty p_j \, E\big(X(t)\,|\,T = \nu,\ R = j\big) \,,$$

where R is the number of reinsurers of the primary company. Each second-level reinsurance company generates an independent chain of reinsurers. Hence,

$$
\begin{aligned}
E\big(X(t)\,|\,T = \nu,\ R = j\big) &= E\Big(\sum_{i=1}^j X_{2,i}(t - \nu)\,|\,T = \nu,\ R = j\Big) \\
&= j\,\mu(t - \nu) \,,
\end{aligned}
$$

which implies the result. $\square$

The following result is typical for the theory of branching processes.

Proposition 7.2 *If distribution F is continuous, then the following statements hold true.*

1. *If $\mu_R = 1$, then $\mu(t) = 1$ for all $t \geq 0$.*

2. *If $\mu_R > 1$, then*
$$\lim_{t \to \infty} \frac{\mu(t)}{e^{\gamma t}} = \frac{\mu_R - 1}{\gamma \mu_R^2 \, |\hat{l}_T'(\gamma)|} \,,$$
where γ is the unique solution of equation
$$\hat{l}_T(y) := \int_0^\infty e^{-xy} \, dF(x) = \mu_R^{-1} \,.$$

3. *If $\mu_R < 1$ and there exists a positive solution γ of equation*

$$\hat{m}_T(y) := \int_0^\infty e^{yx} \, dF(x) = \mu_R^{-1} \,,$$

then

$$\lim_{t \to \infty} \frac{\mu(t)}{e^{-\gamma t}} = \frac{1 - \mu_R}{\gamma \mu_R^2 \, |\hat{m}_T'(\gamma)|} \,,$$

otherwise this limit is equal to zero.

Proof We prove here only the first statement. If $\mu_R = 1$, then $\mu(t) \equiv 1$ is a solution of equation (7.1). Thus, we only need to establish the uniqueness of this solution to

$$\mu(t) = \overline{F}(t) + \int_0^t \mu(t - \nu) \, dF(\nu) \,.$$

Suppose that both $\mu_1(t)$ and $\mu_2(t)$ are solutions of this equation. Then

$$\mu_1(t) - \mu_2(t) = \int_0^t \left(\mu_1(t - \nu) - \mu_2(t - \nu) \right) dF(\nu)$$

if and only if

$$\mu_1(t) - \mu_2(t) = (\mu_1 - \mu_2) * F(t) \,.$$

Further

$$
\begin{aligned}
|\mu_1(t) - \mu_2(t)| &= |(\mu_1 - \mu_2) * F(t)| = |(\mu_1 - \mu_2) * F * F(t)| \\
&= \ldots \\
&= |(\mu_1 - \mu_2) * F^{*n}(t)| \le F^{*n}(t) \sup_{\nu \in [0,t]} |\mu_1(\nu) - \mu_2(\nu)| \,.
\end{aligned}
$$

Now, since $F^{*n}(t) \le \left[F(t) \right]^n$, then

$$\lim_{n \to \infty} F^{*n}(t) = 0 \,,$$

which proves the claim. $\square$

Worked Example 7.3 *Describe the asymptotic behavior of $\mu(t)$ if the risk transfer time has an exponential distribution.*

Solution We have

$$\hat{l}_T(y) := \int_0^\infty e^{-xy} \, dF(x) = \int_0^\infty e^{-xy} \lambda e^{-\lambda x} \, dx = \frac{\lambda}{\lambda + y}$$

and

$$\hat{m}_T(y) := \int_0^\infty e^{yx} \, dF(x) = \int_0^\infty e^{yx} \lambda e^{-\lambda x} \, dx = \frac{\lambda}{\lambda - y} \,,$$

so that their derivatives are

$$\hat{l}'_T(y) = -\frac{\lambda}{(\lambda + y)^2} \quad \text{and} \quad \hat{m}'_T(y) = -\frac{\lambda}{(\lambda - y)^2}.$$

Since

$$\gamma = \lambda\left(\mu_R - 1\right) \quad \text{is a solution to} \quad \hat{l}_T(y) = \frac{1}{\mu_R},$$

and

$$\gamma = \lambda\left(1 - \mu_R\right) \quad \text{is a solution to} \quad \hat{m}_T(y) = \frac{1}{\mu_R},$$

then we deduce that

1) if $\mu_R = 1$, then $\mu(t) = 1$ for all $t \geq 0$;

2) if $\mu_R > 1$, then

$$\lim_{t \to \infty} \frac{\mu(t)}{e^{\lambda(\mu_R - 1)t}} = 1 \qquad \text{(exponential growth)};$$

3) if $\mu_R < 1$, then

$$\lim_{t \to \infty} \frac{\mu(t)}{e^{\lambda(\mu_R - 1)t}} = 1 \qquad \text{(exponential decay)}.$$

In this case, we can also find an exact expression for $\mu(t)$. Substitute $F(y) = 1 - e^{-\lambda y}$ into equation (7.1):

$$\mu(t) = e^{-\lambda t} + \mu_R \int_0^t \mu(t - \nu)\,\lambda\,e^{-\lambda \nu}\,d\nu.$$

Differentiating in t and integrating by parts, we obtain

$$
\begin{aligned}
\mu'(t) &= -\lambda e^{-\lambda t} + \mu_R\,\mu(0)\,\lambda\,e^{-\lambda t} - \mu_R \int_0^t \lambda\,e^{-\lambda \nu}\,d\mu(t - \nu) \\
&= -\lambda e^{-\lambda t} + \mu_R\,\lambda\,\mu(t) - \lambda\,\mu_R \int_0^t \mu(t - \nu)\,\lambda\,e^{-\lambda \nu}\,d\nu \\
&= \mu(t)\,\lambda\,(\mu_R - 1).
\end{aligned}
$$

The Cauchy problem

$$\mu'(t) = \mu(t)\,\lambda\,(\mu_R - 1), \quad t \geq 0, \quad \mu(0) = 1,$$

has the unique solution

$$\mu(t) = e^{\lambda(\mu_R - 1)t},$$

which implies the asymptotic behavior as described above. $\square$

Remark 7.1

The structure of the traditional insurance market provides reasonable protection to insurance companies against "moderately" large risks. There are events (*catastrophes*) that can give rise to giant claims, when the total claim amount can be comparable with the total premium income. Some catastrophes may cause losses that are comparable with the capacity of the whole of the insurance industry.

Risk securitization is one of the possible ways of dealing with this situation. It involves the introduction of *insurance securities*: catastrophe (CAT) bonds, forwards, futures, options, and so forth as derivative instruments in catastrophe reinsurance.

7.3   Elements of traditional life insurance

Life insurance clearly deals with various types of uncertainties, for example, the uncertainty of future lifetimes, variable interest rates, and so forth. Thus, it is natural that stochastic methods are widely used in life insurance mathematics. In this section, we discuss some survival models as one of the key ingredients of the stochastic approach.

We introduce a random variable T representing the future lifetime of a newborn individual; that is, T is the time elapsed between birth and death. The distribution function of T is

$$F(x) = P(\{\omega : T \le x\}), \quad x \ge 0.$$

Define the *survival function* as

$$s(x) = 1 - F(x) = P(\{\omega : T > x\}), \quad x \ge 0.$$

In practice, one usually introduces the *limiting age* (i.e., the age beyond which survival is supposed to be impossible). Traditionally, it is denoted by ω. In order to avoid ambiguities, we will use letter ϖ instead. Thus, we have that $0 \le T \le \varpi < \infty$. Clearly, function $F(x)$ is increasing and continuous.

Next, we define a random variable $T(x)$ to be the future lifetime of an individual of age x. Obviously, $T(0) = T$.

There is standard actuarial notation for probabilities in survival models: $_t p_x$ denotes the probability that an individual of age x survives to age $x + t$. Again, in order to avoid ambiguities, we will write $p_x(t)$ instead. Also $q_x(t) := 1 - p_x(t)$, and $p_x(1) := p_x$, $q_x(1) := q_x$.

From the definition of conditional expectation, we have

$$p_x(t) = P(\{\omega : T(x) > t\}) = P(\{\omega : T > x + t | T > x\})$$

$$= \frac{p_0(x+t)}{p_0(x)} = \frac{s(x+t)}{s(x)},$$

and

$$q_x(t) = 1 - \frac{p_0(x+t)}{p_0(x)} = 1 - \frac{s(x+t)}{s(x)}.$$

One of the most widely used actuarial representations of the survival model is the *life table* (or *mortality table*). Suppose that l_0 is the number of newborn individuals, and let random variable $L(x)$ represent the number of individuals surviving to age x. The life table consists of set of expected values of $L(x)$:

$$l_x = E(L(x)) = l_0 \, s(x)$$

for all $0 \leq x \leq \varpi$.

The following relations hold true:

$$l_1 = l_0 (1 - q_0) = l_0 \, p_0,$$
$$l_2 = l_1 (1 - q_0) = l_0 (1 - q_0(2)) = l_0 \, p_0 \, p_1,$$
$$\ldots$$
$$l_x = l_{x-1} (1 - q_{x-1}) = l_0 (1 - q_0(x)) = \prod_{y=0}^{x-1} p_y = l_0 \, p_0(x).$$

Example 7.2

1. The probability that an individual of age 20 survives to the age of 100 is

$$p_{20}(80) = \frac{s(100)}{s(20)} = \frac{l_{100}}{l_{20}}.$$

2. The probability that an individual of age 20 dies before the age of 70 is

$$q_{20}(80) = \frac{s(80) - s(70)}{s(20)} = 1 - \frac{l_{70}}{l_{20}}.$$

3. The probability that an individual of age 20 survives to the age of 80 but dies before the age of 90 is

$$\frac{s(80) - s(90)}{s(20)} = \frac{l_{80} - l_{90}}{l_{20}}.$$

Introduce the notion of the *force of mortality* at age x as

$$\mu_x = \lim_{h \to 0^+} \frac{P(\{\omega : T \leq x + h \,|\, T > x\})}{h}, \qquad 0 \leq x < \varpi.$$

The following laws for μ_x are widely used in actuarial theory and practice:

- Gompertz's formula: $\mu_x = B\,c^x$,

- Makeham's formula: $\mu_x = A + B\,c^x$.

Now we obtain an expression for density of the distribution function of $T(x)$:

$$f_x(t) = \frac{d}{dt}P\big(\{\omega:\ T(x) \le t\}\big)$$

$$= \lim_{h \to 0^+} \frac{P\big(\{\omega:\ T(x) \le t + h\}\big) - P\big(\{\omega:\ T(x) \le t\}\big)}{h}$$

$$= \lim_{h \to 0^+} \frac{P\big(\{\omega:\ T \le x + t + h \mid T > x\}\big) - P\big(\{\omega:\ T \le x + t \mid T > x\}\big)}{h}$$

$$= \lim_{h \to 0^+} \left[\frac{P\big(\{\omega:\ T \le x + t + h\}\big) - P\big(\{\omega:\ T \le x\}\big)}{s(x)\,h} \right.$$

$$\left. - \frac{P\big(\{\omega:\ T \le x + t\}\big) - P\big(\{\omega:\ T \le x\}\big)}{s(x)\,h} \right]$$

$$= \lim_{h \to 0^+} \frac{P\big(\{\omega:\ T \le x + t + h\}\big) - P\big(\{\omega:\ T \le x + t\}\big)}{s(x)\,h}$$

$$= \frac{s(x+t)}{s(x)} \lim_{h \to 0^+} \frac{P\big(\{\omega:\ T \le x + t + h\}\big) - P\big(\{\omega:\ T \le x + t\}\big)}{s(x+t)\,h}$$

$$= p_x(t) \lim_{h \to 0^+} \frac{P\big(\{\omega:\ T \le x + t + h \mid T > x + t\}\big)}{h}$$

$$= p_x(t)\,\mu_{x+t}, \quad 0 \le t \le \varpi - x.$$

Further,

$$q_x(t) \equiv \int_0^t \frac{d}{ds}\,q_x(s)\,ds = \int_0^t f_x(s)\,ds = \int_0^t p_x(s)\,\mu_{x+s}\,ds,$$

hence,

$$\frac{\partial}{\partial s}p_x(s) \equiv -\frac{\partial}{\partial s}q_x(s) = -p_x(s)\,\mu_{x+s}.$$

Solving this differential equation for $p_x(t)$ with the initial condition $p_x(0) = 1$, we obtain

$$p_x(t) = \exp\left\{ -\int_0^t \mu_{x+s}\,ds \right\}.$$

These expressions for $q_x(t)$ and $p_x(t)$ are widely used for premium calculations in standard life insurance contracts.

We also introduce an integer-valued random variable $K(x) := [\![T(x)]\!]$, which obviously represents the number of whole years survived by an individual of age x. The set of its values is $\{0, 1, 2, \ldots, [\![\varpi - x]\!]\}$. We have

$$
\begin{aligned}
P(\{\omega : K(x) = k\}) &= P(\{\omega : k \leq T(x) < k + 1\}) \\
&= P(\{\omega : k < T(x) < k + 1\}) = p_x(k)\, q_{x+k}\,.
\end{aligned}
$$

It is more convenient to use quantities $K(x)$ when using life tables.

A standard life insurance contract assumes payment of b_t at time t. If ν_t is the discount factor, then the present value (at time $t = 0$) of this payment is $z_t = b_t\, \nu_t$. Since the amount of payment b_t is set at the time of contract issue, then without loss of generality we can assume that $b_t = 1$.

First, we consider contracts when benefits are paid upon the death of the insured individual (i.e., *life insured*). Let $Z = b_{T(x)}\, \nu_{T(x)}$, where x is the age of the life insured at the time of contract issue. The equivalence principle is used for premium calculations.

Term life insurance pays a lump-sum benefit upon the death of the life insured within a specified period of time, say, within n-years term, that is,

$$
b_t = \begin{cases} 1, & t \leq n \\ 0, & t > n\,, \end{cases}
$$

$$
\nu_t = \nu^t\,, \quad t \geq 0\,,
$$

$$
Z = \begin{cases} \nu^{T(x)}, & T(x) \leq n \\ 0, & T(x) > n\,. \end{cases}
$$

The net-premium in this case is

$$
\overline{A}^1_{x:\overline{n}|} = E(Z) = E(z_{T(x)}) = \int_0^\infty z_t\, f_x(t)\, dt = \int_0^n \nu^t\, p_x(t)\, \mu_{x+t}\, dt\,.
$$

Whole life insurance pays a lump sum benefit upon the death of the life insured whenever it should occur:

$$
\begin{aligned}
b_t &= 1, & t \geq 0\,, \\
\nu_t &= \nu^t\,, & t \geq 0\,, \\
Z &= \nu^{T(x)}\,, & T(x) \geq 0\,.
\end{aligned}
$$

The net-premium is

$$
\overline{A}_x = E(Z) = \int_0^\infty \nu^t\, p_x(t)\, \mu_{x+t}\, dt\,.
$$

Worked Example 7.4 *Consider* 100 *whole life insurance contracts. Suppose that all life insured are of age* x *and the benefit payment is* 10. *Let discount factor be* $\nu = e^{-\delta} = e^{-0.06}$ *and* $\mu = 0.04$. *Compute the premium that guarantees the probability of solvency at* 0.95.

Solution For an individual contract, we have

$$b_t = 10, \quad t \geq 0,$$
$$\nu_t = \nu^t, \quad t \geq 0,$$
$$Z = 10\,\nu^{T(x)}, \quad T(x) \geq 0.$$

The risk process in this case is $S = \sum_{i=1}^{100} Z_i$. For individual claims, we have that payment amounts in the case of death are

$$\overline{A}_x = \int_0^\infty e^{-\delta t}\, e^{-\mu t}\, \mu \, dt = \frac{\mu}{\mu + \delta}$$

then

$$E(Z) = 10\,\overline{A}_x = 10\,\frac{0.04}{0.1} = 4\,,$$
$$E(Z^2) = 10^2 \int_0^\infty e^{-2\delta}\, e^{-\mu t}\, \mu \, dt = 100\,\frac{0.04}{0.04 + 2 \times 0.06} = 25\,,$$

which also implies that $Var(Z) = 9$.
The premium payment h can be found from the equation

$$P\big(\{\omega: \ S \leq h\}\big) = 0.95\,,$$

which can be written in the form

$$P\left(\left\{\omega: \ \frac{S - E(S)}{\sqrt{Var(S)}} \leq \frac{h - 400}{30}\right\}\right) = 0.95\,.$$

Since random variable $(S - E(S))/\sqrt{Var(S)}$ is normal, we obtain

$$\frac{h - 400}{30} \approx 1.645 \quad \text{and} \quad h \approx 449.35\,.$$

Thus, we have that the premium is higher than the expected claim payment. The corresponding security loading coefficient is

$$\theta = \frac{h - E(S)}{E(S)} \approx 0.1234\,. \ \square$$

Pure endowment insurance pays a lump-sum benefit on survival of the life insured up to the end of a specified period of time, say, up to the end of n years term:

$$b_t = \begin{cases} 0, & t \le n \\ 1, & t > n \end{cases},$$

$$\nu_t = \nu^n, \quad t \ge 0,$$

$$Z = \begin{cases} 0, & T(x) \le n \\ \nu^n, & T(x) > n \end{cases}.$$

Net-premium is

$$A^1_{x:\overline{n}|} = E(Z) = \nu^n \, p_x(n).$$

Endowment insurance pays a lump-sum benefit on death of the life insured within a specified period of time, say, within the n years term, or on survival of the life insured up to the end of this period:

$$b_t = 1, \quad t \ge 0,$$

$$\nu_t = \begin{cases} \nu^t, & t \le n \\ \nu^n, & t > n \end{cases},$$

$$Z = \begin{cases} \nu^{T(x)}, & T(x) \le n \\ \nu^n, & T(x) > n \end{cases}.$$

This contract is obviously a combination of a pure endowment insurance and a term-life insurance:

$$Z_1 = \begin{cases} 0, & T(x) \le n \\ \nu^n, & T(x) > n \end{cases} \quad \text{and} \quad Z_2 = \begin{cases} \nu^{T(x)}, & T(x) \le n \\ 0, & T(x) > n \end{cases},$$

respectively. Therefore, the net-premium is

$$\overline{A}_{x:\overline{n}|} = E(Z) = E(Z_1 + Z_2) = \overline{A}^1_{x:\overline{n}|} + A^1_{x:\overline{n}|}.$$

Deferred whole life insurance pays a lump-sum benefit upon the death of the life insured if it occurs at least, say, m years after issuing the contract:

$$b_t = \begin{cases} 1, & t > m \\ 0, & t \le m \end{cases},$$

$$\nu_t = \nu^t, \quad t > 0,$$

$$Z = \begin{cases} \nu^{T(x)}, & T(x) > m \\ 0, & T(x) \le m \end{cases}.$$

The net-premium in this case is

$$_{m|}\overline{A}_x = E(Z) = \int_m^\infty \nu^t \, p_x(t) \, \mu_{x+t} \, dt \, .$$

Next, we consider contracts with variable amounts of benefit paid upon the death of life insured.

Increasing whole life insurance :

$$b_t = [\![t+1]\!], \quad t \geq 0,$$
$$\nu_t = \nu^t, \quad t \geq 0,$$
$$Z = [\![T(x)+1]\!] \, \nu^{T(x)}, \quad T(x) \geq 0.$$

Net-premium is

$$\left(I\overline{A}\right)_x = E(Z) = \int_0^\infty [\![t+1]\!] \, \nu^t \, p_x(t) \, \mu_{x+t} \, dt \, .$$

Decreasing term-life insurance :

$$b_t = \begin{cases} n - [\![t]\!], & t \leq n \\ 0, & t > n \end{cases},$$

$$\nu_t = \nu^t, \quad t \geq 0,$$

$$Z = \begin{cases} \nu^{T(x)} \, (n - [\![T(x)]\!]), & T(x) \leq n \\ 0, & T(x) > n \end{cases}.$$

Net-premium is

$$\left(D\overline{A}\right)^1_{x:\overline{n}|} = \int_0^n \nu^t \, (n - [\![t]\!]) \, p_x(t) \, \mu_{x+t} \, dt \, .$$

One can consider variations of these contracts in the case when benefits are paid at the end of the year in which death occurred, that is, at time $K(x)+1$. Some of them are presented in the following table. Note that we write k for $K(x)$ here.

Type of insurance	b_{k+1}	z_{k+1}	Premium
Whole life	1	v^{k+1}	A_x
Term life	$\begin{cases} 1, & k \in K_0 \\ 0, & k \in K_1 \end{cases}$	$\begin{cases} v^{k+1}, & k \in K_0 \\ 0, & k \in K_1 \end{cases}$	$A^1_{x:\overline{n}\rvert}$
Endowment insurance	1	$\begin{cases} v^{k+1}, & k \in K_0 \\ v^n, & k \in K_1 \end{cases}$	$A_{x:\overline{n}\rvert}$
Increasing term life	$\begin{cases} n+1, & k \in K_0 \\ 0, & k \in K_1 \end{cases}$	$\begin{cases} (k+1)v^{k+1}, & k \in K_0 \\ 0, & k \in K_1 \end{cases}$	$(IA)^1_{x:\overline{n}\rvert}$
Decreasing term life	$\begin{cases} n-1, & k \in K_0 \\ 0, & k \in K_1 \end{cases}$	$\begin{cases} (n-1)v^{k+1}, & k \in K_0 \\ 0, & k \in K_1 \end{cases}$	$(DA)^1_{x:\overline{n}\rvert}$
Increasing whole life	$k+1, \ k=0,1,\dots$	$(k+1)v^{k+1}$	$(IA)_x$

Here $K_0 = \{0, \dots, n-1\}$ and $K_1 = \{n, n+1, \dots\}$.

7.4 Risk modeling and pricing in innovative life insurance

Traditional insurance contracts considered in the previous section have an essential common feature with financial products that we studied earlier: contingent payments at some future dates. In traditional insurance theory, it is assumed that the amounts of these payments are deterministic and all randomness is due to the uncertainty of future lifetimes. Because of market competition, some insurers (particularly investment companies, hedging funds, merchant bank, etc.) now offer more attractive (from the investment's point of view) "options-type" insurance contracts such that their structure depends on risky financial assets. These ideas gave rise to a new approach of *innovative methods* in insurance, which is usually referred to as *equity-linked life insurance*.

We begin our discussion of such *flexible insurance methods* by revisiting Worked Example 2.3 from Section 2.2, which is concerned with a pure endowment insurance contract in the framework of a binomial (B, S)-market.

Let $(\Omega_1, \mathcal{F}_N^1, \mathbb{F}_1, P_1)$ be a stochastic basis. Consider a binomial (B, S)-market with

$$\Delta B_n = r\, B_{n-1}, \quad B_0 > 0,$$
$$\Delta S_n = \rho_n\, S_{n-1}, \quad S_0 > 0, \ n \le N,$$

where $r \ge 0$ is a constant rate of interest with $-1 < a < r < b$, and

$$\rho_n = \begin{cases} b & \text{with probability } p \in [0,1] \\ a & \text{with probability } q = 1 - p \end{cases}, \quad n = 1, \ldots, N,$$

form a sequence of independent identically distributed random variables.

Suppose that an insurance company issues l_x contracts with policyholders of age x. As before, random variable $T(x)$ represents the future lifetime of an individual of age x and $p_x(t) = P(\{\omega : T(x) > t\})$.

Introduce a process

$$N_t^x = \sum_{i=1}^{l_x} I_{\{\omega:\ T_i(x) \le t\}}$$

that counts the number of deaths during the time interval from 0 to t.

Random variables $T_1(x), T_2(x), \ldots, T_{l_x}(x)$ are defined on stochastic basis $(\Omega_2, \mathcal{F}_N^2, \mathbb{F}_2, P_2)$, where $\mathcal{F}_n^2 = \sigma(N_k^x, \ k \le n)$, $n \le N$.

Thus, we have two sources of randomness: the future lifetime of life insured and the prices of assets of the financial market. It is natural to assume that these sources of randomness are independent. Hence, formally we have two probability spaces. One of them describes the dynamics of the market, and the other describes lifetimes of the life insured. The following stochastic basis

$$\left(\Omega_1 \otimes \Omega_2, \mathcal{F}_N^1 \otimes \mathcal{F}_N^2, \mathbb{F}_1 \otimes \mathbb{F}_2, P_1 \otimes P_2\right),$$

naturally corresponds to the problem in consideration.

Consider a pure endowment insurance contract that pays a lump sum f_N upon survival of the life insured to time N. The total discounted amount of claims at time N is given by

$$\sum_{i=1}^{l_x} \frac{f_N}{B_N}\, I_{\{\omega:\ T_i(x) > N\}}.$$

Consider the case when $f_N = \max\{S_N, K\}$, where K is the guaranteed minimal payment. We wish to price this contingent claim, that is, to calculate premium $U_x(N)$. One approach consists of applying the equivalence principle for the risk-neutral probability $P_1^* \otimes P_2$. Since S and T are independent, we have

$$
U_x(N) = \frac{1}{l_x} E^* \left(\sum_{i=1}^{l_x} \frac{f_N}{B_N} I_{\{\omega:\ T_i > N\}} \right) = p_x(N) E^* \left(\frac{K + (S_N - K)^+}{B_N} \right)
$$

$$
= p_x(N) \frac{K}{(1+r)^N} + p_x(N) \left[S_0\, B(k_0, N, \tilde{p}) - \frac{K}{(1+r)^N}\, B(k_0, N, p^*) \right],
$$

where p^* is a risk-neutral probability:

$$
p^* = \frac{r - a}{b - a} \qquad \text{and} \qquad \tilde{p} = \frac{1+b}{1+r} p^*.
$$

Recall (see Section 2.2) that

$$
B(j, N, p) := \sum_{k=j}^{N} \binom{N}{k} p^k (1 - p)^{N-k}
$$

and constant k_0 is defined by

$$
k_0 = \min \left\{ k \leqslant N:\ S_0(1+b)^k(1+a)^{N-k} \geqslant K \right\}
$$

so that

$$
k_0 = \left[\ln \frac{K}{S_0(1+a)^N} \middle/ \ln \frac{1+b}{1+a} \right] + 1 .
$$

Alternatively, one can use hedging in mean square for computing the premium. Suppose that the discounted total amount of claims is

$$
H = \sum_{i=1}^{l_x} Y_i \qquad \text{with} \qquad Y_i = \frac{g(S_N)}{B_N} I_{\{\omega:\ T_i(x) > N\}} ,
$$

where function g determines amount of claim for an individual contract.

It was shown in Section 3.3 that the unique optimal (risk-minimizing) strategy $\hat{\pi} = (\hat{\beta}, \hat{\gamma})$ is given by

$$
\hat{\gamma}_n = \gamma_n^H, \qquad \hat{\beta}_n = V_n^* - \hat{\gamma}_n X_n, \qquad n \leq N ,
$$

where X_n represents the capital of the portfolio, V_n^π represents the discounted capital of the portfolio, and

$$
V_n^* = E^* (H | \mathcal{F}_n), \qquad n \leq N ,
$$

with respect to a risk-neutral probability P^*. Sequence (γ_n^H) and martingale (L_n^H) are uniquely determined by the Kunita-Watanabe decomposition (see Lemma 3.2).

Also we have the price-sequence

$$C_n^{\hat{\pi}} = V_n^{\hat{\pi}} - \sum_{k=1}^{n} \hat{\gamma}_k \, \Delta X_k = E^*(H) + L_n^H \,,$$

and the risk-sequence

$$R_n^{\pi} = E^* \left(\left(L_N^H - L_n^H \right)^2 \middle| \mathcal{F}_n \right).$$

Note that this strategy $\hat{\pi} = (\hat{\beta}, \hat{\gamma})$ is not self-financing, but it is self-financing in the mean.

As an illustration, we consider a *lognormal* model of a financial market:

$$S_n = S_0 \, e^{h_1 + \ldots + h_n} \,, \quad h_i = \mu + \sigma \, \varepsilon_n \,, \quad \text{and} \quad B_n = B_0 \, (1 + r)^n \,, \qquad (7.2)$$

where ε_n are independent identically distributed standard normal random variables.

Let

$$g(S_N) = \max \{ S_N, K \} = K + (S_N - K)^+ \,,$$

where K is a constant.

Denote

$$h_k^* = \mu - \delta + \sigma \, \varepsilon_k \quad \text{and} \quad S_k^* = \frac{S_k}{B_k} \,,$$

where $\delta = \ln(1 + r)$. The discounting factor is $\nu = 1/(1 + r)$ and $N_t^x = \sum_{k=1}^{l_x} I_{\{\omega: \, T_k(x) \le t\}}$.

From properties of expectations, we have

$$V_t^{\pi} = \left(l_x - N_t^x \right) B_0^{-1} \nu^N \, p_{x+t}(N - t)$$

$$\times \left[K + S_t \, (1 + r)^{N-t} \, \Phi \left(\frac{\ln \left(S_t/K \right) + (N - t) \left(\delta + \sigma^2/2 \right)}{\sigma \sqrt{N - t}} \right) \right.$$

$$\left. - K \, \Phi \left(\frac{\ln \left(S_t/K \right) + (N - t) \left(\delta - \sigma^2/2 \right)}{\sigma \sqrt{N - t}} \right) \right],$$

$$V_0^{\pi} = l_x \, B_0^{-1} \nu^N \, p_x(N)$$

$$\times \left[K + S_0 \, (1 + r)^N \, \Phi \left(\frac{\ln \left(S_0/K \right) + N \left(\delta + \sigma^2/2 \right)}{\sigma \sqrt{N}} \right) \right.$$

$$\left. - K \, \Phi \left(\frac{\ln \left(S_0/K \right) + N \left(\delta - \sigma^2/2 \right)}{\sigma \sqrt{N}} \right) \right].$$

Also

$$\gamma_t^H = \frac{(l_x - N_t^x)\, B_0^{-1}\, \nu^N\, p_{x+t-1}(N - t + 1)}{S_{t-1}^*\left(\exp\{\sigma^2\} - 1\right)}$$

$$\times \left\{ S_{t-1}^* B_0 (1+r)^N \left[\Phi\left(\frac{\ln\left(S_{t-1}/K\right) + \sigma^2 + (N - t + 1)\left(\delta + \sigma^2/2\right)}{\sigma\sqrt{N - t + 1}} \right) \right.\right.$$

$$\left. - \Phi\left(\frac{\ln\left(S_{t-1}/K\right) + (N - t + 1)\left(\delta + \sigma^2/2\right)}{\sigma\sqrt{N - t + 1}} \right) \right]$$

$$+ K \left[\Phi\left(\frac{\ln\left(S_{t-1}/K\right) + (N - t + 1)\left(\delta - \sigma^2/2\right)}{\sigma\sqrt{N - t + 1}} \right) \right.$$

$$\left.\left. - \Phi\left(\frac{\ln\left(S_{t-1}/K\right) + \sigma^2 + (N - t + 1)\left(\delta - \sigma^2/2\right)}{\sigma\sqrt{N - t + 1}} \right) \right] \right\},$$

and

$$\beta_t^H = V_t^\pi - \gamma_t^H S_t^*, \qquad t = 1, 2, \dots, N.$$

Worked Example 7.5 *A one-step model (7.2) with* $l_x = 2$, $N = 1$ *and*

$B_0 = 100$, $S_0 = 100$, $K = 100$, $r = 0.01$, $\mu = 5$, $\sigma = 0.5$, $p_x(1) = 0.999996$.

Solution The contingent claim is

$$H = \frac{\max\{S_1, K\} I_{\{\omega:\, T_1 > 1\}}}{B_1} + \frac{\max\{S_1, K\} I_{\{\omega:\, T_2 > 1\}}}{B_1}.$$

We have

$$\delta \approx 0.00995, \qquad \nu = \frac{100}{101},$$

and

$$V_0^\pi \approx 2.383, \qquad \gamma_1^H = 1.245.$$

Note that, since $\Phi(\infty) = 1$ and $\Phi(-\infty) = 0$, we obtain

$$V_1^\pi = (2 - N_1^x)\frac{1}{B_1}\max\{S_1, K\}.$$

Here $\max\{S_1, K\}$ is the amount of an individual payment, $(2 - N_1^x)$ is the number of survivors, B_1 is the discounting factor, and $\beta_1^H = V_1^\pi - \gamma_1^H S_1^*$.

Note that, since sequence γ^H is predictable, then value of γ_1^H is chosen at time 0, that is, when the value of S_1 is unknown. The value of β_1^H depends on S_1, and therefore, it is a random variable. □

Now we consider a pure endowment insurance contract in the framework of a continuous Black-Scholes model of a (B, S)-market (4.5). Recall that dynamics of asset S are described the following stochastic differential equation

$$dS_t = S_t \left(\mu \, dt + \sigma \, dW_t \right),$$

and for a bank account B, we have

$$dB_t = r \, B_t \, dt, \quad B_0 = 1, \ t \leq T.$$

As in the case of a binomial model, we assume that the Black-Scholes model of a (B, S)-market is defined on a stochastic basis $(\Omega_1, \mathcal{F}_T^1, \mathbb{F}_1, P_1)$, and random variables $T_1(x), T_2(x), \ldots, T_{l_x}(x)$ are defined on a stochastic basis $(\Omega_2, \mathcal{F}_N^2, \mathbb{F}_2, P_2)$.

Then, on the stochastic basis

$$\left(\Omega_1 \otimes \Omega_2, \mathcal{F}_N^1 \otimes \mathcal{F}_N^2, \mathbb{F}_1 \otimes \mathbb{F}_2, P_1 \otimes P_2 \right),$$

we consider a pure endowment insurance contract with the discounted total amount of claims

$$\sum_{k=1}^{l_x} \frac{\max\{S_T, K\}}{B_T} I_{\{\omega: \, T_k(x) > T\}}.$$

To calculate premium $U_x(T)$, we compute the average of the latter sum with respect to probability $P^* = P_1^* \otimes P_2$:

$$
\begin{aligned}
U_x(T) &= \frac{1}{l_x} E^* \left(\sum_{k=1}^{l_x} \frac{\max\{S_T, K\}}{B_T} I_{\{\omega: \, T_k(x) > T\}} \right) \qquad (7.3) \\
&= p_x(T) \, K \, e^{-rT} + p_x(T) \left[S_0 \, \Phi\left(d_+(0)\right) - K \, e^{-rT} \, \Phi\left(d_-(0)\right) \right],
\end{aligned}
$$

where

$$d_\pm(t) = \frac{\ln\left(S_t/K\right) + (T - t)\left(r \pm \sigma^2/2\right)}{\sigma \sqrt{T - t}},$$

and $p_x(T)$ is the probability that an individual of age x survives to age $x + T$.

Formula (7.3) for premium $U_x(T)$ has the following obvious interpretation that is based on the structure of the payment $\max\{S_T, K\} = K + (S_T - K)^+$. The first term $p_x(T) \, K \, e^{-rT}$ reflects the obligation to pay the guaranteed amount K. Clearly, K is discounted and multiplied by the survival function. The second term takes into account both the risk of surviving and the market risk related to the payment of amount $(S_T - K)^+$. The second risk component is estimated in terms of the price of a European call option. Hence, the Black-Scholes formula is naturally used for calculating $U_x(T)$.

Remark 7.2

1. The discrete Gaussian model of a market gives the same results as hedging in mean square in the Black-Scholes model if the discrete time $t \leq N$ is replaced with the continuous time $t \leq T$.

2. In practice, the premium sometimes is not paid as a lump sum at time 0 but is arranged as a periodic payment. In this case, it is natural to characterize the premium in terms of its density $p(t)$, which can be found from the following equivalence relation:

$$U_x(T) = \int_0^T p(t) e^{-rt} p_x(t) \, dt.$$

3. The notion of *reserve* V_t is an important ingredient of actuarial mathematics. The reserve at time t is defined as the difference between the value of future claims and the value of future premiums:

$$V_t = p_{x+t}(T-t) E^* \left(\max\{S_T, K\} e^{-rt} | \mathcal{F}_t^1 \right)$$
$$- \int_t^T p(u) e^{-r(u-t)} p_{x+t}(u-t) \, du.$$

Assuming that

$$p_x(t) = P_2(\{\omega : T_k(x) > t\}) = \exp\left\{ -\int_0^t \mu_{x+\tau} \, d\tau \right\},$$

and using the Kolmogorov-Itô formula, we obtain the following equation for $V_t \equiv V(t, s)$:

$$\frac{\partial V}{\partial t}(t, s) = p(t) + \left(\mu_{x+t} + r\right) V(t, s) - \frac{1}{2} \sigma^2 s^2 \frac{\partial^2 V}{\partial s^2}(t, s) - r s \frac{\partial V}{\partial s}(t, s). \tag{7.4}$$

Note that since V is a function of the price of asset S, then naturally the latter equation is a generalization of the Black-Scholes equation. Otherwise, it reduces to the well-known Thiele's equation

$$\frac{\partial V}{\partial t} = p(t) + \left(\mu_{x+t} + r\right) V.$$

However, if insurance characteristics $p(t) = \mu_{x+t} = 0$, then equation (7.4) reduces to the Black-Scholes equation. Thus, we can summarize that equation (7.4) for the reserve reflects the presence of both insurance risk and financial risk.

We now consider again a *pure endowment insurance* contract with price given by (7.3). In Section 4.1, we showed that a contingent claim $\max\{S_T, K\}$ can be perfectly hedged. The same hedge can be considered for a mixed claim

$$\max\{S_T, K\} \cdot I_{\{\omega:T(x)>T\}},$$

where random variable $T(x)$ is distributed as random variables $T_k(x)$, $k = 1,\ldots,l_x$, but the corresponding premium $U_x(T)$ is insufficient for perfect hedging since

$$
\begin{aligned}
U_x(T) &= p_x(T)\, E_1^*\big(\max\{S_T, K\}\, e^{-rT}\big) \tag{7.5}\\
&= x_0 < E_1^*\big(\max\{S_T, K\}\, e^{-rT}\big),
\end{aligned}
$$

which can be interpreted as a *budget constraint*. In order to minimize risk associated with such contract, we will use the methodology of *quantile hedging* of Section 4.3. For simplicity, we let $l_x = 1$. Rewriting (7.3) accordingly, we obtain

$$p_x(T)\, E_1^*\left(\frac{(S_T - K)^+}{B_T}\right) = U_x(T) - p_x(T)\,\frac{K}{B_T}. \tag{7.6}$$

Relation (7.6) implies that instead of the initial contract with the budget constraint of type (7.5), we can study an *embedded call option*. Applying the quantile hedging methodology, we can construct the maximal set A^* of successful hedging for the call option $(S_T - K)^+$ with the initial capital $p_x(T)\, E_1^*\big((S_T - K)^+/B_T\big)$. Using the definition of a perfect hedge and Theorem 4.1, we conclude that

$$p_x(T) = \frac{E_1^*\big((S_T - K)^+\, I_{A^*}\big)}{E_1^*\big((S_T - K)^+\big)}. \tag{7.7}$$

Suppose $r = 0$. As in Section 4.2, we consider the following two cases in our further analysis.

Case 1: $\mu \le \sigma^2$. The set $A^* = \{W_T^* < b\}$, where b is a constant and $W_t^* = W_t + \mu t/\sigma$ is a Wiener process under probability P_1^*.

If $p_x(T)$ is known, then we can use equation (7.7) to find b. Using the Black-Scholes formula and the expression for F_T^* from Section 4.3, we rewrite (7.7) in terms of the initial parameters of the model:

$$
\begin{aligned}
p_x(T) &= \frac{S_0\,\Phi\big(y_+(0)\big) - K\,\Phi\big(y_-(0)\big) - S_0\,\Phi\left(\frac{-b+\sigma T}{\sqrt{T}}\right) + K\,\Phi\left(\frac{-b}{\sqrt{T}}\right)}{S_0\,\Phi\big(y_+(0)\big) - K\,\Phi\big(y_-(0)\big)} \\[2mm]
&\qquad\qquad\qquad\qquad\qquad\qquad\qquad\qquad\qquad\qquad\qquad\qquad\qquad (7.8) \\[1mm]
&= 1 - \frac{S_0\,\Phi\left(\frac{-b+\sigma T}{\sqrt{T}}\right) - K\,\Phi\left(\frac{-b}{\sqrt{T}}\right)}{S_0\,\Phi\big(y_+(0)\big) - K\,\Phi\big(y_-(0)\big)},
\end{aligned}
$$

where $y_\pm$ are defined in (4.12). Determining the actuarial parameter $p_x(T)$ from an appropriate life table, we can now construct the corresponding set

A^* of successful hedging. As in Theorem 4.1, the described above procedure gives a solution to a risk minimization problem that corresponds to our equity-linked life insurance contract. However, an insurance company might tolerate a certain *risk level* $\varepsilon \in (0,1)$ such that

$$1 - \varepsilon = P_1(A^*).$$

Since $A^* = \{W_T^* < b\}$, we compute its probability

$$P_1(A^*) = \Phi\left(\frac{b - \mu T/\sigma}{\sqrt{T}}\right)$$

and therefore

$$b = \sqrt{T}\,\Phi^{-1}(1 - \varepsilon) + \frac{\mu}{\sigma}T. \tag{7.9}$$

Note that equation (7.7) can be used for solving a reverse problem. It determines the actuarial parameter $p_x(T)$ that corresponds to a given level ε of shortfall probability. Then using life tables we can find an appropriate age x of insured and an appropriate term T for the corresponding contract.

Case 2: $\mu > \sigma^2$. The set

$$A^* = \{W_T^* < b_1\} \cup \{W_T^* > b_2\},$$

where constants $b_1 < b_2$ can be determined in a way similar to the previous case. The corresponding characteristic equation has the form

$$p_x(T) = 1 - \frac{S_0\left[\Phi\left(\frac{-b_1 + \sigma T}{\sqrt{T}}\right) - \Phi\left(\frac{-b_2 + \sigma T}{\sqrt{T}}\right)\right] - K\left[\Phi\left(\frac{-b_1}{\sqrt{T}}\right) - \Phi\left(\frac{-b_2}{\sqrt{T}}\right)\right]}{S_0\,\Phi(y_+(0)) - K\,\Phi(y_-(0))}.$$

As above, this equation allows us to solve a risk minimization problem that corresponds to our pure endowment life insurance contract with guarantee.

Worked Example 7.6 *Consider a Black-Scholes market and an equity-linked life insurance contract with the following parameters*

$$\mu = 0.08; \ \sigma = 0.3; \ S_0 = 100; \ K = 110; \ T = 1 \text{ and } 5 \text{ (years)}.$$

Conduct the actuarial analysis of risk associated with this contact given the risk level $\varepsilon = 0.01$.

Solution Applying (7.7)–(7.9), we obtain the following survival probabilities

$$p_x(1) = 0.930095 \quad \text{and} \quad p_x(5) = 0.9551055.$$

Using the *Illustrative Life Tables* from the standard actuarial science text by Bowers et al. (1997), we find the suitable age restrictions for these contracts: $x \geq 78$ and $x \geq 53$ (years), respectively.

If the volatility in our model is increased to $\sigma = 0.4$, then the corresponding survival probabilities are

$$p_x(1) = 0.914122 \quad \text{and} \quad p_x(5) = 0.913937 \,,$$

and the corresponding age restrictions for insured are $x \geq 81$ and $x \geq 61$ (years), respectively. Thus, an insurance company has to compensate for such financial risks. One of the ways of reducing the insurance risk in practice involves selling such contracts also to clients with smaller survival probabilities.
□

Chapter 8

Solvency Problem for an Insurance Company: Discrete and Continuous Time Models

8.1 Ruin probability as a measure of solvency of an insurance company

Consider a collective risk model with a *binomial* process $N(t)$ representing the total number of claims up to time t:

$$N(0) = 0, \quad N(t) = \xi_1 + \ldots + \xi_t, \ t = 1, 2, \ldots,$$

where $(\xi_i)_{i=1,2,\ldots}$ is a sequence of independent Bernoulli random variables such that

$$P(\{\omega : \xi_i = 1\}) = q \quad \text{and} \quad P(\{\omega : \xi_i = 0\}) = 1 - q.$$

Sequence of independent identically distributed random variables $(X_i)_{i=1,2,\ldots}$ with values in the set of all natural numbers $\mathbb{N}$, represents the amounts of claims. Denote

$$f_n = P(\{\omega : X_i = n\}), \qquad \tilde{f}(z) = \sum_{n=1}^{\infty} f_n z^n, \qquad \text{and} \qquad \mu = E(X_i)$$

the distribution, the generating function, and the expectation of $(X_i)_{i=1,2,\ldots}$, respectively.

Assuming that sequences $(X_i)_{i=1,2,\ldots}$ and $(\xi_i)_{i=1,2,\ldots}$ are independent, let

$$X(k) = X_1 \xi_1 + \ldots + X_k \xi_k$$

and

$$g_n(k) = P(\{\omega : X(k) = n\}), \quad n = 0, 1, 2, \ldots.$$

Then the sum

$$G_n(k) = \sum_{m=0}^{n} g_n(k), \qquad n = 1, 2, \ldots$$

is the distribution function of $X(k)$, the sum of independent identically distributed random variables $X_l \, \xi_l$, $l = 1, \ldots, k$, with generating functions

$$\sum_{i=0}^{\infty} P(\{\omega : \ X_l \, \xi_l = i\}) \, z^i \ = \ 1 - q + q \sum_{i=1}^{\infty} P(\{\omega : \ X_l = i\}) \, z^i$$

$$= \ 1 - q + q \, \tilde{f}(z) \, .$$

Therefore

$$\tilde{g}(z, k) = \big[1 - q + q \, \tilde{f}(z)\big]^k \, .$$

is the generating function of $X(k)$.

Consider a stochastic sequence

$$R(k) = x + k - X(k), \quad k = 1, 2, \ldots, \quad R(0) = x \in \{0, 1, 2, \ldots\} \, ,$$

which represents the capital of an insurance company, where x is the initial capital, k is premium income (i.e., at each time $k = 1, 2, \ldots$, the company receives the premium of 1). This model is referred to as a *compound binomial model*.

Functions

$$\phi(x, k) = P(\{\omega : \ R(j) \geq 0, \ j = 0, 1, \ldots, k\}) \quad \text{and} \quad \phi(x) = \lim_{k \to \infty} \phi(x, k)$$

are called the *probabilities of non-ruin* (*probabilities of solvency*) on a finite interval $[0, k]$ and infinite interval $[0, \infty)$, respectively.

Clearly, knowing the analytical expressions for this functions, one can estimate the solvency of the company.

To find an expression for $\phi(x, k)$ we assume that the initial capital is $x - 1$, where $x \geq 1$. We also accept that the probability of solvency of a company with negative initial capital is equal to zero. Then, for any integers k and x, we have the following recurrence relation:

$$\phi(x - 1, k) \ = \ E\big(\phi(R(1), k - 1)\big)$$

$$= \ (1 - q) \, \phi(x, k - 1) + q \sum_{y=1}^{x} \phi(x - y, k - 1) \, f_y \, .$$

Further, using the technique of generating functions, we obtain the following expression for the probability of solvency of a company with zero initial capital (for details, see "Mathematical appendix 1" below):

$$\phi(0, k) = \frac{\sum_{m=0}^{k} (k + 1 - m) \, g_m(k + 1)}{(1 - q) \, (k + 1)}, \quad k = 0, 1, \ldots \, .$$

If the initial capital $x = 1, 2, \ldots$, then we have

$$\phi(x, k) = G_{x+k}(k) - (1 - q) \sum_{m=0}^{k-1} \phi(0, k - 1 - m) \, g_{x+m+1}(m) \, ,$$

for $k = 1, 2, \ldots$ (see for details "Mathematical appendix 2").

In the case of the infinite time interval $[0, \infty)$, we use the following formula from "Mathematical appendix 3"

$$\phi(0, k) = \frac{1 - q\,\mu}{1 - q} + \frac{\sum_{m=k+1}^{\infty} \left(1 - G_m(k + 1)\right)}{(1 - q)(k + 1)}.$$

Taking limit as $k \to \infty$, we obtain (see for details "Mathematical appendix 3")

$$\phi(0) = \frac{1 - q\,\mu}{1 - q}.$$

Now we establish a relation between the initial capital and probabilistic characteristics of claims, which guarantees the solvency of an insurance company over the infinite period of time with the probability that corresponds to a chosen (fixed) level of risk $\varepsilon > 0$:

$$\phi(0) \geq 1 - \varepsilon.$$

This implies

$$\mu \leq 1 - \varepsilon + \frac{\varepsilon}{q}.$$

The case when the initial capital is greater than zero is illustrated by the following example.

Worked Example 8.1 *Consider function*

$$\tilde{\phi}(z) = \sum_{x=0}^{\infty} \phi(x)\, z^x.$$

Let $X_i \equiv 2$, then $f(z) = z^2$ and $\mu = 2$.

Find values of the initial capital that guarantee that the probability of insolvency is less than the chosen level of risk.

Solution Note that function $\tilde{\phi}(z)$ can be written in the form (see "Mathematical appendix 4"):

$$\tilde{\phi}(z) = \frac{1}{1 - z}\, \frac{1 - q\,\mu}{1 - q\,\mu\,\tilde{b}(z)},$$

where

$$\tilde{b}(z) = \frac{\tilde{g}(z, 1) - 1}{q\,\mu\,(z - 1)} = \frac{\tilde{f}(z) - 1}{\mu\,(z - 1)}.$$

In our case $\tilde{b}(z) = (1 + z)/2$, hence

$$\tilde{\phi}(z) = \frac{1}{1 - z}\, \frac{q\,(1 - q)^{-1}}{1 - q\,z\,(1 - q)^{-1}}.$$

Expanding $\tilde{\phi}(z)$ in powers of Z, we obtain

$$\phi(x) = 1 - \left(\frac{q}{1-q}\right)^{x+1}.$$

Positivity of the security loading coefficient $1 - q\,\mu > 0$ implies

$$q < \frac{1}{2} \quad \text{and} \quad \frac{q}{1-q} < 1.$$

The following table and Figure 8.1 give probabilities of ruin for four different values of q with accuracy 0.0001.

Initial capital	$q = 0.1$	$q = 0.2$	$q = 0.4$	$q = 0.49$
0	0.1111	0.25	0.6667	0.9608
1	0.0123	0.0625	0.4444	0.9231
2	0.0014	0.0156	0.2963	0.8869
3	0.0002	0.0039	0.1975	0.8521
4	0	0.0010	0.1317	0.8187
5	0	0.0002	0.0878	0.7866
6	0	0.0001	0.0585	0.7558
7	0	0	0.0390	0.7261
8	0	0	0.0260	0.6976
9	0	0	0.0173	0.6703
10	0	0	0.0116	0.6440

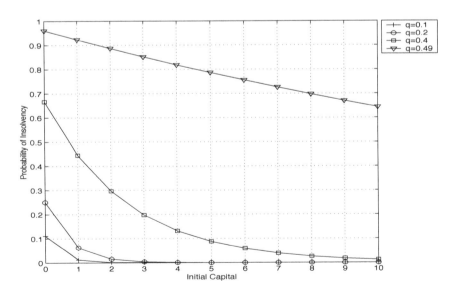

FIGURE 8.1: Probabilities of ruin for four different values of q.

For a given level of risk ε, we solve the following inequality for x:

$$\phi(x) > 1 - \varepsilon.$$

We have

$$x > \frac{\ln(\varepsilon)}{\ln\left(q/(1-q)\right)} - 1.$$

The next table and Figure 8.2 give the minimal values of the initial capital x for various values of q and ε.

q	$\varepsilon = 0.05$	$\varepsilon = 0.03$	$\varepsilon = 0.01$
0.05	1	1	1
0.1	1	1	2
0.15	1	2	2
0.2	2	2	3
0.25	2	3	4
0.3	3	4	5
0.35	4	5	7
0.4	7	8	11
0.45	14	17	22

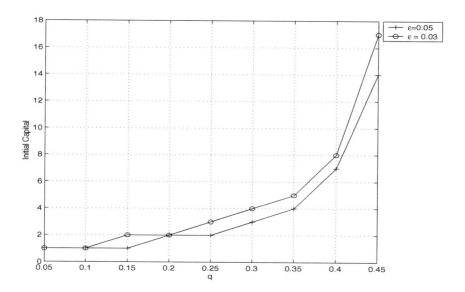

FIGURE 8.2: Minimal values of the initial capital.

□

Let $(\Omega, \mathcal{F}, P)$ be a probability space. Consider a Poisson process $N(t)$, $t \geq 0$, with intensity λ: $N(0) = 0$, $E(N(t)) = \lambda t$, and an independent of N sequence $(X_i)_{i=1,2,...}$ of independent identically distributed random variables with expectation μ and distribution function $F(x)$, $F(0) = 0$.

The claims flow (number of claims received up to time t) is represented by the Poisson process $N(t)$. The amounts of these claims are represented by sequence $(X_i)_{i=1,2,...}$. The premium income of an insurance company is given by $\Pi(t) = ct$, where c is a constant. If x is the initial capital of the company, then the dynamics of the company's capital are given by the *Cramér-Lundberg model*:

$$R(t) = x + ct - \sum_{i=1}^{N(t)} X_i \, .$$

Since $N(t)$ and $(X_i)_{i=1,2,...}$ are independent, then the expectation of the risk process $X(t) = \sum_{i=1}^{N(t)} X_i$ is $E(X(t)) = \lambda t \mu$. Setting the security loading coefficient at

$$\theta = \frac{\Pi(t)}{E(X(t))} - 1 = \frac{c - \lambda \mu}{\lambda \mu},$$

we obtain $c = (1 + \theta) \lambda \mu$.

Now we compute the probability of solvency

$$\phi(x) = P\big(\{\omega : \ R(t) \geq 0, \ R(0) = x, \ t \geq 0\}\big) \, .$$

First, we investigate smoothness of function $\phi(x)$ assuming that the distribution function $F(x)$ has density $f(x)$. Since ruin cannot occur prior time T_1, when the Poisson process N has its first jump, then we can write

$$
\begin{aligned}
\phi(x) &= E\big(\phi(x + cT_1 - x_1)\big) \\
&= \int_0^\infty \lambda e^{-\lambda s} \int_0^{x+cs} \phi(x + cs - y) f(y) \, dy \, ds \, .
\end{aligned}
$$

Making a substitution $q = x - y$, we can rewrite the latter equality in the form

$$\phi(x) = \int_0^\infty \lambda e^{-\lambda s} \int_{-cs}^{x} \phi(q + cs) f(x - q) \, dq \, ds \, .$$

Thus, if $F(y) \in C^n[0, \infty)$, then $\phi(x) \in C^{n-1}[0, \infty)$. In our further discussion, we assume $F(y) \in C^3[0, \infty)$.

Using properties of the Poisson process and the formula for total probability, we obtain

$$
\begin{aligned}
\phi(x) &= \phi(x + c\,\Delta t) \left[1 - \lambda\,\Delta t + o(\Delta t)\right] \\
&\quad + \lambda\,\Delta t \int_0^{x+c\Delta t} \phi(x + c\,\Delta t - y) \, dF(y) + o(\Delta t) \, .
\end{aligned}
$$

By Taylor's formula, we also have

$$
\begin{aligned}
\phi(x) &= \left[\phi(x) + c\,\phi'(x)\,\Delta t\right] \left[1 - \lambda\,\Delta t + o(\Delta t)\right] \\
&\quad + \lambda\,\Delta t \int_0^{x+c\Delta t} \phi(x + c\,\Delta t - y) \, dF(y) + o(\Delta t) \, ,
\end{aligned}
$$

hence

$$\phi(x)\left[\lambda\,\Delta t + o(\Delta t)\right] = c\,\phi'(x)\,\Delta t\left[1 - \lambda\,\Delta t + o(\Delta t)\right]$$
$$+\lambda\,\Delta t\int_0^{x+c\Delta t}\phi(x + c\,\Delta t - y)\,dF(y) + o(\Delta t)\,.$$

Dividing the latter equality by Δt and taking limits as $\Delta t \to 0$, we obtain

$$\phi(x)\,\lambda = c\,\phi'(x) + \lambda\int_0^x \phi(x - y)\,dF(y)\,. \tag{8.1}$$

In the case of an exponential distribution function F, it is not difficult to find an explicit solution of this equation. Indeed, if $F(y) = 1 - e^{-x/\mu}$, then equation (8.1) is reduced to

$$\phi(x)\,\lambda = c\,\phi'(x) + \lambda\int_0^x \phi(x - y)\,\frac{1}{\mu}\,e^{-y/\mu}\,dy\,.$$

Differentiating and integrating by parts, we obtain

$$\begin{aligned}
\lambda\,\phi'(x) &= c\,\phi''(x) + \frac{\lambda}{\mu}\,\phi(0) + \lambda\int_0^x \phi'_x(x - y)\,\frac{1}{\mu}\,e^{-y/\mu}\,dy\\[2mm]
&= c\,\phi''(x) + \frac{\lambda}{\mu}\,\phi(0) - \lambda\int_0^x \frac{1}{\mu}\,e^{-y/\mu}\,d\phi(x - y)\\[2mm]
&= c\,\phi''(x) + \frac{\lambda}{\mu}\,\phi(x) + \lambda\int_0^x \phi(x - y)\,\frac{1}{\mu}\,de^{-y/\mu}\\[2mm]
&= c\,\phi''(x) + \frac{\lambda}{\mu}\,\phi(x) - \frac{1}{\mu}\left[\lambda\int_0^x \phi(x - y)\,\frac{1}{\mu}\,e^{-y/\mu}\,dy\right]\\[2mm]
&= c\,\phi''(x) + \frac{\lambda}{\mu}\,\phi(x) + \frac{c}{\mu}\,\phi'(x) - \frac{\lambda}{\mu}\,\phi(x)\\[2mm]
&= c\,\phi''(x) + \frac{c}{\mu}\,\phi'(x)\,.
\end{aligned}$$

Thus, we arrive at the following differential equation:

$$\phi''(x) + \phi'(x)\left[\frac{1}{\mu} - \frac{\lambda}{c}\right] = 0\,.$$

Its general solution is of the form

$$\phi(x) = B + A\,\exp\left\{x\left[\frac{\lambda}{c} - \frac{1}{\mu}\right]\right\},$$

where A and B are some constants. The inequality

$$\frac{\lambda}{c} < \frac{1}{\mu}$$

can be written as
$$\lambda \mu - c < 0 \,,$$

which reflects the positivity of θ, and therefore $\phi(\infty) = B$.

Unknown constants A and B can be found from the following relations

1. $\phi(\infty) = 1$;

2. Substituting $x = 0$ into equation (8.1):

$$\phi(x)\,\lambda = c\,\phi'(x) + \lambda \int_0^x \phi(x-y)\,dF(y)$$

implies

$$\phi(0)\,\lambda = c\,\phi'(0)\,.$$

Thus, we arrive at the following expression:

$$\phi(x) = 1 - \frac{\lambda\mu}{c}\exp\left\{ x\left[\frac{\lambda}{c} - \frac{1}{\mu}\right]\right\} = 1 - \frac{1}{1+\theta}\exp\left\{ -\frac{\theta\,x}{(1+\theta)\,\mu}\right\}.$$

In general, for an arbitrary distribution function F, it may be difficult to find an explicit expression for ϕ. In this case, one can look for various estimates of $\psi(x) = 1 - \phi(x)$, the probability of ruin. The main result here is usually referred to as the *Cramér-Lundberg inequality*:

$$\psi(x) \le e^{-Rx}\,, \tag{8.2}$$

where R is a positive solution to the equation

$$\lambda + cr = \lambda \int_0^\infty e^{rx}\,dF(x)\,.$$

Note that so far we have dealt with the classical insurance models, where one does not take into account the investment strategies of an insurance company.

Mathematical appendix 1

Consider equation

$$\phi(x-1,k) \;=\; E\big(\phi(X_1,k-1)\big) \tag{8.3}$$

$$=\; (1-q)\,\phi(x,k-1) + q\sum_{y=1}^x \phi(x-y,k-1)\,f_y\,,$$

and

$$\tilde{\phi}_1(z,k) = \sum_{x=0}^\infty \phi(x,k)\,z^x\,,$$

the generating function of $\left(\phi(x, k)\right)_{x=0}^{\infty}$.

Multiplying equation (8.3) by z^x, and summing in x from 1 to ∞, we obtain

$$z \tilde{\phi}_1(z, k) = (1 - q) \left[\tilde{\phi}_1(z, k - 1) - \phi(0, k - 1)\right]$$
$$+ q \tilde{\phi}_1(z, k - 1) \tilde{f}(z)$$

or

$$z \tilde{\phi}_1(z, k) = \tilde{g}(z, 1) \tilde{\phi}_1(z, k - 1) - (1 - q) \phi(0, k - 1). \qquad (8.4)$$

Introduce two auxiliary functions:

$$\tilde{\phi}(z, t) = \sum_{k=0}^{\infty} \tilde{\phi}(z, k) t^k = \sum_{k=0}^{\infty} \sum_{x=0}^{\infty} \phi(z, k) z^x t^k$$

and

$$\tilde{\phi}_0(t) = \sum_{k=0}^{\infty} \phi(0, k) t^k . \qquad (8.5)$$

Multiplying equation (8.4) by t^k, and summing in k from 1 to ∞, we obtain

$$z \tilde{\phi}(z, k) - z \tilde{\phi}_1(z, 0) = t \tilde{g}(z, 1) \tilde{\phi}(z, k) - t (1 - q) \tilde{\phi}_0(t). \qquad (8.6)$$

From the definition of function ϕ, we have that $\phi(x, 0) = 1$ for all $x = 0, 1, 2, \ldots$. Hence

$$\tilde{\phi}_1(z, 0) = \frac{1}{1 - z} \qquad \text{for} \quad |z| < 1.$$

Then equation (8.6) can be written in the form

$$\tilde{\phi}(z, k) \left[z - t \tilde{g}(z, 1)\right] = \frac{z}{1 - z} - t (1 - q) \tilde{\phi}_0(t). \qquad (8.7)$$

Fix t with $|t| < 1$. Consider function

$$F(z) := z - t \tilde{g}(z, 1),$$

then

$$F(0) = 0 \qquad \text{and} \qquad F(1) = 1 - t \left[1 - q + q \sum_{n=1}^{\infty} f_n\right] = 1 - t > 0.$$

Also

$$F'(z) = 1 - t q \sum_{n=1}^{\infty} n f_n z^{n-1} > 1 - q \sum_{n=1}^{\infty} n f_n z^{n-1} > 1 - q \mu > 0.$$

The inequality $1 - q \mu > 0$ is equivalent to positivity of the security loading coefficient, and we assume that it is the case.

Thus, for each fixed t with $|t| < 1$, the equation

$$z = t\, \tilde{g}(z, 1) \tag{8.8}$$

has a unique root $z = z(t) \in (0, 1)$. Therefore, function $z(t)$, $|t| < 1$, is a solution to (8.8).

Now, for any analytic function h with $h(0) = 0$, we have

$$h\big(z(t)\big) = \sum_{n=1}^{\infty} \frac{t^n}{n!} \frac{d^{n-1}}{ds^{n-1}} \Big[h'(s)\,\big(\tilde{g}(s,1)\big)^n\Big]\Big|_{s=0}, \tag{8.9}$$

where $z(t)$ is a solution of (8.8). Note that $\big(\tilde{g}(s,1)\big)^n = \tilde{g}(s,n)$.

If $h(z) = z$, then the solution to (8.8) has the form

$$z(t) = \sum_{n=1}^{\infty} \frac{t^n}{n!}\, g(n-1, n),$$

where

$$g(n-1, n) = \frac{1}{(n-1)!} \frac{d^{n-1}}{ds^{n-1}}\, \tilde{g}(s,n)\Big|_{s=0}.$$

Substituting $h(z) = z/(1-z)$ into (8.9), we obtain

$$\frac{z(t)}{1-z(t)} = \sum_{n=1}^{\infty} \frac{t^n}{n!} \frac{d^{n-1}}{ds^{n-1}} \frac{\tilde{g}(s,n)}{(1-s)^2}\Big|_{s=0}. \tag{8.10}$$

For s with $|s| < 1$, we have

$$\frac{\tilde{g}(s,n)}{(1-s)} \equiv \frac{\sum_{k=0}^{\infty} g_k(n)\, s^k}{(1-s)}$$

$$= \Big[g_0(n) + g_1(n)\, s + g_2(n)\, s^2 + \ldots\Big] \times \Big[1 + s + s^2 + \ldots\Big]$$

$$= g_0(n) + s\,\big[g_1(n) + g_0(n)\big] + s^2\,\big[g_2(n) + g_1(n) + g_0(n)\big] + \ldots$$

$$+ s^k\,\big[g_k(n) + g_{k-1}(n) + \ldots + g_0(n)\big] + \ldots,$$

so the coefficient in front of s^k is

$$\sum_{m=0}^{k} g_m(n) = G_k(n).$$

Similarly, for s with $|s| < 1$, we obtain that the coefficient in front of s^k in the expansion

$$\frac{\tilde{g}(s,n)}{(1-s)^2} \equiv \frac{\tilde{g}(s,n)\,(1-s)^{-1}}{(1-s)}$$

is equal $\sum_{m=0}^{k} G_m(n)$.

Thus, we can write (8.10) in the form

$$\frac{z(t)}{1 - z(t)} = \sum_{n=1}^{\infty} \left[\sum_{m=0}^{n-1} G_m(n) \right] \frac{t^n}{n!}. \tag{8.11}$$

If we substitute $z = z(t)$ in (8.7), then the left-hand side of this equation vanishes, so we can find an expression for $\tilde{\phi}_0(t)$:

$$\tilde{\phi}_0(t) = \frac{z(t)}{t(1-q)(1-z(t))},$$

which in view of (8.11) becomes

$$\tilde{\phi}_0(t) = \frac{1}{1-q} \sum_{k=0}^{\infty} \frac{t^k}{k+1} \left[\sum_{m=0}^{k} G_m(k+1) \right].$$

Since representation of $\tilde{\phi}_0(t)$ in form (8.5) is unique, then

$$\phi_0(0, k) = \frac{\sum_{m=0}^{k} G_m(k+1)}{(1-q)(k+1)}, \qquad k = 0, 1, \ldots. \tag{8.12}$$

Finally, taking into account that

$$\sum_{m=0}^{k} G_m(k+1) = \sum_{m=0}^{k} (k+1-m) g_m(k+1),$$

we write

$$\phi_0(0, k) = \frac{\sum_{m=0}^{k} (k+1-m) g_m(k+1)}{(1-q)(k+1)}, \qquad k = 0, 1, \ldots.$$

Mathematical appendix 2

In the case when the initial capital is greater than zero, equation (8.7) implies

$$\tilde{\phi}(z, t) = \left(\frac{1}{1-z} - \frac{1-q}{z} t \tilde{\phi}_0(t) \right) \Big/ \left(1 - t \frac{\tilde{g}(z, 1)}{z} \right). \tag{8.13}$$

To represent the right-hand side of this equality as a series in powers of t, we write

$$\left(\frac{1}{1-z} \right) \Big/ \left(1 - t \frac{\tilde{g}(z, 1)}{z} \right) = \frac{1}{1-z} \left(1 + t \frac{\tilde{g}(z, 1)}{z} + t^2 \frac{\tilde{g}^2(z, 1)}{z^2} + \ldots \right)$$

$$= \frac{1}{1-z} \left(1 + t \frac{\tilde{g}(z, 1)}{z} + t^2 \frac{\tilde{g}(z, 2)}{z^2} + \ldots + t^k \frac{\tilde{g}(z, k)}{z^k} + \ldots \right)$$

$$= \sum_{k=0}^{\infty} t^k \frac{\tilde{g}(z, k)}{(1-z) z^k},$$

and

$$\left(\frac{1-q}{z}t\tilde{\phi}_0(t)\right)\bigg/\left(1-t\frac{\tilde{g}(z,1)}{z}\right) = \sum_{k=0}^{\infty}\frac{1-q}{z}\phi_0(0,k)t^{k+1}\sum_{m=0}^{\infty}t^m\frac{\tilde{g}(z,m)}{z^m}$$

$$= \sum_{l=1}^{\infty}t^l\,a_l\,,$$

where

$$a_l = (1-q)\sum_{m=0}^{l-1}\tilde{g}(z,m)\,\phi_0(l-m-1)\,z^{-m-1}\,.$$

Substituting these in (8.13) we equate the coefficients in front of t^k, $k \geq 1$:

$$\tilde{\phi}_1(z,k) = \frac{\tilde{g}(z,k)}{(1-z)\,z^k} - (1-q)\sum_{m=0}^{k-1}\tilde{g}(z,m)\,\phi_0(k-m-1)\,z^{-m-1}$$

or

$$z^k\,\tilde{\phi}_1(z,k) = \frac{\tilde{g}(z,k)}{(1-z)} - (1-q)\sum_{m=0}^{k-1}\tilde{g}(z,m)\,\phi_0(k-m-1)\,z^{k-m-1}\,. \quad (8.14)$$

If $k = 0$, then (8.13) reduces to

$$\tilde{\phi}_1(z,0) = \sum_{x=0}^{\infty}\phi(x,0)\,z^x = \sum_{x=0}^{\infty}z^x = \frac{1}{1-z}\,.$$

Noting that

$$z^k\,\tilde{\phi}_1(z,k) = \sum_{j=0}^{\infty}\phi(j,k)\,z^{j+k}\,,$$

and

$$\frac{\tilde{g}(z,k)}{(1-z)} = \sum_{i=0}^{\infty}g_i(k)\,z^i\sum_{j=0}^{\infty}z^j$$

$$= \left[g_0(k) + g_1(k)\,z + g_2(k)\,z^2 + \ldots\right]\times\left[1 + z + z^2 + \ldots\right]$$

$$= g_0(k) + z\left[g_1(k) + g_0(k)\right] + z^2\left[g_2(k) + g_1(k) + g_0(k)\right] + \ldots$$

$$= \sum_{i=0}^{\infty}G_i(k)\,z^i\,,$$

and

$$(1-q) \sum_{m=0}^{k-1} \tilde{g}(z,m) \, \phi_0(k-m-1) \, z^{k-m-1}$$

$$= (1-q) \sum_{m=0}^{k-1} \phi(0,k-m-1) \sum_{j=0}^{\infty} z^{j+k-m-1} \, g_j(m),$$

we can rewrite (8.13) in the form

$$\sum_{j=0}^{\infty} z^{j+k} \, \phi(j,k) = \sum_{i=0}^{\infty} G_i(k) \, z^i$$

$$-(1-q) \sum_{m=0}^{k-1} \phi(0,k-m-1) \sum_{j=0}^{\infty} z^{j+k-m-1} \, g_j(m).$$

Changing summation indices to $i = j + k$ in the first sum and to $i = j + k - 1 - m$ in the last sum, we obtain

$$\sum_{i=0}^{\infty} z^i \, \phi(i-k,k) = \sum_{i=0}^{\infty} G_i(k) \, z^i$$

$$-(1-q) \sum_{m=0}^{k-1} \phi(0,k-m-1) \sum_{i=k-1-m}^{\infty} z^i \, g_{i+m+1-k}(m).$$

We rearrange the last term in the latter relation:

$$\sum_{i=0}^{\infty} z^i \, \phi(i-k,k) = \sum_{i=0}^{\infty} G_i(k) \, z^i$$

$$-(1-q) \sum_{m=0}^{k-1} \sum_{i=k-1-m}^{k-1} z^i \, \phi(0,k-m-1) \, g_{i+m+1-k}(m)$$

$$-(1-q) \sum_{m=0}^{k-1} \sum_{i=k}^{\infty} z^i \, \phi(0,k-m-1) \, g_{i+m+1-k}(m),$$

and change the order of summation:

$$\sum_{i=0}^{\infty} z^i \, \phi(i-k,k) = \sum_{i=0}^{\infty} G_i(k) \, z^i$$

$$-(1-q) \sum_{i=0}^{k-1} z^i \sum_{m=k-1-i}^{k-1} \phi(0,k-m-1) \, g_{i+m+1-k}(m)$$

$$-(1-q) \sum_{i=k}^{\infty} z^i \sum_{m=0}^{k-1} \phi(0,k-m-1) \, g_{i+m+1-k}(m).$$

Equating coefficients in front of z^i, we have

$$\phi(i-k,k) = G_i(k) - (1-q) \sum_{m=0}^{k-1} \phi(0, k-m-1) \, g_{i+m+1-k}(m),$$

for $i \geq k \geq 1$. In other words, for $x = 0, 1, \ldots$ and $k = 1, 2, \ldots$

$$\phi(x,k) = G_{x+k}(k) - (1-q) \sum_{m=0}^{k-1} \phi(0, k-m-1) \, g_{x+m+1}(m).$$

Mathematical appendix 3

Equation (8.12) implies

$$\phi(0,k)$$
$$= \frac{\sum_{m=0}^{k} \left[1 - (1 - G_m(k+1)) \right]}{(1-q)(k+1)} = \frac{k+1 - \sum_{m=0}^{k} (1 - G_m(k+1))}{(1-q)(k+1)}$$
$$= \frac{k+1 - (k+1)\,q\,\mu + \sum_{m=k+1}^{\infty} (1 - G_m(k+1))}{(1-q)(k+1)}.$$

Here we used the relation

$$\sum_{m=0}^{\infty} (1 - G_m(k+1)) = (1 - G_0(k+1)) + (1 - G_2(k+1)) + \ldots$$
$$= P(\{\omega : X(k+1) > 0\}) + P(\{\omega : X(k+1) > 1\}) + \ldots$$
$$= \sum_{j=1}^{\infty} \sum_{i=j}^{\infty} P(\{\omega : X(k+1) = i\})$$
$$= E(X(k+1)) = (k+1)\,q\,\mu.$$

This latter relation also implies the convergence of the series

$$\sum_{m=0}^{\infty} (1 - G_m(k+1))$$

since the sequence of its partial sums is monotonically increasing, and it is bounded from above by $(k+1)\,q\,\mu$.

Thus, the probability of non-ruin on $[0, k]$ has the following analytical form

$$\phi(0,k) = \frac{1 - q\,\mu}{1-q} + \frac{\sum_{m=k+1}^{\infty} (1 - G_m(k+1))}{(1-q)(k+1)}.$$

An expression for the probability of non-ruin on an infinite interval can be obtained directly from (8.1) by passing to the limit as $k \to \infty$:

$$\phi(j) = (1-q)\,\phi(j+1) + q\,E\big(\phi(j+1-X_1)\big), \quad j = 0, 1, 2, \ldots,$$

or

$$\phi(j+1) - \phi(j) = q\left[\phi(j+1) - E\big(\phi(j+1-X_1)\big)\right], \quad j = 0, 1, 2, \ldots.$$

Summing in j from 0 to $k - 1$, we obtain

$$\phi(k) - \phi(0) = q\left[\sum_{j=1}^{k}\phi(j) - E\Big(\sum_{j=1}^{k}\phi(j-X_1)\Big)\right], \quad k = 1, 2, \ldots$$

or

$$\phi(k) - (1-q)\,\phi(0) = q\left[\sum_{j=1}^{k}\phi(j) - E\Big(\sum_{j=1}^{k}\phi(j-X_1)\Big)\right], \quad k = 1, 2, \ldots. \quad (8.15)$$

Introduce function

$$1_+(j) := \begin{cases} 1, & j = 0, 1, 2, \ldots \\[2mm] 0, & j = -1, -2, \ldots \end{cases}.$$

For a pair of integer-valued functions f and g, we define their convolution:

$$(f * g)(j) := \sum_{i=-\infty}^{\infty} f(j-i)\,g(i).$$

If $f(i) = g(i) = 0$ for $i = -1, -2, \ldots$, then

$$(f * g)(j) = \sum_{i=0}^{j} f(j-i)\,g(i).$$

Now, since

$$\sum_{j=0}^{k}\phi(j) = (\phi * 1_+)(k),$$

$$\sum_{j=1}^{k}\phi(j - X_1) = \sum_{j=0}^{k}\phi(j - X_1) = (\phi * 1_+)(k - X_1),$$

then we can rewrite equation (8.15) in the form

$$\begin{aligned} \phi(k) - (1 - q)\,\phi(0) &= q\left[(\phi * 1_+)(k) - E\Big((\phi * 1_+)(k - X_1)\Big)\right] \quad (8.16)\\ &= q\left[(\phi * 1_+)(k) - (\phi * 1_+ * f)(k)\right]\\ &\qquad k = 1, 2, \ldots; f(n) = f_n. \end{aligned}$$

Since $f(0) = 0$, then (8.16) also holds for $k = 0$. Now we can extend (8.16) to all integers k:

$$\phi(k) - (1 - q)\,\phi(0)\,1_+(k) = q\left[(\phi * 1_+)(k) - (\phi * 1_+ * f)(k)\right]. \tag{8.17}$$

Introduce function

$$\delta(j) := \begin{cases} 1, & j = 0 \\ 0, & j \neq 0 \end{cases}.$$

Then (8.17) can be written in the form

$$\phi(k) * \left[\delta(k) - q\left[1_+(k) * (\delta(k) - f(k))\right]\right] = c\,1_+(k),$$

where $c = (1 - q)\,\phi(0)$.

A solution to this equation can be written in the form of the following Neumann series:

$$\phi(k) = c\sum_{n=0}^{\infty} q^n \left[(\delta(k) - f(k))^{*n} * 1_+^{*(n+k)}(k)\right],$$

where $g^{*0} = \delta$, $g^{*n} = g^{*(n-1)} * g$, $n = 1, 2, \ldots$.

If $k \to \infty$, then (8.16) gives

$$1 - (1 - q)\,\phi(0) = q\sum_{j=-\infty}^{\infty}\left[1_+(j) - (1_+ * f)(j)\right]$$

$$= q\sum_{j=0}^{\infty}\left[1 - P(\{\omega : X_1 \leq j\})\right] = q\,\mu.$$

Hence,

$$\phi(0) = \frac{1 - q\,\mu}{1 - q}.$$

Mathematical appendix 4

Introduce function

$$\tilde{\phi}(z) := \sum_{x=0}^{\infty} \phi(x)\,z^x.$$

Taking into account

$$\lim_{t \nearrow 1}(t - 1)\,\tilde{\phi}(z, t) = \lim_{t \nearrow 1}\sum_{k=0}^{\infty}\tilde{\phi}_1(z, k)\,t^k\,(1 - t)$$

$$= \lim_{t \nearrow 1}\left[\tilde{\phi}_1(z, 0)\,(1 - t) + \tilde{\phi}_1(z, 1)\,t\,(1 - t) + \ldots + \tilde{\phi}_1(z, k)\,t^k\,(1 - t) + \ldots\right]$$

$$= \lim_{t \nearrow 1}\left[\tilde{\phi}_1(z, 0) + t\left(\tilde{\phi}_1(z, 1) - \tilde{\phi}_1(z, 0)\right) + \ldots\right]$$

$$= \tilde{\phi}_1(z, \infty) \equiv \tilde{\phi}(z)$$

and equation (8.13), we obtain

$$
\begin{aligned}
\tilde{\phi}(z) &= \lim_{t \nearrow 1} (t-1)\,\tilde{\phi}(z,t) = -\frac{1-q}{z - \tilde{g}\,\tilde{\phi}(z,1)} \lim_{t \nearrow 1}(t-1)\,\tilde{\phi}_0(t) \\
&= \frac{1 - q\,\mu}{\tilde{g}(z,1) - z} = \frac{1}{1-z}\,\frac{1 - q\,\mu}{1 - q\,\mu\,\tilde{b}(z)},
\end{aligned}
$$

where

$$
b(z) = \frac{\tilde{g}(z,1) - 1}{q\,\mu\,(z-1)} = \frac{\tilde{f}(z) - 1}{\mu\,(z-1)}.
$$

Also note that

$$
\phi(x) = \frac{d^x \tilde{\phi}(z)}{dz^x} \left. \frac{\tilde{\phi}(z)}{x!} \right|_{z=0}.
$$

8.2 Solvency of an insurance company and investment portfolios

As in Chapter 2, we consider a binomial (B, S)-market. The dynamics of this market are described by equations

$$
\begin{aligned}
\Delta B_n &= r\,B_{n-1}, \quad B_0 > 0, \\
\Delta S_n &= \rho_n\,S_{n-1}, \quad S_0 > 0, \ n \le N,
\end{aligned}
$$

where $r \ge 0$ is a constant rate of interest with $-1 < a < r < b$, and profitabilities

$$
\rho_n = \begin{cases} b & \text{with probability } p \in [0,1] \\ a & \text{with probability } q = 1 - p \end{cases}, \quad n = 1, \dots, N,
$$

form a sequence of independent identically distributed random variables.

Suppose that an insurance company with the initial capital $x = R_0$ forms an investment portfolio (β_1, γ_1) at time $n = 0$, so that

$$
R_0 = \beta_1\,B_0 + \gamma_1\,S_0.
$$

At time $n = 1$, the capital of the company is

$$
R_1 = \beta_1\,B_1 + \gamma_1\,S_1 + c - Z_1,
$$

where c is the premium income and Z_1 is a non-negative random variable representing total claims payments during this time period. This capital is reinvested into portfolio (β_2, γ_2):

$$
R_1 = \beta_2\,B_1 + \gamma_2\,S_1.
$$

At any time n, we have

$$R_n = \beta_n B_n + \gamma_n S_n + c - Z_n,$$

where predictable sequence $\pi = (\beta_n, \gamma_n)_{n \geq 0}$ is an *investment strategy* and Z_n is a non-negative random variable representing total claims payments during the time step from $n - 1$ to n. The distribution function of Z_n is denoted $F_{Z_n} \equiv F_Z$. It is assumed that sequence $(Z_n)_{n \geq 0}$ of independent identically distributed random variables is also independent of the dynamics of market assets B and S.

Thus, the dynamics of the capital of the insurance company have the form

$$
\begin{aligned}
R_{n+1} &= \beta_n B_{n+1} + \gamma_n S_{n+1} + c - Z_{n+1} \\
&= R_n (1 + r) + \gamma_n S_n (\rho_{n+1} - r) + c - Z_{n+1}.
\end{aligned}
$$

As we discussed in the previous section, the probability of ruin (or insolvency)

$$P(\{\omega: \ R_n < 0 \quad \text{for some} \quad n \geq 0\})$$

is one of the typical measures used in the insurance risk management. Now we study this measure taking into account the investment strategies of an insurance company.

We start with the case when a company invests only in the non-risky asset B. In this case,

$$R_{n+1} = R_n (1 + r) + c - Z_{n+1}.$$

First, we compute the probability of ruin over the finite time interval $[0, k]$:

$$\psi_k(R_0) = P(\{\omega: \ R_n < 0 \quad \text{for some} \quad n \leq k\}).$$

Note that ψ is an increasing function of k and R_0.

The probability of ruin after one time step is given by

$$
\begin{aligned}
\psi_1(R_0) &= P(\{\omega: \ R_1 < 0\}) = P(\{\omega: \ R_0 (1 + r) + c - z_1 < 0\}) \\
&= P(\{\omega: \ z_1 > R_0 (1 + r) + c\}) = 1 - F_z(R_0(1 + r) + c).
\end{aligned}
$$

The probability of ruin after two steps is

$$\psi_2(R_0) \;=\; P\Big(\{\omega:\; R_1 < 0\} \cup \{\omega:\; R_1 > 0,\; R_2 < 0\}\Big)$$

$$=\; P(\{\omega:\; R_1 < 0\}) + P(\{\omega:\; R_1 > 0,\; R_2 < 0\})$$

$$=\; \psi_1(R_0) + \int_{\{\omega:\; R_1 > 0,\, R_2 < 0\}} dF_{Z_1}\, dF_{Z_2}$$

$$=\; \psi_1(R_0) + \int_0^{R_0(1+r)+c} \int_{R_1(1+r)+c}^{\infty} dF_{Z_2}\, dF_{Z_1}$$

$$=\; \psi_1(R_0) + \int_0^{R_0(1+r)+c} \psi_1(R_1)\, dF_{Z_1}$$

$$=\; \psi_1(R_0) + \int_0^{R_0(1+r)+c} \psi_1\big(R_0\,(1+r) + c - Z_1\big)\, dF_{Z_1}\,.$$

And after three steps:

$$\psi_3(R_0)$$

$$= P\Big(\{\omega:\; R_1 < 0\} \cup \{\omega:\; R_1 > 0,\, R_2 < 0\}$$

$$\cup\{\omega:\; R_1 > 0,\, R_2 > 0,\, R_3 < 0\}\Big)$$

$$= \psi_1(R_0) + \int_0^{\{\omega:\; R_1 > 0,\, R_2 < 0\}} dF_{Z_1}\, dF_{Z_2}$$

$$+ \int_{\{\omega:\; R_1 > 0,\, R_2 > 0,\, R_3 < 0\}} dF_{Z_1}\, dF_{Z_2}\, dF_{Z_3}$$

$$= \psi_1(R_0) + \int_0^{R_0(1+r)+c} \int_{R_1(1+r)+c}^{\infty} dF_{Z_2}\, dF_{Z_1}$$

$$+ \int_0^{R_0(1+r)+c} \int_0^{R_1(1+r)+c} \int_{R_2(1+r)+c}^{\infty} dF_{Z_3}\, dF_{Z_2}\, dF_{Z_1}$$

$$= \psi_1(R_0) + \int_0^{R_0(1+r)+c} \psi_1(R_1)\, dF_{Z_1}$$

$$+ \int_0^{R_0(1+r)+c} \int_0^{R_1(1+r)+c} \psi_1(R_2)\, dF_{Z_2}\, dF_{Z_1}$$

$$= \psi_1(R_0)$$

$$+ \int_0^{R_0(1+r)+c} \left[\psi_1(R_1) + \int_0^{R_1(1+r)+c} \psi_1\Big(R_1(1+r)+c-Z_2\Big) dF_{Z_2} \right] dF_{Z_1}$$

$$= \psi_1(R_0) + \int_0^{R_0(1+r)+c} \psi_2(R_1)\, dF_{Z_1}$$

$$= \psi_1(R_0) + \int_0^{R_0(1+r)+c} \psi_2(R_0(1+r)+c-Z_1)\, dF_{Z_1}\,.$$

Using mathematical induction, we obtain that the probability of ruin after $k+1$ steps is

$$\psi_{k+1}(R_0) = 1 - F_Z\big(R_0(1+r)+c\big) + \int_0^{R_0(1+r)+c} \psi_k(R_0(1+r)+c-y)\, dF_y\,,$$

with

$$\psi_1(R_0) = 1 - F_Z\big(R_0(1+r)+c\big)\,.$$

The *probability of solvency* over the time period $[0, n]$ is

$$\phi_n(x) = P\big(\{\omega:\ R_1 > 0,\ R_2 > 0, \ldots, R_n > 0,\}\big)\,.$$

For illustration, we consider a particular example of

$$F_Z(x) \equiv P\big(\{\omega:\ Z_i \le x\}\big) = 1 - e^{-\lambda x}\,.$$

The capital of the company can be represented in the form

$$R_n = R_{n-1}(1+r) + c - Z_n$$

$$= (1+r)\left[R_{n-2}(1+r) + c - Z_{n-1}\right] + c - Z_n$$

$$= R_{n-2}(1+r)^2 + c\left[1 + (1+r)\right] - Z_{n-1}(1+r) - Z_n$$

$$= \ldots$$

$$= R_0(1+r)^n + c\left[1 + (1+r) + (1+r)^2 + \ldots + (1+r)^{n-1}\right]$$

$$- Z_1(1+r)^{n-1} - Z_2(1+r)^{n-2} - \ldots - Z_{n-1}(1+r) - Z_n$$

$$= R_0(1+r)^n + c\frac{(1+r)^n - 1}{r} - S_n,$$

where

$$S_n = Z_1(1+r)^{n-1} + Z_2(1+r)^{n-2} + \ldots + Z_{n-1}(1+r) + Z_n.$$

Note that

$$\phi_n(x) = \int_D \lambda^n e^{-\lambda(z_1 + \ldots + z_n)} \, dz_1 \ldots dz_n,$$

where

$$D = \left\{ 0 < z_1 < R_0(1+r) + c, \right.$$

$$0 < z_k < R_0(1+r)^k + c\frac{(1+r)^k - 1}{r} - z_1(1+r)^{k-1} - \ldots - z_{k-1}(1+r),$$

$$\left. k = 2, \ldots, n \right\}.$$

The integral equation for the probability of ruin has the form

$$\psi_{k+1}(x) = e^{-\lambda[x(1+r)+c]} + \int_0^{x(1+r)+c} \psi_k\big(x(1+r) + c - y\big)\lambda e^{-\lambda y} \, dy$$

with

$$\psi_1(x) = e^{-\lambda[x(1+r)+c]}.$$

Compute the probability of ruin after two steps:

$$\psi_2(x) \;=\; \psi_1(x) + \int_0^{x(1+r)+c} \psi_1\big(x(1+r)+c-y\big)\,\lambda e^{-\lambda y}\,dy$$

$$=\; \psi_1(x) + \int_0^{x(1+r)+c} e^{-\lambda\big([x(1+r)+c-y](1+r)+c-y\big)}\,\lambda e^{-\lambda y}\,dy$$

$$=\; \psi_1(x) + e^{-\lambda\big(x(1+r)^2+c(2+r)\big)} \int_0^{x(1+r)+c} \lambda\,e^{yr}\,dy$$

$$=\; e^{-\lambda(x(1+r)+c)} + e^{-\lambda\big(x(1+r)^2+c(2+r)\big)}\left.\frac{e^{\lambda r}}{r}\right|_0^{x(1+r)+c}$$

$$=\; e^{-\lambda(x(1+r)+c)}\left(1+\frac{e^{-\lambda c}}{r}\right) - \frac{e^{-\lambda\big(x(1+r)^2+c(1+(1+r))\big)}}{r}.$$

For probability of ruin after infinite number of steps, we have (see "Mathematical appendix 5" for details)

$$\psi_\infty(x)$$

$$= b\left[e^{-\lambda(x(1+r)+c)} + \sum_{m=2}^{\infty}(-1)^{m-1}\frac{e^{-\lambda\big(x(1+r)^m+c(1+(1+r)+\ldots+(1+r)^{m-1})\big)}}{r[(1+r)^2-1]\times\ldots\times[(1+r)^{m-1}-1]}\right],$$

where

$$b = \left(1 - \sum_{m=1}^{\infty}(-1)^{m-1}\frac{e^{-\lambda c\big(1+(1+r)+\ldots+(1+r)^{m-1}\big)}}{r\,[(1+r)^2-1]\times\ldots\times[(1+r)^m-1]}\right)^{-1}.$$

If the rate of interest $r=0$, then the equation for the probability of ruin has the form

$$\tilde{\psi}_{k+1}(x) = \tilde{\psi}_1(x) + \int_0^{x+c} \tilde{\psi}_k(x+c-y)\,dF(y)$$

with

$$\tilde{\psi}_1(x) = 1 - F(x+c).$$

In the case of the exponential distribution function $F(y) = 1 - e^{-\lambda y}$, we obtain

$$\tilde{\psi}_1(x) = e^{-\lambda(x+c)},$$

and

$$\tilde{\psi}_2(x) \;=\; \tilde{\psi}_1(x) + \int_0^{x+c} e^{-\lambda(x+2c-y)}\,\lambda e^{-\lambda y}\,dy$$

$$=\; e^{-\lambda(x+c)} + e^{-\lambda(x+2c)}\,\lambda\,(x+c).$$

Note that these formulas can be also obtained by passing to the limit in expressions for ψ_1 and ψ_2:

$$\lim_{r \to 0} \psi_1(x) = e^{-\lambda(x+c)},$$

$$\lim_{r \to 0} \psi_2(x) = e^{-\lambda(x+c)} - \lim_{r \to 0} \frac{e^{-\lambda\left(x(1+r)^2 + c(2+r)\right)} - e^{-\lambda c}}{r}$$

$$= e^{-\lambda(x+c)} - e^{-\lambda\left(x(1+r)^2 + c(2+r)\right)} \Big|_{r=0}$$

$$= e^{-\lambda(x+c)} + \lambda\,(x+2c)\,e^{-\lambda(x+c)} = \widetilde{\psi}_2(x).$$

Next, we consider the case when an insurance company invests in both risky and non-risky assets. By

$$\alpha_n = \frac{\gamma_{n+1}\,S_n}{R_n},$$

we denote the proportion of the risky asset in the investment portfolio. Let us consider a class of strategies with constant proportion $\alpha_n \equiv \alpha$. In the case of the exponential distribution function F, we will obtain an estimate from above for function ψ_∞, and hence, for ψ_k since

$$\psi_1(x) < \psi_2(x) < \ldots < \psi_k(x) < \ldots < \psi_\infty(x).$$

Note that γ_{n+1} is the number of units of asset S that a company buys at time n after collecting premium c and making claim payment Z_n, so that its capital is R_n.

The dynamics of the capital are given by

$$\begin{aligned}
R_{n+1} &= R_n\,(1+r) + \gamma_n\,S_n\,(\rho_n - r) + c - Z_{n+1} \\
&= R_n\,(1 + r + \alpha\,(\rho_n - r)) + c - Z_{n+1}.
\end{aligned}$$

Hence, the probability of ruin after one step is

$$\begin{aligned}
\psi_1(R_0) &= P(\{\omega:\ R_1 < 0\}) \\
&= P(\{\omega:\ R_0\,[1 + r + \alpha\,(\rho_1 - r)] + c - Z_1 < 0\}) \\
&= 1 - F_Z\Big(R_0\,[1 + r + \alpha\,(\rho_1 - r)] + c\Big) \\
&= 1 - p\,F_Z\Big(R_0\,[1 + r + \alpha\,(b - r)] + c\Big) \\
&\qquad\qquad - q\,F_Z\Big(R_0\,[1 + r + \alpha\,(a - r)] + c\Big).
\end{aligned}$$

As in the previous case, we obtain the following integral equation

$$\psi_{k+1}(R_0) = 1 - p\, F_Z\Big(R_0\,[1 + r + \alpha\,(b - r)] + c\Big)$$
$$- q\, F_Z\Big(R_0\,[1 + r + \alpha\,(a - r)] + c\Big)$$
$$+ p \int_0^{R_0\,[1+r+\alpha\,(b-r)]+c} \psi_k\Big(R_0\,[1 + r + \alpha\,(b - r)] + c - y\Big)\, dF_Z(y)$$
$$+ q \int_0^{R_0\,[1+r+\alpha\,(a-r)]+c} \psi_k\Big(R_0\,[1 + r + \alpha\,(a - r)] + c - y\Big)\, dF_Z(y).$$

For the exponential claims distribution function $F(y) = 1 - e^{-\lambda y}$, we have the following estimate:

$$\psi_\infty(x) \le \psi_1(x)\left[1 - e^{-\lambda c}\,\frac{r + p\alpha\,(b - r) + q\alpha\,(a - r)}{[r + \alpha\,(b - r)]\,[r + \alpha\,(b - r)]}\right]^{-1}$$

under condition that

$$\frac{q\alpha\,(b - r) + p\alpha\,(a - r) + r}{e^{\lambda c}\,[r + \alpha\,(b - r)]\,[r + \alpha\,(b - r)] - r - q\alpha\,(b - r) - p\alpha\,(a - r)} > 0.$$

In particular, for $\alpha = 0$ (i.e., when investing in non-risky asset only), we have

$$\psi_\infty(x) \le \psi_1(x)\left[1 - \frac{e^{-\lambda c}}{r}\right]^{-1}$$

under condition $r > e^{-\lambda c}$.

If $\alpha = 1$ (i.e., if investing in risky asset only), then

$$\psi_\infty(x) \le \psi_1(x)\left[1 - e^{-\lambda c}\,\frac{pb + qa}{ab}\right]^{-1}$$

under condition

$$\frac{qb + pa}{ba\,e^{\lambda c} - qb - pa} > 0.$$

We can give the following interpretation of these estimates. Clearly, for all k and x, we have

$$\psi_1(x) < \psi_2(x) < \ldots < \psi_k(x) < \ldots < \psi_\infty(x).$$

Hence,

$$\psi_\infty(x) < \overline{C}\,\psi_1(x),$$

where $\overline{C}$ is independent of x; that is, the probability of ruin after infinite number of time steps can be estimated in terms of the the probability of ruin after one step. Note that, because of our additional assumptions, constant $\overline{C}$

is always positive. And since $\psi_\infty(x)$ is less or equal to 1, then these estimates are satisfactory if

$$\overline{C}\,\psi_1(x) < 1\,,$$

which holds true for sufficiently big initial capital x.

Worked Example 8.2 *Let $r = 0.2$, $c = 1$, $\lambda = 2$, $\alpha = 1$. Given values of the initial capital: 0, 0.1, 0.2, 0.5, 1, 1.5, 3, compute values of $\psi_1(x)$, $\psi_2(x)$ and upper estimate for ψ_∞ with accuracy 0.0001.*

Solution The results are given in the following table and Figure 8.3.

Initial capital	Lower bound	Upper bound	$\psi_2(x)$
0	0.1353	0.4186	0.1655
0.1	0.1065	0.3293	0.1325
0.2	0.0837	0.2590	0.1059
0.5	0.0408	0.1261	0.0538
1	0.0123	0.03797	0.0171
1.5	0.0037	0.0114	0.0054
3	0.0001	0.0003	0.0002

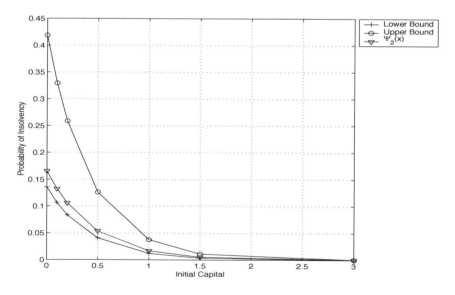

FIGURE 8.3: Probability of ruin.

□

Consider a generalization of the Cramér-Lundberg model when it is assumed that an insurance company has an opportunity to invest in the framework of the Black-Scholes model of a (B, S)-market. Recall that the dynamics of the risky asset in this model are described by the following stochastic differential equation:

$$dS_t = S_t(\mu\,dt + \sigma\,dW_t)\,, \qquad S_0 > 0\,.$$

Then the capital of the company can be written in the form:

$$R(t) = x + \mu\int_0^t R(s)\,ds + \sigma\int_0^t R(s)\,dW_s + ct - \sum_{k=1}^{N(t)} X_k\,.$$

In this case, the probability of solvency ϕ satisfies the following integro-differential equation:

$$\frac{1}{2}\sigma^2 x^2\,\phi''(x) + (\mu\,x + c)\,\phi'(x) - \lambda\,\phi(x) + \lambda\int_0^x \phi(x-y)\,dF(y) = 0\,.$$

Analyzing the behavior of function ϕ as $x \to \infty$ in the case of exponential distribution function of claims

$$X_k \sim F(x) = 1 - e^{-x/\alpha} \qquad x > 0\,,$$

leads to the following result. If the profitability μ of asset S is greater than $\sigma^2/2$, where σ is the volatility of the market, then the probability of ruin $\psi(x) = 1 - \phi(x)$ converges to zero according to the following power law (not exponentially!):

$$\psi(x) = \mathcal{O}\big(x^{1-2\mu/\sigma^2}\big)\,.$$

If asset S is not profitable enough ($\mu < \sigma^2/2$), then for any initial capital $x > 0$, the probability of ruin $\psi(x) = 1$.

Mathematical appendix 5

We will look for a solution of the form

$$\psi_k(x)$$

$$= e^{-\lambda(x(1+r)+c)}\,b_1^k + \sum_{m=2}^{k} b_m^k\,\frac{e^{-\lambda\big(x(1+r)^m + c(1+(1+r)+\dots+(1+r)^{m-1})\big)}}{r\,[(1+r)^2 - 1]\times\dots\times[(1+r)^{m-1} - 1]}\,,$$

where (b_m^k) is a two-parameter sequence independent of x. Here parameter k corresponds to function ψ_k and parameter m corresponds to factor

$$\frac{e^{-\lambda\big(x(1+r)^m + c(1+(1+r)+\dots+(1+r)^{m-1})\big)}}{r\,[(1+r)^2 - 1]\times\dots\times[(1+r)^{m-1} - 1]}\,.$$

Expressions for probabilities of ruin after one and two time steps imply that

$$b_1^1 = 1\,,$$

$$b_1^2 = 1 + \frac{e^{-\lambda c}}{r}\,b_2^2 = -1\,.$$

It is convenient to write sequence (b_m^k) in the form of a triangular table:

$$
\begin{array}{ccccc}
b_1^1 & & & & \\
b_1^2 & b_2^2 & & & \\
b_1^3 & b_2^3 & b_3^3 & & \\
b_1^4 & b_2^4 & b_3^4 & b_4^4 & \\
\cdots & \cdots & \cdots & \cdots & \cdots
\end{array}
$$

From the recurrence equation, we have

$$\psi_k(x(1+r)+c-y)$$

$$= \sum_{m=2}^{k} b_k^m \, \frac{e^{-\lambda\left(x(1+r)^{m+r}+c(1+(1+r)+\ldots+(1+r)^{m-1})-y\,(1+r)^m\right)}}{r\left[(1+r)^2-1\right]\times\ldots\times\left[(1+r)^{m-1}-1\right]}$$

$$+ b_k^1 \, e^{-\lambda[x(1+r)^2+c(1+(1+r))]},$$

and

$$\int_0^{x(1+r)+c} \psi_k(x(1+r)+c-y)\,\lambda e^{-\lambda y}\,dy$$

$$= \sum_{m=2}^{k} b_k^m \, \frac{e^{-\lambda\left(x(1+r)^{m+1}+c(1+(1+r)+\ldots+(1+r)^m)\right)}}{r\left[(1+r)^2-1\right]\times\ldots\times\left[(1+r)^{m-1}-1\right]}$$

$$\times \int_0^{x(1+r)+c} \lambda e^{\lambda y\,[(1+r)^m-1]}\,dy$$

$$+ b_k^1 \, e^{-\lambda[x(1+r)^2+c(1+(1+r))]} \int_0^{x(1+r)+c} \lambda e^{\lambda y\,r}\,dy$$

$$= \sum_{m=2}^{k} b_k^m \, \frac{e^{-\lambda\left(x(1+r)^{m+1}+c(1+(1+r)+\ldots+(1+r)^m)\right)}}{r\left[(1+r)^2-1\right]\times\ldots\times\left[(1+r)^{m-1}-1\right]}$$

$$\times \frac{e^{\lambda\,[(1+r)^m-1]}}{(1+r)^m-1}\Bigg|_0^{x(1+r)+c}$$

$$+b_k^1\, e^{-\lambda[x(1+r)^2+c(1+(1+r))]}\, \frac{e^{\lambda r}}{r}\bigg|_0^{x\,(1+r)+c}$$

$$=\sum_{m=2}^{k} b_k^m\, \frac{e^{-\lambda\left(x(1+r)+c(1+(1+r)+\ldots+(1+r)^{m-1}+c)\right)}}{r\,[(1+r)^2-1]\times\ldots\times[(1+r)^m-1]}$$

$$-\sum_{m=2}^{k} b_k^m\, \frac{e^{-\lambda\left(x(1+r)^{m+1}+c(1+(1+r)+\ldots+(1+r)^m)\right)}}{r\,[(1+r)^2-1]\times\ldots\times[(1+r)^m-1]}$$

$$+b_k^1\, \frac{e^{-\lambda[x(1+r)+c+c]}}{r}+b_k^1\, \frac{e^{-\lambda[x(1+r)^2+c(1+(1+r))]}}{r}\,.$$

Thus,

$$\psi_{k+1}(x)=e^{-\lambda(x(1+r)+c)}$$

$$+\sum_{m=1}^{k} b_k^m\, e^{-\lambda(x(1+r)+c)}\, \frac{e^{-\lambda c(1+(1+r)+\ldots+(1+r)^{m-1})}}{r\,[(1+r)^2-1]\times\ldots\times[(1+r)^m-1]}$$

$$-\sum_{m=1}^{k} b_k^m\, \frac{e^{-\lambda\left(x(1+r)^{m+1}+c(1+(1+r)+\ldots+(1+r)^m)\right)}}{r\,[(1+r)^2-1]\times\ldots\times[(1+r)^m-1]}$$

$$=e^{-\lambda(x(1+r)+c)}\left[1+\sum_{m=1}^{k} b_k^m\, \frac{e^{-\lambda c(1+(1+r)+\ldots+(1+r)^{m-1})}}{r\,[(1+r)^2-1]\times\ldots\times[(1+r)^m-1]}\right]$$

$$-\sum_{m=1}^{k+1} b_k^{m-1}\, \frac{e^{-\lambda\left(x(1+r)^m+c(1+(1+r)+\ldots+(1+r)^{m-1})\right)}}{r\,[(1+r)^2-1]\times\ldots\times[(1+r)^{m-1}-1]}\,,$$

which implies

$$b_{k+1}^1 \;=\; 1+\sum_{m=1}^{k} b_k^m\, \frac{e^{-\lambda c(1+(1+r)+\ldots+(1+r)^{m-1})}}{r\,[(1+r)^2-1]\times\ldots\times[(1+r)^m-1]}\,,$$

$$b_{k+1}^m \;=\; -b_k^{m-1}.$$

So sequence (b_m^k) has the following structure:

$$b_1$$

$$b_2 \quad -b_1$$

$$b_3 \quad -b_2 \quad b_1$$

$$b_4 \quad -b_3 \quad b_2 \quad -b_1$$

$$\cdots \quad \cdots \quad \cdots \quad \cdots \quad \cdots$$

where we introduced the notation: $b_i := b_i^1$ with $b_1 = 1$,

$$b_{k+1} = 1 + \sum_{m=1}^{k} (-1)^{m-1} b_{k-m+1} \frac{e^{-\lambda c(1+(1+r)+...+(1+r)^{m-1})}}{r\left[(1+r)^2 - 1\right] \times ... \times \left[(1+r)^m - 1\right]}.$$

Properties of

$$0 < \psi_1(x) < \psi_2(x) < ... < \psi_k(x) < ...$$

imply that sequence $(b_i)_{i=1}^{\infty}$ is positive and increasing.
Condition

$$\frac{e^{-\lambda c}}{r} < 1, \quad \text{i.e.,} \quad P(\{\omega: Z_1 > c\}) < r \quad \text{or} \quad c > \frac{-\ln r}{\lambda},$$

is sufficient for boundedness of $(b_i)_{i=1}^{\infty}$ and therefore for existence of finite $b = \lim_{i \to \infty} b_i$.

Then passing to the limit in

$$\psi_k(x) = e^{-\lambda(x(1+r)+c)} b_1^k + \sum_{m=2}^{k} b_m^k \frac{e^{-\lambda\left(x(1+r)^m + c(1+(1+r)+...+(1+r)^{m-1})\right)}}{r\left[(1+r)^2 - 1\right] \times ... \times \left[(1+r)^{m-1} - 1\right]},$$

we obtain

$$\psi_{\infty}(x)$$
$$= b\left[e^{-\lambda(x(1+r)+c)} + \sum_{m=2}^{\infty} (-1)^{m-1} \frac{e^{-\lambda\left(x(1+r)^m + c(1+(1+r)+...+(1+r)^{m-1})\right)}}{r\left[(1+r)^2 - 1\right] \times ... \times \left[(1+r)^{m-1} - 1\right]}\right],$$

where

$$b = \left(1 - \sum_{m=1}^{\infty} (-1)^{m-1} \frac{e^{-\lambda c\left(1+(1+r)+...+(1+r)^{m-1}\right)}}{r\left[(1+r)^2 - 1\right] \times ... \times \left[(1+r)^m - 1\right]}\right)^{-1}.$$

8.3   Solvency problem in a generalized Cramér-Lundberg model

As we discussed earlier, solvency of an insurance company is a natural characterization of its exposure to risk, and the traditional actuarial measure of such exposure is the probability of ruin (or insolvency).

Representing the capital R of the insurance company as the difference between the premium process Π and risk process X, we can define the probability of solvency on finite and infinite intervals as

$$
\begin{aligned}
\varphi(t,x) &= P(\{\omega: R(s) > 0 \quad \text{for all} \quad s \le t\}), & R(0) = x, \\
\varphi(x) &= P(\{\omega: R(t) > 0 \quad \text{for all} \quad t \ge 0\}), & R(0) = x,
\end{aligned}
$$

respectively.

Suppose that

$$
E\big(\Pi(t)\big) > E\big(X(t)\big),
$$

that is, *pure income* is positive.

This section is devoted to generalizations of the Cramér-Lundberg model. In particular, we study the probability of ruin as a measure of exposure to risk in situations when the premium process Π has more complex structure than in the original model. We also will take into account various factors of financial and insurance markets.

First, we consider a case when premiums are received at some random times and their amounts are also random. The capital of the company has the form

$$
R(t) = x + \sum_{i=1}^{N_1(t)} c_i - \sum_{i=1}^{N(t)} X_i,
$$

where N_1 and N are independent Poisson processes with intensities λ_1 and λ, respectively, and (c_i) and (X_i) are sequences of independent random variables with distribution functions $G(\cdot)$ and $F(\cdot)$, respectively.

Hence,

$$
\begin{aligned}
\lambda_1\, t\, E(c_i) &= E\left(\sum_{i=1}^{N_1(t)} c_i\right) = E\big(\Pi(t)\big) > E\big(X(t)\big) \\
&= E\left(\sum_{i=1}^{N(t)} X_i\right) = \lambda\, t\, E(X_i).
\end{aligned}
$$

Therefore, the condition of positivity of pure income is reduced to inequality

$$
\lambda_1\, E(c_i) > \lambda\, E(X_i).
$$

Under these assumptions, we have that the probability of solvency satisfies the inequality

$$\varphi(x) \geq 1 - e^{-Rx},$$

where constant R is a solution of the *characteristic equation*

$$\lambda_1 \left[E\left(e^{-Rc_i}\right) - 1 \right] + \lambda \left[E\left(e^{-RX_i}\right) - 1 \right] = 0.$$

For exponential distribution functions $G(\cdot)$ (premium process) and $F(\cdot)$ (risk process), we have the following result.

Proposition 8.1 *If*

$$G(x) \;=\; P(\{\omega: \; c_i \leq x\}) = 1 - e^{-bx},$$

$$F(x) \;=\; P(\{\omega: \; X_i \leq x\}) = 1 - e^{-ax},$$
$$a > 0, \; b > 0, \; x > 0,$$

then we have an exact expression

$$\varphi(x) = 1 - \frac{(a+b)\lambda}{(\lambda_1 + \lambda)a} \exp\left\{ \frac{\lambda b - \lambda_1 a}{\lambda_1 + \lambda} x \right\}.$$

Proof Using the independence of $\Pi(t)$ and $X(t)$, we have

$$E\left(e^{-R\left[\Pi(t)-X(t)\right]}\right)$$

$$= \left(\sum_{k=0}^{\infty} e^{-\lambda_1 t} \frac{(\lambda_1 t)^k}{k!} E\left(e^{-R\sum_{i=1}^{k} c_i}\right) \right) \left(\sum_{k=0}^{\infty} e^{-\lambda t} \frac{(\lambda t)^k}{k!} E\left(e^{R\sum_{i=1}^{k} X_i}\right) \right)$$

$$= \exp\left\{ \lambda_1 t \left[E\left(e^{-Rc_i}\right) - 1 \right] + \lambda t \left[E\left(e^{RX_i}\right) - 1 \right] \right\},$$

for any constant R. Now let R be a solution of the characteristic equation, then for $t > s$ we have

$$E\left(e^{-R\left[\Pi(t)-X(t)\right]} \Big| \mathcal{F}_s\right)$$

$$= e^{-R\left[\Pi(s)-X(s)\right]} E\left(e^{-R\left[\Pi(t)-\Pi(s)-X(t)+X(s)\right]} \Big| \mathcal{F}_s\right)$$

$$= e^{-R\left[\Pi(s)-X(s)\right]} E\left(e^{-R\left[\Pi(t-s)-X(t-s)\right]}\right) = e^{-R\left[\Pi(s)-X(s)\right]},$$

where $\mathcal{F}_t$ is a σ-algebra generated by processes $\Pi(s)$ and $X(s)$ up to time t. Hence, the process

$$Y_R(t) = e^{-R\left[\Pi(t)-X(t)\right]}$$

is a martingale with the initial condition $Y_R(0) = 1$.

Consider the ruin time

$$\tau = \inf\{t \geq 0 : R(t) < 0\}.$$

Since the average value of a martingale is constant, we obtain

$$
\begin{aligned}
1 &= E\big(Y_R(\tau \wedge t)\big) \geq E\big(Y_R(\tau \wedge t)\, I_{\{\omega:\ \tau \leq t\}}\big) \\
&= E\big(e^{-R\,[\Pi(\tau) - X(\tau)]}\, I_{\{\omega:\ \tau \leq t\}}\big) > e^{Rx}\, P\big(\{\omega :\ \tau \leq t\}\big),
\end{aligned}
$$

where we also used the fact that $\Pi(\tau) = X(\tau)$ for $\tau < \infty$. Passing to the limit as $t \to \infty$, we obtain

$$\varphi(x) \geq 1 - e^{-Rx}.$$

Note that, if distribution functions $G(\cdot)$ and $F(\cdot)$ are exponential, then the condition of positivity of pure income has the form

$$\frac{\lambda_1}{b} > \frac{\lambda}{a},$$

and the characteristic equation is

$$\lambda_1\left[\frac{b}{b+R} - 1\right] + \lambda\left[\frac{a}{a-R} - 1\right] = 0.$$

Hence, constant R is either equal to zero or to

$$\frac{\lambda_1\, a - \lambda\, b}{\lambda_1 + \lambda}.$$

Thus, we have that either $\varphi(\infty) = 1$ or

$$\varphi(x) > 1 - \exp\left\{\frac{\lambda\, b - \lambda_1\, a}{\lambda_1 + \lambda}\, x\right\}.$$

One can use the formula for total probability to obtain the following integral equation for $\varphi(x)$:

$$(\lambda + \lambda_1)\, \varphi(x) = \lambda_1 \int_0^\infty \varphi(x + v)\, b\, e^{-bv}\, dv + \lambda \int_0^x \varphi(x - u)\, a\, e^{-au}\, du.$$

Changing variables $v_1 = v + x$, $u_1 = x - u$, we can write $\varphi(x)$ in the form

$$\varphi(x) = \frac{\lambda_1}{\lambda + \lambda_1} \int_x^\infty \varphi(v_1)\, b\, e^{-b(v_1 - x)}\, dv_1 + \frac{\lambda}{\lambda + \lambda_1} \int_0^x \varphi(u_1)\, a\, e^{-a(x - u_1)}\, du_1,$$

which, in particular, indicates that function φ is differentiable. Also note that

$$\left(\int_0^\infty \varphi(x + v)\, b\, e^{-bv}\, dv\right)' = -b\,\varphi(x) + b \int_0^\infty \varphi(x + v)\, b\, e^{-bv}\, dv,$$

$$\left(\int_0^x \varphi(x - u)\, a\, e^{-au}\, du\right)' = a\,\varphi(x) - a \int_0^x \varphi(x - u)\, a\, e^{-au}\, du.$$

Differentiating the equation for $\varphi(x)$, we have

$$(\lambda + \lambda_1)\,\varphi'(x) + (\lambda_1\,b - \lambda\,a)\,\varphi(x)$$
$$= b\,\lambda_1 \int_0^\infty \varphi(x+v)\,b\,e^{-bv}\,dv - a\,\lambda \int_0^x \varphi(x-u)\,a\,e^{-au}\,du\,.$$

Differentiating the second time, we obtain

$$(\lambda + \lambda_1)\,\varphi''(x) + (\lambda_1\,b - \lambda\,a)\,\varphi'(x) + (\lambda_1\,b^2 + \lambda\,a^2)\,\varphi(x)$$
$$= b^2\,\lambda_1 \int_0^\infty \varphi(x+v)\,b\,e^{-bv}\,dv + a^2\,\lambda \int_0^x \varphi(x-u)\,a\,e^{-au}\,du\,.$$

This implies that

$$\varphi''(x) = \frac{\lambda\,b - \lambda_1\,a}{\lambda + \lambda_1}\,\varphi'(x)\,.$$

This equation has a solution of the form

$$\varphi(x) = C_1 + C_2\,\exp\left\{\frac{\lambda\,b - \lambda_1\,a}{\lambda + \lambda_1}\,x\right\}\,.$$

It is clear from the statement of the problem that

$$C_1 = \varphi(\infty) = 1\,.$$

Substituting this expression into the initial integral equation for $\varphi(x)$, we obtain that for $x = 0$

$$(\lambda + \lambda_1)\,\varphi(0) = \lambda_1 \int_0^\infty \varphi(v)\,b\,e^{-bv}\,dv\,,$$

and hence,

$$C_2 = -\frac{(a+b)\,\lambda}{(\lambda + \lambda_1)\,a}\,,$$

which completes the proof. $\square$

Remark 8.1

Similar results can be obtained for a discrete version of the Cramér-Lundberg model, when $\Pi(t)$ and $X(t)$ are independent compound binomial processes. $\square$

Now we consider a generalization of the Cramér-Lundberg model that takes into account the insurance market competition. Suppose the pool of insurance companies is large enough, and each company has only limited influence on the insurance market. Then it is natural to use Gaussian diffusion for modeling the capital of an insurance company:

$$R(t) = x + \Pi(t) - X(t) + \sigma\,W_t\,,$$

where

$$\Pi(t) = \mu\, t + \sum_{i=1}^{N_1(t)} c_i \,, \quad \mu > 0 \,,$$

is the premium process,

$$X(t) = \sum_{i=1}^{N(t)} X_i \,,$$

is the risk process, W_t is a standard Wiener process and $\sigma \geq 0$.

It is assumed that all processes $\Pi(t)$, $X(t)$, and W_t are independent, and the condition of positivity of income,

$$\mu + \lambda_1\, E(c_i) > \lambda\, E(X_i) \,,$$

holds true.

In this case, the probability of solvency again satisfies the estimate

$$\varphi(x) \geq 1 - e^{-R\,x} \,,$$

where R is a solution of the *characteristic equation*

$$-R\,\mu + \sigma^2\, R^2 + \lambda_1 \left[\int_0^\infty e^{-R\,v}\, dG(v) - 1 \right] + \lambda \left[\int_0^\infty e^{R\,y}\, dF(y) - 1 \right] = 0 \,.$$

Another generalization of the Cramér-Lundberg model takes into account the fact that insurance companies are active participants of the financial market. Earlier we discussed several discrete models of this type. Now we consider a version of Cramér-Lundberg model in the framework of the following Black-Scholes market:

$$dB_t = r\, B_t\, dt \,, \quad B_0 = 1 \,,$$
$$dS_t = S_t \left(\mu\, dt + \sigma\, dW_t \right) \,, \quad S_0 > 0 \,.$$

Suppose that the initial capital of an insurance company is x, and the capital of the investment portfolio $\pi = (\beta, \gamma)$ is

$$R(t) = \beta_t\, B_t + \gamma_t\, S_t \,,$$

and its dynamics are described by

$$dR(t) = \beta_t\, dB_t + \gamma_t\, dS_t + B_t\, d\beta_t + S_t\, d\gamma_t \,.$$

If $\Pi(t) = \sum_{i=1}^{N_1(t)} c_i$ is the premium process and $X(t) = \sum_{i=1}^{N(t)} X_i$ is the risk process, then the following constraint

$$B_t\, d\beta_t + S_t\, d\gamma_t = \sum_{i=N_1(t)}^{N_1(t+dt)} c_i - \sum_{i=N(t)}^{N(t+dt)} X_i \,,$$

is natural for the class of admissible strategies. It means that the redistribution of capital in the portfolio happens because of premium and claim flows.

Suppose that *all capital is invested into a bank account*, then its dynamics are described by equation

$$R(t) = x + \int_0^t r\,R(s)\,ds + \sum_{i=1}^{N_1(t)} c_i - \sum_{i=1}^{N(t)} X_i\,.$$

Its solution has the form

$$R(t) = e^{rt}\left[x + \sum_{i=1}^{N_1(t)} c_i\,e^{-r\sigma_i} - \sum_{i=1}^{N(t)} X_i\,e^{-r\tau_i}\right],$$

where σ_i are jumps of process $N_1(t)$ and τ_i are jumps of $N(t)$.

Since random variable

$$\tau = \inf\{t \geq 0 : R(t) < 0\}$$

represents the ruin time, then the probability of solvency

$$\varphi(x) = P(\{\omega : \tau = \infty\})$$

is established in the following theorem.

Theorem 8.1 *Suppose that all capital of an insurance company is invested in a bank account, then the probability of the company's solvency satisfies the integro-differential equation*

$$r\,x\,\varphi'(x) - (\lambda_1 + \lambda)\,\varphi(x) + \lambda\int_0^x \varphi(x - y)\,dF(y) + \lambda_1 \int_0^\infty \varphi(x + \nu)\,dG(\nu) = 0\,.$$

Proof Since for a fixed $R(t) = x$ the further evolution of the process depends neither on t nor on its history, then using the equation for $R(t)$, we can write for a small time interval Δt:

$$
\begin{aligned}
\varphi(x) &= \left[1 - (\lambda_1 + \lambda)\right]\varphi(x + r\,x\,\Delta t) + \lambda_1 \int_0^\infty \varphi(x + r\,x\,\Delta t + \nu)\,dG(\nu) \\
&\quad + \lambda \int_0^x \varphi(x + r\,\Delta t - u)\,dF(u) + o(\Delta t)\,.
\end{aligned}
$$

Since by Taylor's formula we have

$$\varphi(x + r\,x\,\Delta t) = \varphi(x) + r\,\Delta t\,\varphi'(x) + o(\Delta t)\,,$$

then dividing the latter equality by Δt and taking limits as $\Delta \to \infty$ proves the claim. $\square$

To estimate the probability of solvency on a finite time interval, we consider the discounted capital $\tilde{R}(t) = R(t)\,e^{-rt}$.

Clearly, for any finite interval, we have

$$P(\{\omega: \ \tilde{R}(s) \geq 0 \quad \text{for all} \quad 0 \leq s \leq t\})$$
$$= P(\{\omega: \ R(s) \geq 0 \quad \text{for all} \quad 0 \leq s \leq t\}) = \varphi(x,t),$$

since processes $\tilde{R}(t)$ and $R(t)$ are positive multiples of each other.

Then we have the following estimate from below.

Theorem 8.2 *For all R such that*

$$f(R,t)$$
$$= \exp \left\{ \int_0^t \left[\lambda_1 + \lambda - \lambda_1\, E\left(\exp\{-Rc_i e^{-rs}\}\right) - \lambda E\left(\exp\{RX_i e^{-rs}\}\right)\right] ds \right\}$$

$$< \infty,$$

and for all $t \geq 0$, the process $e^{-R\tilde{R}(t)}/f(R,t)$ is a martingale and

$$\varphi(x,t) \geq 1 - f(R,t)\, e^{-Rx}.$$

Proof Denote $g(\tilde{R}(t),t) = e^{-R\tilde{R}(t)}$, and compute

$$
\begin{aligned}
E\big(g(x,t+\Delta t)\big) \ &= \ \big[1 - (\lambda_1 + \lambda)\,\Delta t\big]\, E\big(g(x,t)\big) \\
&\quad + \lambda_1\, \Delta t \int_0^\infty g(x + \nu e^{-rt}, t)\, dG(\nu) \\
&\quad + \lambda\, \Delta t \int_0^\infty g(x - u e^{-rt}, t)\, dF(u) + o(\Delta t).
\end{aligned}
$$

Hence, we obtain the following integro-differential equation:

$$\frac{\partial}{\partial t} E\big(g(x,t)\big) + (\lambda_1 + \lambda)\, g(x,t)$$
$$= \lambda_1 \int_0^\infty g(x + \nu e^{-rt}, t)\, dG(\nu) + \lambda \int_0^\infty g(x - u e^{-rt}, t)\, dF(u).$$

Let us find a solution of the form $E\big(g(x,t)\big) = b(t)\, e^{-Rx}$. We obtain that $b(t)$ satisfies the equation

$$b'(t) = b(t)\big[-\lambda_1 - \lambda + \lambda_1\, E\left(\exp\{-R c_i\, e^{-rt}\}\right) + \lambda E\left(\exp\{R X_i\, e^{-rt}\}\right)\big]$$

with the initial condition $b(0) = 1$.

Further

$$
\begin{aligned}
E\left(\frac{e^{-R\tilde{R}(t)}}{f(R,t)} \Big| \mathcal{F}_s \right) \ &= \ \frac{e^{-R\tilde{R}(s)}}{f(R,s)} E\left(e^{-R(\tilde{R}(t)-\tilde{R}(s))} \Big| \mathcal{F}_s \right) \frac{f(R,s)}{f(R,t)} \\
&= \ \frac{e^{-R\tilde{R}(s)}}{f(R,s)} E\left(e^{-R(\tilde{R}(t)-\tilde{R}(s))} \right) \frac{f(R,s)}{f(R,t)},
\end{aligned}
$$

where the latter equality holds true because of the independence of increments of $\tilde{R}(t)$. Also note that random variables $\tilde{R}(t) - \tilde{R}(s)$ and $\tilde{R}(t-s)\,e^{-rs}$ have the same distribution function. Hence,

$$E\left(e^{-R(\tilde{R}(t)-\tilde{R}(s))}\right) = E\left(e^{-rs\tilde{R}(t-s)}\right)$$

$$= e^{-(\lambda_1+\lambda)\,(t-s)}\,\exp\left\{\int_0^{t-s}\left[\lambda_1\,E\left(\exp\{-R\,c_i\,e^{-r(s+l)}\}\right)\right.\right.$$

$$\left.\left. + \lambda\,E\left(\exp\{R\,X_i\,e^{-r(s+l)}\}\right)\right]dl\right\}$$

$$= e^{-(\lambda_1+\lambda)\,(t-s)}\,\exp\left\{\int_s^t\left[\lambda_1\,E\left(\exp\{-R\,c_i\,e^{-rl}\}\right)\right.\right.$$

$$\left.\left. + \lambda\,E\left(\exp\{R\,X_i\,e^{-rl}\}\right)\right]dl\right\}$$

$$= \frac{f(R,t)}{f(R,s)},$$

and therefore the process $e^{-R\tilde{R}(t)}/f(R,t)$ is a martingale.

Using martingale properties, we obtain

$$1 = \frac{E\left(e^{-R\tilde{R}(t)}\right)}{f(R,t)} = \frac{E\left(e^{-R\tilde{R}(t\wedge\tau)}\right)}{f(R,t\wedge\tau)} \geq \frac{E\left(e^{-R\tilde{R}(t\wedge\tau)}\,I_{\{w:\ \tau\le t\}}\right)}{f(R,t\wedge\tau)}$$

$$\geq \frac{e^{Rx}\,P\left(\{w:\ \tau\le t\}\right)}{f(R,t)},$$

which proves the result. □

Now suppose that all capital of an insurance company is invested in stock. The dynamics of prices of stock S are described by the Black-Scholes model (with $\beta_0 = 0$). In this case, the capital of the insurance company satisfies the equation

$$R(t) = \mu\int_0^t R(s)\,ds + \sigma\int_0^t R(s)\,dW_s + \sum_{i=1}^{N_1(t)} c_i - \sum_{i=1}^{N(t)} X_i\,.$$

We have the following result.

Theorem 8.3 *Suppose that all capital of an insurance company is invested in stock, then the probability of company's solvency satisfies the integro-differential equation*

$$\frac{\sigma^2}{2}x^2\,\varphi''(x) \quad + \quad \mu\,x\,\varphi'(x) - (\lambda_1 + \lambda)\,\varphi(x) \tag{8.18}$$

$$+ \lambda\int_0^x \varphi(x-y)\,dF(y) + \lambda_1\int_0^\infty \varphi(x+v)\,dG(v) = 0\,,$$

which in the case of

$$G(\nu) = 1 - e^{-b\nu} \qquad and \qquad F(u)1 - e^{-au},$$

can be reduced to a third-order ordinary differential equation. For $\mu > \sigma^2/2$, we have the following asymptotic behavior:

$$\varphi(x) = K_1 + x^{1-2\mu/\sigma^2} \left(K_2 + o(1) \right).$$

Proof As in Theorem 8.1, we can write

$$
\begin{aligned}
\varphi(x) \quad = \quad & \left[1 - (\lambda_1 + \lambda) \right] \varphi(x + \mu\,x\,\Delta t + \sigma\,x\,\Delta t) \\
& + \lambda_1 \int_0^\infty \varphi(x + \mu\,x\,\Delta t + \sigma\,x\,\Delta t + \nu)\,dG(\nu) \\
& + \lambda \int_0^{x+\mu\,x\,\Delta t + \sigma\,x\,\Delta t} \varphi(x + \mu\,x\,\Delta t + \sigma\,x\,\Delta t - u)\,dF(u) + o(\Delta t).
\end{aligned}
$$

Using the Kolmogorov-Itô formula, we obtain

$$
\begin{aligned}
\varphi(x) \quad = \quad & \left[1 - (\lambda_1 + \lambda)\,\Delta t \right] \left(\varphi(x) + \mu\,x\,\varphi'(x)\,\Delta t + \frac{\sigma^2}{2}\,x^2\,\varphi''(x)\,\Delta t \right) \\
& + \lambda_1\,\Delta t \int_0^\infty \varphi(x + \nu)\,dG(\nu) + \lambda\,\Delta t \int_0^x \varphi(x - u)\,dF(u),
\end{aligned}
$$

which implies (8.18).

Now consider the case of

$$G(\nu) = 1 - e^{-b\nu} \qquad and \qquad F(u) = 1 - e^{-au}.$$

Equation (8.18) becomes

$$(\lambda_1 + \lambda)\,\varphi(x) - \mu\,x\,\varphi'(x) - \frac{\sigma^2}{2}\,x^2\,\varphi''(x) = \lambda_1 I_1 + \lambda I, \qquad (8.19)$$

where

$$I_1 = \int_0^\infty \varphi(x + \nu)\,b\,e^{-b\nu}\,d\nu \qquad and \qquad I = \int_0^x \varphi(x - u)\,a\,e^{-au}\,du.$$

Since

$$I_1' = -b\,\varphi(x) + b\,I_1 \qquad and \qquad I' = a\,\varphi(x) - a\,I,$$

then differentiating (8.19), we obtain

$$(\lambda_1 + \lambda)\,\varphi'(x) - \mu\,\varphi'(x) \quad - \quad \mu\,x\,\varphi''(x) - \sigma^2\,x\,\varphi''(x) - \frac{\sigma^2}{2}\,x^2\,\varphi^{(3)}(x)$$

$$= a\,\lambda\,\varphi(x) - a\,\lambda I - b\,\lambda_1\,\varphi(x) + b\,\lambda_1 I_1$$

or

$$(\lambda_1 + \lambda - \mu)\,\varphi'(x) \quad - \quad (\mu + \sigma^2)\,x\,\varphi''(x) - \frac{\sigma^2}{2}\,x^2\,\varphi^{(3)}(x) \tag{8.20}$$

$$= (a\,\lambda - b\,\lambda_1)\,\varphi(x) - a\,\lambda\,I + b\,\lambda_1\,I_1\,.$$

Further differentiation gives

$$(\lambda_1 + \lambda \quad - \quad \mu)\,\varphi''(x) - (\mu + \sigma^2)\,\varphi''(x)$$

$$-(\mu + \sigma^2)\,x\,\varphi^{(3)}(x) - \sigma^2\,x\,\varphi^{(3)}(x) - \frac{\sigma^2}{2}\,x^2\,\varphi^{(4)}(x)$$

$$= (a\,\lambda - b\,\lambda_1)\,\varphi'(x) - (b\,\lambda_1^2 + a\,\lambda^2)\,\varphi(x) + a^2\,\lambda\,I + b^2\,\lambda_1\,I_1$$

or

$$(\lambda_1 + \lambda - 2\,\mu \quad - \quad \sigma^2)\,\varphi''(x) - (\mu + 2\,\sigma^2)\,x\,\varphi^{(3)}(x) - \frac{\sigma^2}{2}\,x^2\,\varphi^{(4)}(x) \tag{8.21}$$

$$= (a\,\lambda - b\,\lambda_1)\,\varphi'(x) - (b\,\lambda_1^2 + a\,\lambda^2)\,\varphi(x) + a^2\,\lambda\,I + b^2\,\lambda_1\,I_1\,.$$

Now we multiply equation (8.19) by $(a - b)$, equation (8.20) by ab, and add both to equation (8.21):

$$\varphi^{(4)}(x) \quad + \quad \left[(a - b) + \frac{2\,(\mu + 2\,\sigma^2)}{\sigma^2\,x}\right]\varphi^{(3)}(x)$$

$$+ \left[-a\,b + \frac{2\,(\mu + \sigma^2)(a - b)}{\sigma^2\,x} - \frac{2\,(\lambda_1 + \lambda - 2\,\mu - \sigma^2)}{\sigma^2\,x^2}\right]\varphi''(x)$$

$$+ \left[-\frac{2\,a\,b\,\mu}{\sigma^2\,x} - \frac{2\,(\lambda_1\,a - \lambda\,b + \mu\,(b - a))}{\sigma^2\,x^2}\right]\varphi'(x) = 0\,.$$

Making substitution $G = \varphi'$, we obtain

$$G^{(3)}(x) \quad + \quad \left[(a - b) + \frac{2\,(\mu + 2\,\sigma^2)}{\sigma^2\,x}\right]G^{(2)}(x)$$

$$+ \left[-a\,b + \frac{2\,(\mu + \sigma^2)(a - b)}{\sigma^2\,x} - \frac{2\,(\lambda_1 + \lambda - 2\,\mu - \sigma^2)}{\sigma^2\,x^2}\right]G'(x)$$

$$+ \left[-\frac{2\,a\,b\,\mu}{\sigma^2\,x} - \frac{2\,(\lambda_1\,a - \lambda\,b + \mu\,(b - a))}{\sigma^2\,x^2}\right]G(x) = 0\,.$$

We can use standard methods of theory of ordinary differential equations to find the asymptotic behavior of a solution of the latter equation as $x \to \infty$. We use the substitution $G(x) = e^{\tau x} G_1(x)$, where τ is chosen so that the constant coefficient in front of $G_1(x)$ vanishes. This implies that τ satisfies the equation

$$\tau^3 + \tau^2 (a - b) - \tau\, a\, b = 0 \,,$$

which has solutions $\tau = 0$, $-a$, b. The case of $\tau = b$ is not suitable as $\varphi(x)$ is a bounded function. If $\tau = 0$, then the equation stays unchanged. Next, we use the substitution $G_1(x) = x^r\, G_2(x)$ with r such that the coefficient in front of $G_2(x)/x$ is zero. Hence, r satisfies the equation

$$-r\, a\, b - \frac{2\, a\, b\, \mu}{\sigma^2} = 0 \,,$$

which implies $r = -2\,\mu/\sigma^2$.

Thus, we find a solution in the form of the series

$$G_2(x) = \sum_{k=0}^{\infty} \frac{c_k}{x^k} \,,$$

which, in general, may be divergent, but it gives us the following asymptotic representation:

$$G_2(x) = c_0 + o(1) \,.$$

In this case,

$$\varphi'(x) = x^{-2\,\mu/\sigma^2} \left(c_0 + o(1) \right) .$$

If $\tau = -a$, then

$$\varphi'(x) = o\!\left(x^{-2\,\mu/\sigma^2} \right) . \quad \square$$

Note that this theorem reiterates the following important observation: if an insurance company has investments in risky assets of the financial market, then the asymptotic behavior of the probability of its solvency in general cannot be exponential as it was in the standard Cramér-Lundberg model.

Appendix A

Problems

A.1 Probability theory and elements of stochastic analysis

Problem A.1.1 *Consider probability space $([0,1], B(0,1), m)$, where m is the Lebesgue measure. Find $E(\xi|\eta)$ for random variables*

$$\xi(\omega) = 2\omega^2 \text{ and } \eta(\omega) = \begin{cases} 0, & \omega \in [0, 1/3] \\ 2, & \omega \in (1/3, 2/3] \\ 1, & \omega \in (2/3, 1] \end{cases}.$$

Problem A.1.2 *Consider a sequence of independent random variables $(\xi_n)_{n=1,\ldots,N}$ such that each ξ_n takes two values: $e = 2.71828\ldots$ and $-2e$ with probabilities $(2e+1)/3e$ and $(e-1)/3e$, respectively. Define*

$$X_n = \prod_{k=1}^{n} \xi_k, \quad \mathcal{F}_n = \sigma\{\xi_1, \ldots, \xi_n\}.$$

Is sequence $(X_n, \mathcal{F}_n)_{n=1,\ldots,N}$ a martingale?

Problem A.1.3 *Let $(\xi_n)_{n=1,\ldots,N}$ be a sequence of independent random variables taking values -1 and $+1$ with probabilities 0.5. Define $X_n = \sum_{k=1}^{n} \xi_k$. Show that $(X_n^2 - n)_{n \geq 1}$ is a martingale with respect to the natural filtration $\mathcal{F}_n = \sigma\{\xi_1, \ldots, \xi_n\}$.*

Problem A.1.4 *Let $(X_n)_{n \geq 1}$ be a sequence of identically distributed independent random variables with the density function given by the formula*

$$f_{a,b}(x) = \begin{cases} a, & \text{if } 0 \leq x < 1, \\ \exp\{-bx\}, & \text{if } x > 1 \end{cases}.$$

Define new stochastic sequences $Y_n = \sum_{i=1}^{n} X_i$ and $Z_n = \prod_{i=1}^{n} X_i$. Are there positive parameters a, b such that $(Y_n)_{n \geq 1}$ and/or $(Z_n)_{n \geq 1}$ are martingales with respect to filtration $\mathcal{F}_n = \sigma(X_1, \ldots, X_n)$?

265

Problem A.1.5 *Let* $(P_n)_{n \geq 1}$ *and* $(Q_n)_{n \geq 1}$ *be two sequences of probability measures such that* $Q_n \ll P_n$ *for all* n. *Consider measures* $\mu = \sum_{n=1}^{\infty} \alpha_n P_n$,

where $(\alpha_n)_{n \geq 1}$ *is a sequence of positive numbers, and* $\nu = \sum_{n=1}^{\infty} \beta_n Q_n$, *where* $(\beta_n)_{n \geq 1}$ *is a sequence of non-negative numbers. Show that* $\nu \ll \mu$.

Problem A.1.6 *Consider a geometrical Brownian motion* $\exp\{at + bW_t\}$. *Find its expected value and variance.*

Problem A.1.7 *Let* $(\varepsilon_i)_{i=1,\ldots,N}$ *be a sequence of independent random variables taking values* -1 *and* $+1$ *with probabilities* 0.5 *and define*

$$X_n = (-1)^n \cos\left(\pi \sum_{k=1}^{n} \varepsilon_k\right).$$

Show that $(X_n)_{n=1,\ldots,N}$ *is a martingale with respect to filtration* $\mathcal{F}_n = \sigma(\varepsilon_1, \ldots, \varepsilon_N)$ *on a given probability space* $(\Omega, \mathcal{F}, P)$.

Hint: verify martingale property using the formula $\cos(a+b) = \cos(a)\cos(b) - \sin(a)\sin(b)$.

Problem A.1.8 *Let* $(X_n)_{n \geq 1}$ *be a sequence of independent identically distributed random variables with the density*

$$f(x) = \begin{cases} 1, & x \in \left[-\dfrac{1}{2}, \dfrac{1}{2}\right] \\ 0, & otherwise \end{cases}.$$

Define a new stochastic sequence $Y_n = \sum_{i=1}^{n} X_i$.

(a) *Is it a martingale with respect to filtration* $\mathcal{F}_n = \sigma(X_1, \ldots, X_n)$?

Will the answer change for densities

(b)

$$f(x) = \begin{cases} 1, & x \in \left[-\dfrac{1}{3}, \dfrac{2}{3}\right] \\ 0, & otherwise \end{cases}$$

      and

(c)

$$f(x) = \begin{cases} 1, & x \in \left[-\dfrac{2}{3}, \dfrac{1}{3}\right] \\ 0, & otherwise \end{cases} ?$$

Problem A.1.9 *Check whether the following processes are martingales with respect to a filtration generated by Wiener process* (W_t).

(a) $X_t = W_t^3 - 3tW_t$,

(b) $X_t = W_t + 108t$,

(c) $X_t = t^2 W_t - 2 \int_0^t sW_s \, ds$,

(d) $X_t = \exp\left\{ \frac{t}{2} \right\} \cdot \sin(W_t)$.

Problem A.1.10 *Find a stochastic differential for the process*

$$X_t = \left(\sqrt{15} + \frac{1}{2} W_t \right)^2.$$

Problem A.1.11 *Consider the process* $dX_t = -\alpha X_t dt + \sigma d W_t$ *determined by positive numbers* α, σ, *and some initial value* $X_0 \in \mathbb{R}$. *Prove that*

$$X_t = \exp\{-\alpha t\} \left(X_0 + \sigma \int_0^t \exp\{\alpha t\} d W_t \right)$$

and

$$Var \, X_t \to \frac{\sigma^2}{2\alpha} \quad \text{as} \quad t \to \infty.$$

Problem A.1.12 *Prove that for every random variable* $\xi \geq 0$ *on a probability space* $(\Omega, \mathcal{F}, P)$ *there exists a sequence of simple random variables* ξ_n *such that* $\xi_n(\omega) \to \xi(\omega)$ *for all* $\omega \in \Omega$.

Problem A.1.13 *Show that for a random variable* ξ *on a probability space* $(\Omega, \mathcal{F}, P)$ *the family of sets* $\{\omega : \xi(\omega) \in B\}|_{B \in Borel \, algebra}$ *is a* σ-*algebra.*

Problem A.1.14 *Assume that a non-negative random variable* Z *with* $E(Z) = 1$ *is defined on a probability space* $(\Omega, \mathcal{F}, P)$. *Show that* $P^*(A) = E(ZI_A)$ *defines a new probability.*

Problem A.1.15 *Let* $(\xi_k)_{k \geq 1}$ *be a sequence of independent random variables on a probability space* $(\Omega, \mathcal{F}, P)$ *and suppose* $E(\xi_k) = 1$ *for all* k. *Prove that*

$$\left(\prod_{k=1}^n \xi_k \right)_{n \geq 1}$$

is a martingale with respect to the natural filtration $(\mathcal{F}_n)_{n=1,2,\dots}$ *generated by this sequence.*

Problem A.1.16 *Assume* $(\Omega, \mathcal{F}, (\mathcal{F}_n)_{n=0,1,\dots}, P)$ *is a stochastic basis and a stochastic sequence* $X = (X_n, \mathcal{F}_n)$ *satisfies the linear stochastic differential equation* $\Delta X_n = X_{n-1} \Delta U_n$, *where* U *is a stochastic sequence with* $U_0 = 0$ *and* $X_0 = 1$. *Show that* X *admits the representation*

$$X_n = \prod_{k=1}^n (1 + \Delta U_k), \quad n = 1, 2, \dots .$$

Problem A.1.17 *Assume $M = (M_n)_{n \geq 0}$ and $N = (N_n)_{n \geq 0}$ are martingales on a stochastic basis $(\Omega, \mathcal{F}, (\mathcal{F}_n)_{n=0,1,\dots}, P)$. Show that their product $(M_n \cdot N_n)_{n \geq 0}$ is a martingale if and only if $\langle M, N \rangle_n = 0$.*

Problem A.1.18 *Consider a non-homogeneous stochastic differential equation*

$$\Delta X_n = \Delta N_n + X_{n-1} \Delta U_n,$$

where $(N_n)_{n \geq 0}$ is a stochastic sequence such that $X_0 = N_0$. Show that the solution of this equation admits the following representation:

$$X_n = \varepsilon_n^N(U) = \varepsilon_n(U)\left[N_0 + \sum_{k=1}^{n} \frac{\Delta N_k}{\varepsilon_k(U)}\right],$$

where $\varepsilon_n(U) = \prod_{k=1}^{n}(1+U_k)$. In particular, we have $X_n = X_0 \varepsilon_n(U)$ if $\{N_n\}_{n \geq 0}$ is a constant sequence.

Problem A.1.19 *Show that stochastic exponents satisfy to the following properties:*

(1) $\dfrac{1}{\varepsilon_n(U)} = \varepsilon_n(-U^*)$, *where* $\Delta U_n^* = \dfrac{\Delta U_n}{1 + \Delta U_n}$, $\Delta U_n \neq -1$;

(2) $\varepsilon_n(U)$ *is a martingale if and only if U is a martingale;*

(3) $\varepsilon_n(U) = 0$ *for all $n \geq \tau_0 = \inf\{k : \varepsilon_k(U) = 0\}$;*

(4) $\varepsilon_n(U)\varepsilon_n(V) = \varepsilon_n(U + V + [U, V])$, *where* $[U, V]_n = \sum_{k=1}^{n} \Delta U_k \Delta V_k$.

Problem A.1.20 *Consider a Bernoulli random variable ρ taking values b and a with probabilities p and $1 - p$, $p \in (0, 1)$, respectively. Show that*

$$\mu = E(\rho) = p(b - a) + a$$

and

$$\sigma^2 = Var(\rho) = (b - a)^2 p(1 - p).$$

Problem A.1.21 *Let $(\Omega, \mathcal{F}, (\mathcal{F}_n)_{n=0,1,\dots}, P)$ be a stochastic basis and $\widetilde{Z}$ be a positive random variable with $E(\widetilde{Z}) = 1$. Define a stochastic sequence $\widetilde{Z}_n = E(\widetilde{Z}|\mathcal{F}_n)$. Prove that it is a martingale with respect to probability $\widetilde{P}$ defined by $\widetilde{P}(A) = E(\widetilde{Z} \cdot I_A)$.*

Problem A.1.22 *Assume* $(M_n)_{n=0,1,...}$ *is a martingale on a stochastic basis* $(\Omega, \mathcal{F}, (\mathcal{F}_n)_{n=0,1,...}, P)$. *Show that* $(M_n)^2_{n=0,1,...}$ *is a submartingale and* $M_n^2 = m_n + \langle M, M \rangle_n$, *where* $(m_n)_{n=0,1,...}$ *is a martingale and* $\langle M, M \rangle_n = \sum_{k=1}^{n} E\big((M_k - M_{k-1})^2 | \mathcal{F}_{k-1}\big)$ *is a predictable sequence. (It is called the quadratic characteristic of M.)*

Problem A.1.23 *Let* ξ *be a standard normal random variable on some* $(\Omega, \mathcal{F}, P)$ *and a, b, K are constants. Prove that*

$$E\Big(a \cdot \exp\{b\xi - 0.5b^2\} - K\Big)^+ = a\Phi\left(\frac{\ln\left(\frac{a}{K}\right) + \frac{b^2}{2}}{b}\right) - K\Phi\left(\frac{\ln\left(\frac{a}{K}\right) - \frac{b^2}{2}}{b}\right),$$

where

$$\Phi(x) = \frac{1}{\sqrt{2\pi}} \int_{-\infty}^{x} \exp\left\{-\frac{y^2}{2}\right\} dy.$$

Problem A.1.24 *Let* $(\Omega, \mathcal{F}, (\mathcal{F}_n)_{n=0,1,...}, P)$ *be a stochastic basis. Assume that there exists another probability* $\widetilde{P}$ *such that* $P_n \sim \widetilde{P}_n$, *where* P_n *and* $\widetilde{P}_n$ *are restrictions of* P *and* $\widetilde{P}$ *on* $\mathcal{F}_n$. *Denote the corresponding density* $\widetilde{Z}_n = d\widetilde{P}_n/dP_n$. *Prove the following formula for changing the probability under the conditional expectation sign: for a fixed* $N \in \mathbb{Z}_+$ *and for any integrable* $\mathcal{F}_N$-*measurable random variable* Y *we have*

$$\widetilde{Z}_{N-1}\widetilde{E}(Y|\mathcal{F}_{N-1}) = E(Y\widetilde{Z}_N|\mathcal{F}_{N-1}) \quad (P-\text{a.s.}) \text{ and } (\widetilde{P}-\text{a.s.}).$$

Problem A.1.25 *Suppose that stochastic sequences* (U_n) *and* $(\widehat{U}_n)$ *are such that the usual exponential of* (U_n) *coincides with the stochastic exponential of* $(\widehat{U}_n)$:

$$\exp\{U_n\} = \varepsilon_n(\widehat{U}_n), \quad n \in \mathbb{Z}_+.$$

Express sequence $(\widehat{U}_n)$ *in terms of* (U_n) *without using the stochastic exponential.*

Problem A.1.26 *Suppose that a positive numerical sequence* $(\alpha_n)_{0 \leq n \leq N}$ *and a stochastic sequence* $(V_n)_{0 \leq n \leq N}$ *are such that sequence*

$$A_n = \sum_{k=1}^{n} E\big(e^{\alpha_k \Delta V_k} - 1 | \mathcal{F}_{k-1}\big), \quad n \leq N$$

is well defined. Prove that stochastic sequence

$$Z_n = \exp\left\{\sum_{k=1}^{n} \alpha_k \Delta V_k\right\} \varepsilon_n^{-1}(A), \ Z_0 = 1,$$

is a martingale with respect to the same filtration $(\mathcal{F}_n)_{n \leq N}$.

Defining the probability $\widetilde{P}(A) = E(Z_N I_A)$ on $\mathcal{F}_N$, prove that, for independent increments ΔV_n, this probability has the form

$$\widetilde{P}(A) = E\left(I_A \frac{\exp\{\alpha_N \Delta V_N\}}{E(\exp\{\alpha_N \Delta V_N\})}\right).$$

Problem A.1.27 *Investigate the martingale property of the stochastic sequence* $(Y_n, \mathcal{F}_n)$*, where*

$$Y_n = \alpha X_n^2 + \beta X_n + \gamma,$$

α, β, γ *are real numbers and* (X_n) *is a martingale on a stochastic basis* $(\Omega, \mathcal{F}, (\mathcal{F}_n)_{n=0,1,\ldots}, P)$*.*

Problem A.1.28 *Let* $(X_n)_{n \geq 1}$ *be a sequence of identically distributed independent random variables with the density function*

$$f(x) = \begin{cases} \dfrac{1}{8} & \text{if } x \in [0, 2), \\ y & \text{if } x \in [2, 4), \\ 0 & \text{outside of the interval } [0, 4]. \end{cases}$$

(a) *Determine the value of* y *and then calculate* $\mu = E(X_n)$ *and* $\sigma^2 = Var(X_n)$*.*

(b) *Define new sequences* $Y_n = \displaystyle\sum_{k=1}^{n} X_k$*, and* $Z_n = \displaystyle\prod_{k=1}^{n} X_k$ *and investigate their martingale properties with respect to filtration* $\mathcal{F}_n = \sigma(X_1, \ldots, X_n)$*. If they are not martingales, show how to modify these sequences in order to obtain martingales.*

Problem A.1.29 *Consider a sequence of independent random variables* $(\xi_n)_{n=1,2,\ldots,N}$ *such that each* ξ_n *takes values* $\pi = 3.14\ldots$ *and* $-\pi/2$ *with probabilities* $\dfrac{2 + \pi}{3\pi}$ *and* $\dfrac{2\pi - 2}{3\pi}$*, respectively. Define* $X_n = \displaystyle\prod_{k=1}^{n} \xi_k$ *and let* $\mathcal{F}_n = \sigma\{\xi_1, \ldots, \xi_n\}$*. Is the sequence* $(X_n)_{n=1,\ldots,N}$ *a martingale with respect to filtration* $(\mathcal{F}_n)_{n=1,\ldots,N}$*?*

A.2   General questions on financial markets

Problem A.2.1 *Consider a single-period* (B, S)*-market on a probability space* $(\Omega, \mathcal{F}, P)$*, where* $\Omega = \{\omega_1, \omega_2, \omega_3\}$*. Let* $B_0 = 1$*,* $S_0 = 100$*,* $r = 0.1$*,* $S_1(\omega_1) = 60$*,* $S_1(\omega_2) = 120$*,* $S_1(\omega_3) = 180$*. Find a risk-neutral probability.*

Problem A.2.2 *Consider a single-period (B, S)-market with $B_0 = 1$, $S_0 = 150$, $r = 0.2$. Suppose price S_1 can take 3 values: $60, 120,$ or 180. Is such a market incomplete? If the answer is positive, describe the set of martingale probabilities.*

Problem A.2.3 *Let (r_n) be a predictable sequence and (ρ_n) be a stochastic sequence in a binomial model and let Z_n be the local density of a martingale probability P^* with respect to original probability P. Find a general formula connecting sequences (r_n) and (ρ_n) with the local density Z_n.*

Problem A.2.4 *Suppose that (B, S)-market is determined by a d-dimensional sequence $(S_n)_{n \leq N}$ of prices and $B_n \equiv 1$. Prove that, in general, there is no martingale probability in the two "infinite" markets with $d = \infty$ and $N = \infty$, respectively.*

Problem A.2.5 *Let $(\Omega, \mathcal{F}, (\mathcal{F}_n)_{n \in \mathbb{Z}_+}, P)$ be a stochastic basis. Let $\mathcal{F}_n^N = \mathcal{F}_n$ for $n \leq N$ and let $P^N = P|_{\mathcal{F}_N^N}$ for each stochastic basis $(\Omega, \mathcal{F}_N^N, (\mathcal{F}_n^N)_{n \leq N}, P^N)$, $N = 1, 2, \ldots$. Consider a no-arbitrage (B^N, S^N)-market, where $B_n^N \equiv 1$, S_n^N is $\mathcal{F}_n^N$-measurable, and P^{*N} is a probability with respect to which S^N is a martingale. Then a portfolio π^N, which is determined by a predictable sequence $(\gamma_n^N)_{n \leq N}$, has the values*

$$X_n^N = X_0^N + \sum_{k=1}^{n} \gamma_k^N \Delta S_k^N, \quad n \leq N.$$

Prove that

(a) *X_n^N and $X_n^N Z_n^N$ are martingales with respect to P^{*N} and P^N, where $Z_n^N = d P_n^{*N} / d P_n^N$;*

(b) *$(Z_N, \mathcal{F}_N, P)_{N \geq 1} = (Z_N^N, \mathcal{F}_N^N, P)_{N \geq 1}$ is a martingale, $Z_\infty = \lim_{N \to \infty} Z_N$, and $0 \leq E(Z_N) \leq 1$;*

(c) *the condition $P\{Z_\infty > 0\} = 1$ is sufficient for the sequence of strategies (π^N) to be an arbitrage sequence asymptotically, i.e., $X_0^N \to 0$ (P-a.s.) as $N \to \infty$; $X_N^N \geq -C_N$, where $C_N \downarrow 0$ and for some $\varepsilon > 0$ $\lim_{N \to \infty} \sup P^N \{X_N^N \geq \varepsilon\} > 0$.*

Problem A.2.6 *Consider standard European call and put options on a no-arbitrage (B, S)-market. Prove that, if $N_2 \geq N_1$, then $C(N_2) \geq C(N_1)$ and $P(N_2) \geq P(N_1)$, respectively.*

Problem A.2.7 *Let $C = C(N, S_0, K)$ be the fair price of a standard European call option, where N is the exercise time, S_0 is the initial price of the stock, and K is the strike price. Prove that $C = C(N, S_0, K)$ has the following properties:*

(a) $C(N, S_0, K)$ *is a monotone function of* S_0 *and* K;

(b) $C(N, S_0, K)$ *is a convex function of* S_0 *and* K;

(c) $C(N, \lambda S_0, \lambda K) = \lambda C(N, S_0, K)$ *for* $\lambda > 0$.

Problem A.2.8 *An investor buys two European put options with the strike price \$40 and one European call option with the strike price \$50 on the same stock S, all options with the same maturity date N. The total price of these options is \$10. Write down the gain-loss function and discuss the possible outcomes.*

Problem A.2.9 *Suppose that an analysis of the market data suggests that the price of a certain asset S will increase by 2% in one month time with probability p, or will decrease by 1% with probability $1 - p$. Find all values of p such that an investment in this asset will be on average more profitable than an investment in a bank account with the effective monthly interest rate of 1%.*

Problem A.2.10 *Let the rate of interest be $r \geq 0$ and suppose that price of an asset S has the following dynamics:*

Ω	$n = 0$	$n = 1$	$n = 2$
ω_1	$S_0 = 10$	$S_1 = 12$	$S_2 = 15$
ω_2	$S_0 = 10$	$S_1 = 12$	$S_2 = 10$
ω_3	$S_0 = 10$	$S_1 = 6$	$S_2 = 10$
ω_4	$S_0 = 10$	$S_1 = 6$	$S_2 = 3$

1. *Find the expression for a risk-neutral probability.*

2. *Find all values of $r \geq 0$ that admit the existence of a risk-neutral probability.*

3. *Consider an American call option with the sequence of claims*

$$f_0 = \left(S_0 - 9\right)^+, \qquad f_1 = \left(S_1 - 9\right)^+, \qquad f_2 = \left(S_2 - 10\right)^+.$$

Price this option; find the minimal hedge and stopping times for $r = 0$.

Problem A.2.11 *Consider a single-period (B, S)-market with $B_0 = 1$, $S_0 = 10$, $r = 0.2$, and*

$$S_1(\omega_1) = 6, \quad S_1(\omega_2) = 12, \quad S_1(\omega_3) = 18.$$

Find a risk-neutral probability P^*.

Problem A.2.12 *Consider a single-period* (B, S)-*market with a non-risky asset* B *and two risky assets* S^1 *and* S^2, *where*

$$B_0 = 1, \qquad r = 0.2,$$
$$S_0^1 = 150, \quad S_1^1(\omega_1) = 200, \quad S_1^1(\omega_2) = 190, \quad S_1^1(\omega_3) = 170,$$
$$S_0^2 = 200, \quad S_1^2(\omega_1) = 270, \quad S_1^2(\omega_2) = 250, \quad S_1^2(\omega_3) = 230.$$

Find a risk-neutral probability P^*. *If it does not exist, find an arbitrage strategy.*

Problem A.2.13 *Consider a single-period* (B, S)-*market with* $B_0 = 1$, $S_0 = 100$, $r = 0$, *and*

$$S_1(\omega_1) = 80, \quad S_1(\omega_2) = 90, \quad S_1(\omega_3) = 180.$$

Is there a hedging strategy for a European call option with $f_1 = (S_1 - 100)^+$?

Problem A.2.14 *Consider a single-period* (B, S)-*market with* $B_0 = 1$, $S_0 = 200$, *and*
$$S_1(\omega_1) = 150, \quad S_1(\omega_2) = 190, \quad S_1(\omega_3) = 250.$$
Find all values of r *that admit the existence of a risk-neutral probability* P^*.

Problem A.2.15 *Consider a single-period* (B, S)-*market with* $B_0 = 1$, $S_0 = 10$, $r = 0.2$, *and* S_1 *that takes 3 values* $S_1(\omega_1) = 6$, $S_1(\omega_2) = 12$, $S_1(\omega_3) = 18$. *Find a risk-neutral probability* P^* *or describe the set of such probabilities.*

Problem A.2.16 *Suppose the monthly price evolution of stock* S *is given by* $\Delta S_n = S_{n-1}\rho_n$, $n = 1, 2, \ldots$, *where returns* ρ_n *are independent identically distributed random variables taking values* 0.2 *and* -0.1 *with probabilities* 0.4 *and* 0.6, *respectively. Given* $S_0 = \$300$, *find the predicted mean price of* S *for the next 3 months.*

Problem A.2.17 *Suppose the joint distribution of stock returns is given in the table*

$\rho \backslash \tilde{\rho}$	-1	0	1
-1	$1/8$	$1/12$	$7/24$
1	$5/24$	$1/6$	$1/8$

Determine their individual (marginal) distributions, expected values, and prediction of $\tilde{\rho}$ *given* ρ.

Problem A.2.18 *A joint distribution of risky asset returns* ρ *and* $\tilde{\rho}$ *is given in the table*

$\rho \backslash \tilde{\rho}$	-0.2	0	0.3
-0.1	0.3	0.2	0.1
0.2	0.1	0.2	0.1

Determine their individual (marginal) distributions, expected values, and prediction of $\tilde{\rho}$ given ρ.

Problem A.2.19 *Suppose the joint distribution of risky asset returns is given in the table*

$\rho \backslash \tilde{\rho}$	0	1	2	3	4
0	0.108	0.100	0.079	0.026	0.026
1	0.066	0.132	0.079	0.050	0.029
2	0.063	0.055	0.066	0.008	0.018
3	0.016	0.032	0.008	0.010	0.013
4	0.008	0.003	0.005	0.000	0.000

Find their individual (marginal) distributions, expected values, and variances. What is the distribution of $\tilde{\rho}$ conditional on $\rho = 1$?

Problem A.2.20 *The joint distribution of profitabilities α and β is given in the following table*

$\alpha \backslash \beta$	−0.1	0	0.1
−0.2	0.1	0	0.4
0.1	0.3	0.1	0.1

Find their individual distributions, average of β, and the conditional expectation $E(\beta|\alpha)$.

A.3   Binomial model

Problem A.3.1 *Consider a 1-step binomial (B, S)-market with $B_0 = 1$, $S_0 = 100$, $r = 0.1$, and*

$$S_1 = \begin{cases} 130 \text{ with probability } 0.4 \\ 80 \text{ with probability } 0.6 \end{cases}.$$

Consider a contingent claim $f_1 = S_1 - \min\{S_0, S_1\}$. Find:

(a) $E\left(\dfrac{f_1}{B_1}\right)$, *the heuristic price of f_1;*

(b) *Replicating portfolio and the initial capital of this strategy;*

(c) *Fair price using a martingale probability.*

Problem A.3.2 *Consider a 1-step binomial* (B, S)*-market with* $B_0 = 1$, $S_0 = 100$, $r = 0.2$, *and*

$$S_1 = \begin{cases} 150 \text{ with probability } 0.4 \\ 70 \text{ with probability } 0.6 \end{cases}.$$

Consider a contingent claim $f_1 = S_1 - \min\{S_0, S_1\}$. *Find*

(a) *Heuristic price of* f_1.

(b) *Replicating portfolio and the initial capital of this strategy.*

(c) *Fair price using a martingale probability.*

Problem A.3.3 *Consider a 1-step binomial* (B, S)*-market with* $B_0 = 1$, $S_0 = 200$, $r = 0.2$, $a = -0.4$, $b = 0.6$, *and a contingent claim (look-back call option)* $f_1 = (S_1 - K_1)^+$, *where* $K_1 = \min\{S_0, S_1\}$. *Find*

(a) *A risk-neutral probability.*

(b) *Fair price of* f_1.

(c) *Determine also heuristic prices of this option for initial probabilities* $p = 0.8$ *and* $p = 0.3$. *Comparing the results with* (b) *can you explain why this simplistic approach can not be regarded as an appropriate principle for option pricing?*

Problem A.3.4 *Consider a 1-step binomial* (B, S)*-market with* $B_0 = 1$, $S_0 = 100$, $r = 0.2$, *and*

$$S_1 = \begin{cases} 150 \text{ with probability } 0.7 \\ 80 \text{ with probability } 0.3 \end{cases}.$$

Assume that an investor has an initial capital $X = 200$. *He is going to maximize the terminal capital using expected logarithmic utility maximization. Find the optimal strategy and its terminal wealth.*

Problem A.3.5 *Consider a 1-step binomial* (B, S)*-market with* $B_0 = 1$, $S_0 = 100$, $r = 0.1$, *and return*

$$\rho_1 = \begin{cases} 0.8 \text{ with probability } 0.7 \\ -0.6 \text{ with probability } 0.3 \end{cases}.$$

Consider an option with pay-off $f_1 = \min\{110, S_1\}$. *Find*

(a) *Heuristic price of* f_1;

(b) *Fair price using a martingale probability;*

(c) *Arbitrage losses/gains if the price of the contract is chosen to be the heuristic price.*

Problem A.3.6 *Consider a 1-step binomial* (B, S)*-market with* $B_0 = 1$, $S_0 = 100$, $r = 0.2$, *and*

$$S_1 = \begin{cases} 150 \text{ with probability } 0.7 \\ 80 \text{ with probability } 0.3 \end{cases}.$$

An investor is going to maximize his terminal capital using expected logarith-mic utility maximization. Assume there is an option $f_1 = \max\{110, S_1\} - \min\{110, S_1\}$ *in the market. Is it possible to replicate this option using the optimal investment strategy of the investment problem?*

Problem A.3.7 *Consider a 1-step binomial* (B, S)*-market with* $B_0 = 1$, $S_0 = \pi/2$, $r = 0.1$, *and risky asset profitability taking values* $a = -1/3$ *and* $b = 0.5$. *Determine*

(a) *Risk-neutral probability in the market;*

(b) *Fair price of the option with pay-off* $f = |\sin S_1|$.

Problem A.3.8 *Consider a 2-step binomial* (B, S)*-market with* $B_0 = 1$, $S_0 = \pi$, $r = 0.1$, *and the rate of stock return (profitability)* ρ_1 *and* ρ_2 *are equal to*

$$\begin{cases} 0.5 \text{ with probability } 0.6 \\ -0.5 \text{ with probability } 0.4 \end{cases}.$$

Consider an option with the pay-off $f_2 = |\cos S_2|$ *and determine*

(a) *Heuristic price of* f_2;

(b) *Fair price using a martingale probability;*

(c) *Arbitrage losses/gains if the price of the contract is chosen to be the heuristic price.*

Problem A.3.9 *Consider 1- and 2-step binomial* (B, S)*-markets with* $B_0 = 1$, $S_0 = 120$, $r = 10\%$. *The risky asset return is*

$$\rho = \begin{cases} 0.6 \text{ with probability } 0.7 \\ -0.6 \text{ with probability } 0.3 \end{cases}.$$

Assume there are options $f_i = \max\{110, S_i\} - \min\{110, S_i\}$, $i = 1, 2$, *in the market. In both cases, find heuristic and fair prices. Is it possible to repli-cate these options using the optimal investment strategy maximizing expected logarithmic utility?*

Problem A.3.10 *Consider a binomial market with* $S_0 = 150$, $S_1 = \begin{cases} 240 \\ 120 \end{cases}$, $B_0 = 1$, $r = 0.2$. *Determine a martingale probability* p^* *and parameters* a, b. *Considering a 2-step market find fair prices of the following contingent claims* $f = (S_2 - 150)^+$ *and* $f = \sqrt{(150 - S_2)^+}$.

Problem A.3.11 *Consider a 1-step binomial market with $B_0 = 1$, $S_0 = 150$, $r = 0.2$. Assume S_1 takes values 240 and 120 with probabilities 0.8 and 0.2, respectively. Find an optimal proportion and optimal strategy that maximizes the expected logarithmic utility. Find its terminal capital for the initial value $200.*

Problem A.3.12 *Consider a 1-step binomial market with $B_0 = 1$, $S_0 = 150$, $r = 0.2$. Assume S_1 takes values 240 and 120 with probabilities 0.8 and 0.2, respectively. Consider the following three options in the market: $(S_1 - 150)^+$, $\max\{S_0, S_1\} - S_0$ and $S_1 - \min\{S_0, S_1\}$. Show that the optimal investment strategy, which maximizes the expected logarithmic utility, is not equal to replicating strategies of these options.*

Problem A.3.13 *Consider a binomial market with $S_0 = 200$, $B_0 = 1$, $r = 0.2$, and S_1 takes values 320 and 120. Determine a martingale probability p^* and parameters a, b. Considering a 2-step market, find fair prices of the following contingent claims $f = (S_2 - 210)^+$ and $f = \sqrt{S_2}$.*

Problem A.3.14 *Consider the following three options in the market with $S_0 = 200$, $r = 0.2$, $a = -0.2$, $b = 0.5$: $(S_1 - 100)^+$, $\max\{S_0, S_1\} - S_0$, and $S_1 - \min\{S_0, S_1\}$. Show that the investment strategy maximizing expected logarithmic utility is not equal to replicating strategies of these options.*

Problem A.3.15

(a) *Consider a 1-step binomial (B, S)-market with $B_0 = 1$, $S_0 = 100$, $r = 0.1$, and the rate of return*

$$\rho_1 = \begin{cases} 0.4 \text{ with probability } 0.7 \\ -0.2 \text{ with probability } 0.3 \end{cases}.$$

Find the price of call option with strike price $100, the price of put option with the same strike price, optimal strategy for the logarithmic utility.

(b) *How these prices are changed if*

$$\rho_1 = \begin{cases} 0.8 \text{ with probability } 0.7 \\ -0.4 \text{ with probability } 0.3 \end{cases} ?$$

Compare your results and give the corresponding explanations.

Problem A.3.16 *Consider two models*

(a) *with $B_0 = 1$, $S_0 = 100$, $r = 0.1$, and the rate of stock return*

$$\rho_1 = \begin{cases} 0.4 \text{ with probability } 0.7 \\ -0.2 \text{ with probability } 0.3 \end{cases},$$

(b) *with $B_0 = 1$, $S_0 = 100$, $r = 0.1$, and the rate of stock return*

$$\rho_1 = \begin{cases} 0.8 \text{ with probability } 0.7 \\ -0.4 \text{ with probability } 0.3 \end{cases}.$$

Taking initial prices of options to be different from fair prices (for example, heuristic prices), calculate possible arbitrage profits/losses.

Problem A.3.17 *Find an interval of non-arbitrage prices for a call option with strike price \$90 in a binomial (B^1, B^2, S)-market with $r^1 = 0.1$, $r^2 = 0.15$, $S_0 = 100$, and*

$$S_1 = \begin{cases} 120 \text{ with probability } 0.4 \\ 80 \text{ with probability } 0.6 \end{cases}.$$

Problem A.3.18 *Prove that $\sum\limits_{k=1}^{n}(\rho_k - r)$ is a martingale with respect to a martingale probability p^*, where (ρ_n) is a sequence of independent random variables such that*

$$\rho_n = \begin{cases} b \text{ with probability } p^* \\ a \text{ with probability } 1 - p^* \end{cases},$$

and

$$p^* = \frac{r - a}{b - a}, \quad -1 < a < r < b.$$

Problem A.3.19 *Consider a binomial market with interest rate r and rate of stock return*

$$\rho = \begin{cases} b \text{ with probability } p \\ a \text{ with probability } 1 - p \end{cases}.$$

Show that

$$1 - \frac{\mu - r}{\sigma^2}(b - \mu) = \frac{p^*}{p}, \quad 1 - \frac{\mu - r}{\sigma^2}(a - \mu) = \frac{1 - p^*}{1 - p},$$

where

$$p^* = \frac{r - a}{b - a}, \quad E(\rho) = \mu, \quad Var(\rho) = \sigma^2.$$

Problem A.3.20 *Consider a binomial (B^1, B^2, S)-market with two interest rates r^1 and r^2, and the rate of return ρ_n. Show that $\Delta X_n^\pi(x)$ is represented as*

$$\Delta X_n^\pi(x) = X_{n-1}^\pi(x)\left[(1 + \alpha_n)^+ r^1 - (1 - \alpha_n)^- r^2 + \alpha^n \rho^n\right],$$

$X_0^\pi = x > 0$, where $\alpha^n = \dfrac{\gamma_n S_{n-1}}{X_{n-1}^\pi}$ is a proportion of risky asset in portfolio π.

Problem A.3.21 *For* (B^1, B^2, S)-*market and its auxiliary* (B^d, S)-*market suppose* $X_0^{\pi(\alpha)} = X_0^{\pi(\alpha,d)}$, *then* $X_n^{\pi(\alpha)} = X_n^{\pi(\alpha,d)}$ *for all* $n \le N$ *if and only if* $(r^2 - r^1 - d)(1 - \alpha_n)^- + d(1 - \alpha_n)^+ = 0$.

Problem A.3.22 *For a one-period binomial symmetric model of a* (B, S)-*market let us consider an American option with contingent claim* $f_n = \beta^n (S_n - 1)^+$, $\beta \in (0, 1)$, $n = 0, 1$. *Prove that:*

(a) *for* $S_0 = 1$, *the fair price* $C(1)$ *is equal to* $\alpha \beta p^* (\lambda - 1)$, *and the exercise time* τ^* *is equal to 1, where*

$$\lambda > 1, \ \alpha = \frac{1}{1+r}, \ p^* = \frac{r - a}{b - a} = \frac{r - (\lambda^{-1} - 1)}{\lambda - 1 - (\lambda^{-1} - 1)};$$

(b) *for* $S_0 = \lambda$ *and* $\beta \in \left(\frac{\lambda - 1}{\lambda - \alpha}, 1 \right)$, *the fair price* $C(1)$ *is equal to* $\beta(\lambda - \alpha)$, *and the exercise time* τ^* *is equal to 0.*

Problem A.3.23 *Consider a one-period binomial symmetric model of a* (B, S)-*market and an American option with payoff function* $f_n = \beta^n (S_n - 1)^+$, $\beta \in (0, 1)$, $n = 0, 1$. *Assume* $S_0 = \lambda^k$ *with* $k > 1$ *or* $k \le -1$ *and show that the fair price* $C(1)$ *is equal to* $\max \left\{ \lambda^k - 1, \beta(\lambda^k - \alpha) \right\}$ *or 0, respectively. Further, show that the exercise time* τ^* *is equal to* $\min \left\{ 0 \le m \le 1 : S_m \in [\lambda^{k_1^* - m}, \infty) \right\}$, *where*

$$k_0^* = -\infty, \ k_1^* = \max \left\{ 0, \log_\lambda \frac{1 - \alpha}{1 - \beta} \right\}.$$

Problem A.3.24 *Give a pure probabilistic derivation of the Cox-Ross-Rubinstein formula assuming that stock prices go up with probability* p^* *and go down with probability* $(1 - p^*)$.

Problem A.3.25 *Consider a binomial* (B, S)-*market. Suppose we are given the following values of its parameters:*

$$a = -0.4, \quad b = 0.6, \quad r = 0.2, \quad B_0 = 1, \quad S_0 = 200.$$

Find the price and the minimal hedge of a "look-back" European call option with the contingent claim

$$f_2 = (S_2 - K_2)^+, \quad \text{where} \quad K_2 = \min\{S_0, S_1, S_2\}.$$

Problem A.3.26 *Consider a single-period binomial* (B, S)-*market with* $B_0 = 1$, $S_0 = 300$, $r = 0.1$, *and*

$$S_1 = \begin{cases} 350 & \text{with probability } 0.6 \\ 250 & \text{with probability } 0.4 \end{cases}.$$

Use the logarithmic utility function to find an optimal strategy with the initial capital 200.

Problem A.3.27 *Consider a single-period binomial (B, S)-market with with $B_0 = 1$, $S_0 = 100$, $r = 0.2$ and and*

$$S_1 = \begin{cases} 150 & \text{with probability } 0.7 \\ 80 & \text{with probability } 0.3 \end{cases}.$$

Use the logarithmic utility function to find an optimal strategy with the initial capital 200.

Problem A.3.28 *Consider a single-period binomial (B, S)-market with $B_0 = 1$, $S_0 = 300$, $r = 0.1$, and*

$$S_1 = \begin{cases} 350 & \text{with probability } 0.6 \\ 250 & \text{with probability } 0.4 \end{cases}.$$

Use the logarithmic utility function to find the optimal strategy with the initial capital $x = 200$.

Problem A.3.29 *Consider a single-period binomial (B, S)-market with $B_0 = 1$, $S_0 = 100$, $r = 0.2$ and*

$$S_1 = \begin{cases} 120 & \text{with probability } 0.4 \\ 80 & \text{with probability } 0.6 \end{cases}.$$

For a contingent claim $f_1 = S_1 - \min\{S_0, S_1\}$ find

(a) *its heuristic price;*

(b) *its fair price;*

(c) *a replicating portfolio and its initial capital.*

Problem A.3.30 *Consider a single-period binomial (B, S)-market with interest rate r and risky asset return*

$$\rho_1 = \begin{cases} b & \text{with probability } p \\ a & \text{with probability } 1 - p \end{cases}, \quad a < r < b, \ \mu = E(\rho_1).$$

Let x be an initial capital, α_0 and α_1 be proportions of non-risky and risky assets in investment portfolio with rate of return $\rho^{(\alpha_0, \alpha_1)}(x) = \alpha_0 r + \alpha_1 \rho_1$. Maximize the expected return $\alpha_0 r + \alpha_1 \mu$ over all α_0, α_1 such that $\alpha_0 + \alpha_1 \leq 1$, $x \geq 0$, under initial constraint $P(\rho^{(\alpha_0, \alpha_1)}(x) < r) \leq \alpha$, where $\alpha \in (0, 1)$.

Problem A.3.31 *Consider the following binomial (B, S, τ)-market: $B_0 = 1$, $S_0 > 0$, $1 + r(\tau) = e^{r\tau}$, $1 + b(\tau) = e^{\sigma \sqrt{(\tau)}}$, $1 + a(\tau) = e^{-\sigma \sqrt{\tau}}$, $r \geq 0$, $\sigma > 0$, where τ is the length of intervals of a subdivision of given time interval $[0, \tau]$. Consider a European call option with strike K and maturity $M = \llbracket T/\tau \rrbracket$ on* this (B, S, τ)-*market and denote* $C_{M,\tau}$ *its fair price given by the Cox-Ross-Rubinstein formula. Prove that as* $M \to \infty$ $(\tau \to \infty)$, $C_{M,\tau}$ *converges to the Black-Scholes price*

$$C = S_0 \Phi\left(\frac{\ln \frac{S_0}{K} + (r + \frac{\sigma^2}{2})T}{\sigma\sqrt{T}} \right) - Ke^{-rT}\left(\frac{\ln \frac{S_0}{K} + (r - \frac{\sigma^2}{2})T}{\sigma\sqrt{T}} \right),$$

$$\Phi(x) = \frac{1}{\sqrt{2\pi}} \int_{-\infty}^{x} e^{-y^2/2}\,dy,$$

of call option with strike price K *and maturity* T, *and the rate of convergence is at least* $M^{-1/2}$.

A.4 The Black-Scholes model

Problem A.4.1 *Consider Black-Scholes models with interest rates 2% and 4%, and volatility 20%. Assuming that initial stock and bank account values are 100 and 1, respectively, determine fair price of call/put options with strike price 110 and maturities* $T = 108/365, 215/365$. *Comparing these prices, is it possible to say something about their behavior as functions of interest rate? Give similar comparison of call prices for a dividend paying stock with dividend rates 1% and 2%.*

Problem A.4.2 *Consider the Black and Scholes model:* $dS_t(\mu\,dt + \sigma\,dW_t)$ *with interest rate* r. *Derive the following heuristic formula for a call option with strike price* K:

$$C_T^{heuristic}(\mu) = e^{(\mu-r)T} S_0 \Phi\big(d_+(\mu, T)\big) - Ke^{-rT}\Phi\big(d_-(\mu, T)\big)$$

where

$$d_\pm(\mu, T) = \frac{\ln(S_0/K) + (\mu \pm \sigma^2/2)T}{\sigma\sqrt{T}},$$

$\Phi(\cdot)$ *is a standard normal distribution function and* S_0 *is the initial price of stock.*

Problem A.4.3 *Consider a Black-Scholes market with parameters* $r = 0.02$, $S_0 = 100$, $\mu = r$, $\sigma = 0.2$. *Find prices of call and put options with* $K = 110$, $T = \dfrac{108}{365}$, *and* $T = \dfrac{215}{365}$. *Determine also the prices of call and put options for other values of volatility:* $\sigma = 0.1$, $\sigma = 0.4$, *and give an explanation of existing differences in prices.*

Problem A.4.4 *Consider a Black-Scholes market with parameters* $\rho = \mu = 2\%$, $S_0 = 100$, $\sigma = 0.2$.

(a) *Find prices of call and put options with strike price $110 and maturities*
$$T_1 = \frac{108}{365} \text{ and } T_2 = \frac{215}{365}.$$

(b) *Compare these prices with the corresponding prices (only for call option with maturity $T_2 = 215/365$) in the model with a proportional dividend 10%.*

(c) *How these results will change for the market volatility 80% (only for call option with maturity $T_2 = 215/365$)?*

(d) *How these results will change for the strike $80?*

Problem A.4.5 *Assume that $(\varepsilon_n)_{n=1,2,...,N}$ is a sequence of independent standard normal random variables on a probability space $(\Omega, \mathcal{F}, P)$, and (μ_n) and (σ_n) are two sequences of real numbers, $\sigma_n > 0$. Show that*

$$E\left(\exp\left\{ -\frac{\mu_k}{\sigma_k}\varepsilon_k - \frac{1}{2}\left(\frac{\mu_k}{\sigma_k}\right)^2 \right\} \right) = 1$$

and hence

$$\widetilde{Z}_N = \exp\left\{ -\sum_{k=1}^{N} \frac{\mu_k}{\sigma_k}\varepsilon_k - \frac{1}{2}\sum_{k=1}^{N}\left(\frac{\mu_k}{\sigma_k}\right)^2 \right\}$$

defines a new probability $\widetilde{P}$ such that

$$\frac{d\widetilde{P}}{dP} = \widetilde{Z}_N.$$

Problem A.4.6 *Let (μ_n) and (σ_n) be two real-valued sequences and $\sigma_n > 0$, $n = 1, 2, \ldots, N$. Show that*

$$\widetilde{E}\left(e^{i\lambda W_n}\right) = \exp\left\{ -\frac{\lambda^2}{2}\sigma_n^2 \right\}, \quad \lambda_n \geq 0, \quad n = 1, 2, \ldots, N,$$

where
$$W_n = \mu_n + \sigma_n\varepsilon_n,$$

$(\varepsilon_n)_{n=1,2,...}$ *is a sequence of independent standard normal variables and*

$$\frac{d\widetilde{P}_n}{dP} = \widetilde{Z}_n = \exp\left\{ -\sum_{k=1}^{n} \frac{\mu_k}{\sigma_k}\varepsilon_k - \frac{1}{2}\sum_{k=1}^{n}\left(\frac{\mu_k}{\sigma_k}\right)^2 \right\}, \quad n = 1, 2, \ldots, N.$$

Problem A.4.7 *Let (μ_n) and (σ_n) be two real-valued sequences and $\sigma_n > 0$, $n = 1, 2, \ldots, N$ and (ε_n) be a sequence of independent standard normal random variables on a probability space $(\Omega, \mathcal{F}, P)$. Define*

$$\frac{d\widetilde{P}_n}{dP} = \exp\left\{ -\sum_{k=1}^{n} \frac{\mu_k}{\sigma_k}\varepsilon_k - \frac{1}{2}\sum_{k=1}^{n}\left(\frac{\mu_k}{\sigma_k}\right)^2 \right\}.$$

Prove that

$$\tilde{E}\exp\{\tilde{\mu}_n + \sigma_n\varepsilon_n\} = 1,$$

where $\tilde{\mu}_n = \mu_n - \delta_n$ *and*

$$1 + r_n = e^{\delta_n} \text{ if and only if } \tilde{E}\Big(\frac{S_n}{B_n}\Big|\mathcal{F}_{n-1}\Big) = \frac{S_{n-1}}{B_{n-1}}.$$

Here

$$S_n = S_0\exp\Big\{\sum_1^n(\mu_k + \sigma_k\varepsilon_k)\Big\}$$

and

$$B_n = \prod_{k=1}^n(1 + r_k),$$

$B_0 = 1$, $S_0 > 0$, *and* $\mathcal{F}_n = \sigma(\varepsilon_1 \ldots \varepsilon_n)$.

Problem A.4.8 *Consider the Black-Scholes model of a (B, S)-market, and compare the optimal investment strategy with the minimal hedge of an European call option with $f_T = (S_T - K)^+$.*

Problem A.4.9 *In the framework of the Black-Scholes model of a (B, S)-market, consider an investment portfolio π with the initial capital x. Estimate the asymptotic profitability of π:*

$$\limsup_{T\to\infty}\frac{1}{T}\ln E\big(X_T^\pi(x)\big)^\delta, \qquad \delta \in (0, 1].$$

Problem A.4.10 *Let $C = C(S_0, T, K, \sigma, r)$ be the fair price of a call option in the Black-Scholes model. It is a function of S_0 (initial stock price), T (exercise time), K (strike price), σ (volatility), r (interest rate). Prove that*

(a) $C(S_0) \to 0$ *as* $S_0 \to 0$, $C(S_0) \to \infty$ *as* $S_0 \to \infty$;

(b) $C(K) \to S_0$ *as* $K \to 0$, $C(K) \to 0$ *as* $K \to \infty$;

(c) $C(S_0, T, K) \to S_0$ *as* $T \to \infty$, $C(S_0, T, K) \to S_0 - K$ *as* $T \to 0$ $(S_0 > K)$;

(d) $C(S_0, T, K, \sigma, r) \to S_0 - Ke^{-rT}$ *as* $\sigma \to 0$ $(S_0 > Ke^{-rT})$, $C(S_0, T, K, \sigma, r) \to S_0$ *as* $\sigma \to \infty$ $(S_0 > Ke^{-rT})$;

(e) $C(S_0, T, K, \sigma, r) \to S_0$ *as* $r \to \infty$.

Problem A.4.11 *Let $P = P(S_0, T, K, \sigma, r)$ be the fair price of a put option in the Black-Scholes model. Prove the following properties of P as a function of S_0, T, K, σ, r:*

(a) $P(S_0, T) \to Ke^{-rT}$ *as* $S_0 \to 0$, $P(S_0, T) \to 0$ *as* $S_0 \to \infty$;

(b) $P(K) \to 0$ *as* $K \to 0$, $P(K) \to \infty$ *as* $K \to \infty$;

(c) $P(T) \to 0$ as $T \to \infty$, $P(T) \to 0$ as $T \to 0$ $(S_0 > K)$;

(d) $P(\sigma) \to 0$ as $\sigma \to 0$ $(S_0 > Ke^{-rT})$, $P(\sigma) \to Ke^{-rT}$ as $\sigma \to \infty$ $(S_0 > Ke^{-rT})$;

(e) $P(r) \to 0$ as $r \to \infty$.

A.5   Bond market

Problem A.5.1 *Determine the yield to maturity for a two-year bond with payments $20 in the first year and $120 in the second year, if the market value of the bond is $100.*

Problem A.5.2 *Find the market price, yield to maturity and duration of a 4% annual coupon bond, if the term structure of interest rates and bond's face value are given in the following table*

years	1.0	2.0	3.0	face value
Term structure	2.5%	3.5%	4.5%	$100

Problem A.5.3 *Suppose that the term structure of zero-coupon yields has the polynomial form with the following shape parameters:*

(a) $A_0 = 0.08$; $A_1 = 0.02$; $A_2 = -0.003$; $A_3 = 0.0001$;

(b) $A_0 = 0.06$; $A_1 = 0.01$; $A_2 = -0.001$; $A_3 = 0.0001$.

The bond's face value, annual coupon rate and maturity are $1000, 5%, and 4 years, respectively. Find the price of this bond. Then assume that the short rate is increased by 40 basic points, the slope is decreased by 10 basic points and the other parameters are unchanged. Find the percentage change of bond's price.

Problem A.5.4 *Suppose that the term structure of zero-coupon yields has the polynomial form with the following shape parameters:*

(a) $A_0 = 0.08$; $A_1 = 0.02$; $A_2 = -0.003$; $A_3 = 0.0001$;

(b) $A_0 = 0.06$; $A_1 = 0.01$; $A_2 = -0.001$; $A_3 = 0.0001$.

The bond's face value, annual coupon rate, and maturity are $1000, 5%, and 4 years, respectively. Find the term structure of instantaneous forward rates. Then determine the bond's price, the shift in the term structure, and the percentage change in the price.

Problem A.5.5 *Suppose that the term structure of zero-coupon yields has the polynomial form with the following shape parameters:*

(a) $A_0 = 0.08$; $A_1 = 0.02$; $A_2 = -0.003$; $A_3 = 0.0001$;

(b) $A_0 = 0.06$; $A_1 = 0.01$; $A_2 = -0.001$; $A_3 = 0.0001$.

Find the price of a call option on the zero-coupon bond (with maturity $T = 4$ years) with strike $K = \$80$ and the expiration time $T^1 = 1$ year. Hint: use the following pricing formula

$$C = B(0,T)\Phi(d_1) - KB(0,T^1)\Phi(d_2),$$

where

$$d_1 = \frac{\ln\left(\frac{B(0,T)}{B(0,T^1)K}\right) + \frac{V}{2}}{\sqrt{V}}, \quad d_2 = d_1 - \sqrt{V}$$

and $V = 0.25$.

How the pricing formula will change for a face value that is different from $\$1$? Assuming that face value is $\$80$, calculate the price of such option. Compare it with the previous result.

Problem A.5.6 *The market price of a two-year bond at time $t = 0$ is $\$100$. The bond has the following payments: $\$10$ in the first year and $\$118$ in the second year. Determine the yield to maturity for this bond.*

Problem A.5.7 *Build a theoretical yield curve based on the following bonds that exist in the market:*

years	0.5	1.0	1.5	2.0	Price at 0
$B(1)$	104				100
$B(2)$		116			108
$B(3)$	8	8	118		112
$B(4)$	10	10	10	120	130

Problem A.5.8 *We are given the following term structure of interest rates in the market: 3% in the first year, 4% in the second year, and 5% in the third year. Find the market price, yield to maturity, and duration of a three-year bond that pays an annual coupon of 4% and has the nominal value of $\$100$.*

Problem A.5.9 *Consider the following six bonds:*

			(a_1)	(a_2)
B_1	98	1	1	2
B_2	96	2	2	2
B_3	94	3	3	2
B_4	92	4	2	3
B_5	90	5	2	2
B_6	88	6	2	1
	Price	Maturity	Annual Coupon %	Annual Coupon %

(a) *Assuming the continuous compounding, use the bootstrapping method to find the term structure of interest rates in both cases* (a_1) *and* (a_2).

(b) *Repeat part* (a) *first assuming that all coupons are 1% and then assuming that all coupons are 3%. Is there any difference between these two term structures of interest rates?*

(c) *Repeat part* (a) *in the case when all coupons are semiannual. Discuss the difference in the term structure.*

(d) *Is it possible to use the Nelson-Siegel model in cases* (a_1) *and* (a_2)?

Problem A.5.10 *There are two bonds B_1 and B_2 given by*

Face value	Coupon rate %		Maturity
$100	4	Semiannual	2 years
$100	8	Annual	2 years

(a) *Assume that an investor has $800 to invest $400 into each bond B_1 and B_2. Immediately after $t = 0$ the (continuously compounding) interest rate is changed from 5% to 6% p.a. Find D_{port} and C_{port}, and relative changes of the portfolio price under this change of interest rate. Also find the expected and real investment costs at time $t = 2$ and at duration. Make a conclusion about portfolio's immunization property.*

(b) *Repeat part* (a) *in the case when $200 was invested in B_1 and $600 was invested in B_2.*

(c) *Repeat part* (a) *in the case when $600 was invested in B_1 and $200 was invested in B_2.*

Problem A.5.11 *Assume that at time $t = 0$, the annual continuously compounded interest rate is 5%. Consider two bonds B_1 and B_2 with face values $100, annual coupons 8% each, and*

(**case 1**) *maturities 1 and 3 years, respectively;*

(**case 2**) *maturities 2 and 3 years, respectively.*

Assuming that the investment horizon is 3 years and the initial investment is $800, find an immunized portfolio against the following changes of interest rate 5% → 6% (after $t = 0$) → 5% (after $t = 1$).

Problem A.5.12 *Assume that at time $t = 0$, the annual continuously compounded interest rate is 5%. Consider two bonds B_1 and B_2 with face values $100, annual coupons 8% each, and*

(**case 1**) *maturities 1 and 3 years, respectively;*

(**case 2**) *maturities 2 and 3 years, respectively.*

Assuming that the investment horizon is 3 years, the initial investment is $840 and the transaction cost is 0.5%, find an immunized portfolio against the following changes of interest rate $5\% \rightarrow 6\%$ (after $t = 0$) $\rightarrow 7\%$ (after $t = 1$).

Problem A.5.13 *Suppose that effective annual rate of interest is 10%. Find the present value of a 3-year bond with face value $500 and with annual coupon payments of $100.*

A.6   Risk and performance measurement

Problem A.6.1 *Company A, a $4 billion firm, plans to expand within their industry through merger. Their preliminary scan of candidates yielded two firms: B and C, both $4 billion companies. In order to prepare a more thorough analysis of the acquisition candidate, company A collected information about the long-term returns of the two firms. The annual time series of total returns for the last 28 years for each firm are shown in the table below.*

(a) *Suppose that company A mergers with company B. Find Sharpe and Sortino ratios for the new firm. Recall that the Sortino ratio for company X is*

$$SoR_X = \frac{E(R_X) - L}{\sqrt{LPM_{2X}}},$$

where R_X is the return of company X, L is the minimal acceptable return and LPM_{2X} is the second lower partial moment:

$$LPM_{nX} = \frac{1}{T} \sum_{t=1}^{T} \max\{0, L - R_X\}^n$$

with T equal to number of returns.

(b) *Assume that company A mergers with company C. Find Sharpe and Sortino ratios for the new firm.*

(c) *Which merger is more profitable? (Assume, that the minimal acceptable return is equal to the riskless interest rate of 4% compounded annually.)*

Company A	Company B	Company C
0.130	0.195	0.123
0.050	0.100	0.196
0.072	0.085	0.221
-0.003	0.213	0.205
0.324	0.058	-0.106
0.155	0.168	0.152
0.058	-0.034	0.181
0.177	0.103	0.092
0.074	0.298	0.152
-0.016	-0.006	0.108
0.131	0.097	-0.171
0.233	0.063	0.103
0.099	-0.049	0.179
0.071	0.053	0.158
0.177	0.323	0.201
-0.022	0.104	0.111
-0.153	0.126	-0.017
0.184	0.146	0.181
0.116	0.088	-0.018
0.174	-0.013	0.114
0.053	0.213	-0.023
-0.072	0.130	-0.048
-0.031	0.062	0.103
0.078	-0.003	0.137
0.198	0.152	0.000
-0.004	0.114	0.149
0.124	-0.145	0.166
0.048	0.228	0.163

Problem A.6.2 *Company A, a $4 billion firm, plans to expand within their industry through merger. Their preliminary scan of candidates yielded two firms: B and C, both $4 billion companies. In order to prepare a more thorough analysis of the acquisition candidate, company A collected information about the long-term returns of the two firms. The annual time series of total returns for the last 28 years for each firm are shown in the table for Problem A.6.1 above.*

(a) *Suppose that company A mergers with company B. Find Sharpe and Omega ratios for the new firm. Recall that the Omega ratio for company X is*

$$OmR_X = \frac{E(R_X) - R_f}{LPM_{1X}} + 1,$$

where R_X is the return of company X, R_f is a risk-free interest rate, and LPM_{1X} is the first lower partial moment:

$$LPM_{nX} = \frac{1}{T} \sum_{t=1}^{T} \max\{0, L - R_X\}^n,$$

with T equal to number of returns.

(b) *Assume that company A mergers with company C. Find Sharpe and Omega ratios for the new firm.*

(c) *Which merger is more profitable? (Assume, that minimal acceptable return is equal to the riskless interest rate of 4% compounded annually.)*

Problem A.6.3 *Company A, a $4 billion firm, plans to expand within their industry through merger. Their preliminary scan of candidates yielded two firms: B and C, both $4 billion companies. In order to prepare a more thorough analysis of the acquisition candidate, company A collected information about the long-term returns of the two firms. The annual time series of total returns for the last 28 years for each firm are shown in the table for Problem A.6.1 above.*

(a) *Suppose that company A mergers with company B. Find Sharpe and Kappa3 ratios for the new firm. Recall that the Kappa3 ratio for company X is*

$$K3R_X = \frac{E(R_X) - R_f}{\sqrt[3]{LPM_{3X}}},$$

where R_X is the return of company X, R_f is a risk-free interest rate, and LPM_{3X} is the third lower partial moment:

$$LPM_{nX} = \frac{1}{T} \sum_{t=1}^{T} \max\{0, L - R_X\}^n$$

with T equal to number of returns.

(b) *Assume that company A mergers with company C. Find Sharpe and Kappa3 ratios for the new firm.*

(c) *Which merger is more profitable? (Assume, that minimal acceptable return is equal to the riskless interest rate of 4% compounded annually.)*

Problem A.6.4 *Company A, a $1 billion firm, plans to expand within their industry through merger. Their preliminary scan of candidates yielded two firms: B and C, both $4 billion companies. In order to prepare a more thorough analysis of the acquisition candidate, company A collected information about the long-term returns of the two firms. The annual time series of total returns for the last 28 years for each firm are shown in the table below.*

(a) *Suppose that company A mergers with company B. Find Sharpe and Upside Potential ratios for the new firm. Recall that the Upside Potential ratio for company X is*

$$UPR_{XY} = \frac{HPM_{1XY}}{\sqrt{LPM_{2XY}}},$$

HPM_{1X} is the first higher partial moment and LPM_{2X} is the second lower partial moment:

$$HPM_{nX} = \frac{1}{T} \sum_{t=1}^{T} \max\{0, R_X - L\}^n,$$

$$LPM_{nX} = \frac{1}{T} \sum_{t=1}^{T} \max\{0, L - R_X\}^n$$

with T equal to number of returns.

(b) Suppose that company A mergers with company C. Find Sharpe and Upside Potential ratios for the new firm.

(c) Which merger is more profitable? (Assume, that minimal acceptable return is equal to the riskless interest rate of 10% compounded annually.)

Company A	Company B	Company C
0.13	0.091	0.042
0.05	0.148	0.219
0.072	0.127	0.123
−0.003	0.2	0.142
0.324	0.148	0.032
0.155	0.086	0.031
0.058	0.214	0.175
0.177	0.121	0.136
0.074	0.158	0.132
−0.016	0.089	0.054
0.131	0.032	0.098
0.233	−0.004	0.128
0.099	0.085	−0.009
0.071	0.138	0.191
0.177	0.105	0.206
−0.022	0.001	0.063
−0.153	0.127	0.043
0.184	0.077	0.255
0.116	0.107	0.464
0.174	0.153	0.025
0.053	0.049	0.234
−0.072	0.043	0.011
−0.031	0.064	0.159
0.078	0.076	0.064
0.198	0.074	0.06
−0.004	0.135	0.112
0.124	0.06	0.072
0.048	0.135	0.078

Problem A.6.5 *Company A, a $4 billion firm, plans to expand within their industry through merger. Their preliminary scan of candidates yielded two firms: B and C, both $1 billion companies. In order to prepare a more thorough analysis of the acquisition candidate, company A collected information about the long-term returns of the two firms. The annual time series of total returns for the last 28 years for each firm are shown in the table below.*

(a) *Suppose that company A mergers with company B. Find Sharpe and Israelsen ratios for the new firm. Recall that the Israelsen ratio of company X is*

$$
IsR_X =
\begin{cases}
\dfrac{E(R_X) - R_f}{\sigma_X} & \text{if } E(R_X) - R_f > 0 \\[2mm]
\big(E(R_X) - R_f\big) \cdot \sigma_X & \text{if } E(R_X) - R_f < 0
\end{cases}
,
$$

where R_X is the return of company X, R_f is a risk-free interest rate, and σ_X is the standard deviation of returns of company X.

(b) *Suppose that company A mergers with company C. Find Sharpe and Israelsen ratios for the new firm.*

(c) *Which merger is more profitable? (Assume, that minimal acceptable return is equal to the riskless interest rate of 5% compounded annually.)*

Company A	Company B	Company C
0.692	−1.185	−4.905
0.383	−0.331	5.642
−0.098	0.037	−2.591
0.039	0.505	3.777
0.645	0.093	−2.301
−0.167	−0.201	1.453
−0.601	0.236	1.684
1.106	−0.285	2.917
0.296	0.292	−4.81
−0.464	−0.973	−7.544
−0.614	0.622	3.155
−1.798	0.135	0.093
−0.061	0.556	−1.118
−0.647	0.109	−5.594
−1.226	0.086	3.282
0.874	−0.006	−0.866
−1.533	−1.045	3.491
−0.533	−0.178	−3.697
0.521	0.951	0.311
−0.788	0.508	2.358
2.503	−0.465	−2.152
0.329	0.702	−2.295
−0.44	−0.332	2.426
1.244	0.032	5.088
1.448	0.296	−0.327
−0.552	−0.842	3.377
0.757	0.019	0.831
−1.057	0.937	−1.416

Problem A.6.6 *Suppose* 107, 207, 162, 61, 47, 16, −99, 269, 24, 101, 63, 173, −278, 159, 184 *are 15 observations of the return of a one dollar investment in some portfolio. This return has a normal distribution with density*

$$f(x) = \frac{1}{\sigma\sqrt{2\pi}} \exp\left\{-\frac{(x-a)^2}{2\sigma^2}\right\}.$$

Calculate the sample $AVaR_{0.02}$.

Problem A.6.7 *Let* −12, 35, 12, −132, 373, −110, 15, 55, −13, −11, −42, −28, 283, 313, 76 *be observations of the return of some portfolio. Assume that this return has the Gram-Charlier distribution with density*

$$f(x) = \frac{1}{\sigma\sqrt{2\pi}} \exp\left\{-\frac{(x-a)^2}{2\sigma^2}\right\}\left[1 + \frac{\xi}{6}H_3\left(\frac{x-a}{\sigma}\right) + \frac{\kappa-3}{24}H_4\left(\frac{x-a}{\sigma}\right)\right],$$

where H_n *is the n-th order Hermite polynomial. Calculate the sample* $AVaR_{0.02}$.

Problem A.6.8 *Consider 20 observations of the return of some portfolio:*
149, 25, 72, 64, 203, 14, 55, 3, 88, 141, 2, 104, 135, 221, 16, 229, 245, 30, 11,
6. Suppose that this return has the Exponential distribution with density

$$f(x) = ae^{-ax} \quad (x \geq 0).$$

Calculate the sample $AVaR_{0.02}$.

Problem A.6.9 *Consider a portfolio with the initial capital 1000 at time*
$t = 0$. *Suppose that the interest rate is 0.04 p.a. and the dynamics of stock*
price S are given by the following equation

$$\frac{d\,S_t}{S_t} = 0.09\,dt + 0.3\,dW_t.$$

Find an optimal one-year investment strategy that minimizes the Capital-at-
Risk at confidence level $\alpha = 0.05$. *Also find a strategy that maximizes the*
expected return, if the Capital-at-Risk is restricted by 300.

Problem A.6.10 *Consider two mutual funds A and B with twelve monthly*
returns given in the following table

A	B
−0.011099605	0.02
−0.01085255	0.049258728
−0.009761267	−0.008659982
−0.007490409	−0.002298851
−0.00696508	−0.013364055
−0.003605429	0.034096217
0.000779376	0.038392051
0.008108763	−0.062635929
0.009798749	−0.003712297
0.010074997	−0.047973917
0.01347361	0.033757339
0.015737041	−0.014671084

Using two performance measures: Sharpe ratio and Upside Potential ratio,
analyze the effectiveness if their management.

A.7 Elements of insurance and actuarial science

Problem A.7.1 *Consider the Black-Scholes model of a* (B, S)-*market with*
$T = 215/365$, $S_0 = 100$ *and* $\mu = r$. *Calculate the premium for a pure endow-*
ment insurance with the guaranteed minimal payment $K = 80$ *in the cases*
when $r = 0.1$ *or* $r = 0.2$, *and* $\sigma = 0.1$ *or* $\sigma = 0.8$.

Problem A.7.2 *Consider the discrete Gaussian model of a (B,S)-market with $T = 215/365$, $S_0 = 100$, and $\mu = r$. Calculate premium for a pure endowment insurance with the guaranteed minimal payment $K = 80$ in the cases when $r = 0.1$ or $r = 0.2$, and $\sigma = 0.1$ or $\sigma = 0.8$.*

Problem A.7.3 *Consider the binomial model of a (B,S)-market with $S_0 = 100$, $B_0 = 1$, $r = 0.2$, and*

$$\rho = \begin{cases} 0.5 & \text{with probability } 0.4 \\ -0.3 & \text{with probability } 0.6 \end{cases}.$$

Calculate the premium for a pure endowment insurance with the guaranteed minimal payment $K = 100$ in the cases when $N = 1$ and $N = 2$.

Problem A.7.4 *Suppose that an insurance company issues 90 independent identical policies, and suppose that the average amount of claims is $300 with standard deviation $100. Estimate the probability of total claim amount S to be greater than $29000.*

Problem A.7.5 *Suppose that an insurance company issues 100 independent identical policies. Find probabilistic characteristics of an individual claim X given the following statistical data:*

	amount of claim	number of claims
1	0 − 400	2
2	400 − 800	24
3	800 − 1200	32
4	1200 − 1600	21
5	1600 − 2000	10
6	2000 − 2400	6
7	2400 − 2800	3
8	2800 − 3200	1
9	3200 − 3600	1
10	> 3600	0

Problem A.7.6 *Suppose that an insurance company issued 1000 independent identical policies, and as a result, 120 claims were received during the past 12 months. Find the probability of not receiving a claim from an individual policyholder during the next 9 months.*

Problem A.7.7 *Suppose that the following table describes the frequency of receiving claims by an insurance company during one year:*

number of claims	number of policies
0	3288
1	642
2	66
3	4

Find the probability of receiving only one claim from two independent policies during the next year.

Problem A.7.8 *Suppose that an insurance company issued 4000 independent identical policies. Find the expected number of policies that will result in 0, 1, 2, and 3 claims per year if*

(1) *the number of claims from one policy per year has Poisson distribution with parameter 0.1965;*

(2) *the number of claims from one policy per year has binomial distribution with the average 0.1965.*

Problem A.7.9 *Suppose that an insurance company issued 1000 independent identical policies. Further, suppose that the probability of receiving a claim from one policy is 0.5, and that each policy allows no more than one claim to be made. Find the probability of the total number of claims between 470 and 530.*

Problem A.7.10 *An insurance company estimated that the probability of receiving a claim from one policy during one year is 0.01 and the average amount of a claim is $980. Suppose that the company issues 1000 independent identical one-year policies. Find the probability of the total amount of claims to be more than $14850.*

Problem A.7.11 *Suppose that the following table describes q, the frequency of receiving claims by an insurance company during one year:*

number of claims	number of policies
0	3280
1	640
2	64
3	4

Determine a 95% confidence interval for q.

Problem A.7.12 *Consider three policies with claims X_1, X_2, and X_3, respectively. Suppose*

$$P(\{\omega: X_1 = 0\}) = 0.5, \quad P(\{\omega: X_1 = 100\}) = 0.5,$$
$$P(\{\omega: X_2 = 0\}) = 0.8, \quad P(\{\omega: X_2 = 250\}) = 0.2,$$
$$P(\{\omega: X_3 = 0\}) = 0.4, \quad P(\{\omega: X_3 = 100\}) = 0.4,$$
$$P(\{\omega: X_3 = 50\}) = 0.2.$$

Find the most and the least risky policy.

Problem A.7.13 *Consider two independent policies with the following distributions of claims:*

$$P(\{\omega: X_1 = 100\}) = 0.6, \quad P(\{\omega: X_1 = 200\}) = 0.4,$$
$$P(\{\omega: X_2 = 100\}) = 0.7, \quad P(\{\omega: X_2 = 200\}) = 0.3.$$

Suppose that the probability of receiving a claim from the first policy is 0.1 and from the second one is 0.2. Find the distribution of claims for the portfolio formed by these two policies.

Problem A.7.14 *In the framework of the individual risk model, consider a portfolio of 50 independent identical claims. Suppose that premiums are calculated according to the expectation principle with the security loading coefficient 0.1. Assuming that exactly one claim is received from each policyholder, find the probability of solvency in the following cases:*

(a) *each claim has exponential distribution with average 100;*

(b) *each claim has normal distribution with average 100 and variance 400;*

(c) *each claim has uniform distribution in the interval $[70, 130]$.*

Problem A.7.15 *In the framework of a binomial model, consider two insurance companies. Suppose that the claims of the first company are distributed according to the Poisson law with average 2, and that the probability of receiving a claim is equal to 0.1. For the second company, we assume the same probability of receiving a claim and the following distribution of claims: $P(\{\omega: X = 2\}) = 1$. Given that both companies receive a premium of 1 and have zero initial capitals, find the corresponding probabilities of solvency: $\phi(0, 1)$, $\phi(0, 2)$, and $\phi(0)$.*

Problem A.7.16 *Consider the Cramér-Lundberg model with the premium income $\Pi(t) = t$ and with the claims flow represented by a Poisson process with intensity 0.5. Suppose that the average claim amount is 1 with variance 5. Estimate the Cramér-Lundberg coefficient.*

Problem A.7.17 *Consider the Cramér-Lundberg model with the premium income* $\Pi(t) = t$ *and with the claims flow represented by a Poisson process with intensity* 0.5. *Suppose that claim amounts are equal to* 1 *with probability* 1. *Find the Cramér-Lundberg coefficient.*

Problem A.7.18 *Consider* 50 *independent identical insurance policies. Suppose that the average claim received from a policy during a certain time period is* 100 *with variance* 200. *Also suppose that the equivalence principle is used for premiums calculations and that all premiums income is invested in a non-risky asset with the yield rate of* 0.025 *per specified period. Estimate the probability of solvency and the expected profit.*

Problem A.7.19 *Consider* 50 *independent identical insurance policies. Suppose that the average claim received from a policy during a certain time period is* 100 *with variance* 200. *Also suppose that the equivalence principle is used for premiums calculations and that all premiums income is invested in a non-risky asset with the yield rate of* 0.025 *per specified period. Estimate the probability of solvency and the expected profit assuming that there is an opportunity to invest in a risky asset with profitability*

$$\rho = \begin{cases} 0.06 & \text{with probability } 0.5 \\ -0.005 & \text{with probability } 0.5 \end{cases}.$$

Problem A.7.20 *Consider an insurance company such that its annual aggregate claims payment has exponential distribution with the average of* 40000. *Suppose that this company operates in the framework of a* (B, S)*-market, where the profitability of a risky asset is*

$$\rho = \begin{cases} 0.1 & \text{with probability } 0.5 \\ 0.3 & \text{with probability } 0.5 \end{cases},$$

and the rate of interest is 0.2. *Suppose that* $S_0 = 10$, *and that all premium income is invested in a portfolio. Find an investment strategy* $\pi = (\beta, \gamma)$ *that minimizes the probability of ruin.*

Problem A.7.21 *Find the probability that a newborn individual survives to the age of* 30 *if the force of mortality is constant* $\mu_x \equiv \mu = 0.001$.

Problem A.7.22 *Explain why function* $(1+x)^{-2}$ *cannot be used as the force of mortality.*

Problem A.7.23 *Consider the survival function*

$$s(x) = 1 - \frac{x}{100}, \quad 0 \le x \le 100.$$

Find the force of mortality and the probability that a newborn individual survives to the age of 20 *but dies before the age of* 40.

Problem A.7.24 *Consider the Gompertz's model with* $\mu = [\![1.1]\!]^x$. *Find* $p_0(t)$.

Problem A.7.25 *Consider an insurance company with the initial capital of* 250. *Suppose that the company issues* 40 *independent identical insurance policies and that the average claim amount is* 50 *per policy with standard deviation* 40. *Premiums are calculated according to the expectation principle with the security loading coefficient* 0.1. *The company has an option of entering a quota share reinsurance contract with retention function* $h(x) = x/2$. *The reinsurance company calculates its premium according to the expectation principle with the security loading coefficient* 0.15. *Estimate the expected profit and the probability of ruin of the (primary) insurance company in the cases when it purchases the reinsurance contract and when it does not.*

Problem A.7.26 *Suppose that annual aggregate claims payments of an insurance company are uniformly distributed in* $[0, 2000]$. *Consider a stop-loss reinsurance contract with the retention level* 1600. *Compute expectations and variances of aggregate claims payments of both insurance and reinsurance companies.*

Appendix B

Bibliographic Remarks

Chapter 1. We introduce the notions of a financial market and of basic and derivative securities. We discuss the notion of a bank account as a risk-free asset and the related methods of dealing with interest rates. It is illustrated that probabilistic methods are the natural choice of tools for financial modeling. We give a brief introduction to probability theory and stochastic analysis as a foundation for modeling and quantification of risks in finance and insurance ([13], [49], [87], [88], [22], [82], [50]).

Chapter 2. As in the probability theory, where many general ideas and methods are often first explained in a discrete (Bernoulli) case, in financial mathematics binomial markets are considered to be a good starting point in studying such fundamental notions as arbitrage, completeness, hedging, and optimal investment. We use this approach in our book, and this chapter is focused on quantitative analysis of risks related to contingent claims and maximization of utility functions in the framework of the simplest (binomial or Cox-Ross-Rubinstein) model of a market. We also study the asymptotic behavior of binomial markets. In particular, we show that Black-Scholes formula and Black-Scholes equation can be introduced from the Cox-Ross-Rubinstein formula by limit arguments ([1], [20], [30], [34], [44], [62], [76], [88], [42], [19], [27], [43], [65]).

Chapter 3. This chapter begins with a comprehensive study of discrete markets. We discuss two fundamental theorems of financial mathematics, and give a systematic presentation of quantitative methodologies for pricing contingent claims in complete and incomplete markets, in markets with constraints, and in markets with transaction costs. In the setting of discrete time Gaussian markets, we study the discrete time Black-Scholes formula and the methodology of mean-variance hedging ([16], [34], [30], [76], [88], [24], [39], [40], [92], [65], [84], [85], [18], [35], [26]).

Chapter 4. We start with a brief introduction to stochastic analysis in continuous time, including Wiener process, Kolmogorov-Itô formula, martingale representation, Girsanov theorem, and so forth. Then we introduce the Black-Scholes model as a framework for a systematic study of financial risks. The methodology of martingale measures is used here to derive the Black-Scholes formula. We also discuss in detail various extensions of Black-Scholes model and formula, and their applications. This includes pricing contingent claims and optimal investment problems for models with stochastic volatility; with or without taking into account dividends and transaction costs, including the case of insider information. The final part of the chapter is devoted to the topic of imperfect hedging and risk measures, which became pertinent because of developments in the theory and practice of financial risk management, including some advances in the international financial market regulation ([3], [9], [12], [30], [48], [53], [55], [66], [67], [70], [88], [7], [51], [41], [93], [58], [94], [96], [5], [32], [33], [31], [79], [71], [10], [6], [75], [54], [77], [78], [11], [73], [45], [64], [61], [59], [23]).

Chapter 5. This chapter is devoted to models of bond markets and pricing of derivatives in these markets. We begin with a detailed introduction to deterministic models, which helps to develop stochastic models in the second part of the chapter. Vasiček model is one of the key ingredients of this modeling. We also discuss some computational aspects of pricing derivatives in bond markets ([8], [30], [66], [70], [80], [88], [75], [46], [43], [72]).

Chapter 6. First section is devoted to real options that are associated with long-term investment projects. The Bellmann principle is one of the main tools in studying real options. The second section is devoted to technical analysis, which is a very common tool in investigating the qualitative structure of risks. We demonstrate how probabilistic methods can add some quantitative aspects to technical analysis. The final section deals with performance analysis, which is based on risk-adjusted performance measures. We demonstrate applications of this methodology to ranking of managers and to assessments of effectiveness of firms' mergers and acquisitions ([14], [28], [52], [69], [74], [89] [75], [56], [57], [29], [37], [4]).

Handbooks [2], [47] are the standard sources of information on special functions and differential equations that are useful for solving the Bellmann equation, optimal stopping stopping time problem, and so forth.

Chapter 7. Complex binomial and Poisson models are used for modeling the capital of an insurance company. Actuarial criteria for premium calculations are presented. An important type of insurance that is related to combination of risks in insurance and in finance is represented by equity-linked life insurance contracts and by reinsurance with the help of derivative securities. Quantile hedging is used for pricing such contracts. We also discuss the relation between the Black-Scholes equation/formula (finance) and the Thiele differential equations (insurance) ([15], [63], [66], [68], [81], [38], [83], [64], [61], [59], [60], [17], [1], [25], [91]).

Chapter 8. Probability of ruin is used as a measure of solvency of an insurance company. Various estimates of probability of ruin are given, including the celebrated Cramér-Lundberg estimate. We discuss models that take into account an insurance company's financial investment strategies ([15], [21], [36], [63], [66], [81], [90], [95], [83]).

Bibliography

[1] K.K. Aase. On the St. Petersburg paradox. *Scand. Actuar. J.*, (1):69–78, 2001.

[2] M. Abramowitz and I.A. Stegun, editors. *Handbook of Mathematical Functions with Formulas, Graphs, and Mathematical Tables*. Dover Publications, Inc., New York, 1992.

[3] J. Amendinger, P. Imkeller, and M. Schweizer. Additional logarithmic utility of an insider. *Stochastic Process. Appl.*, 75(2):263–286, 1998.

[4] G. Andrade and E. Stafford. Investigating the economic role of mergers. *J. Corporate Finance*, 10(1):1–36, 2004.

[5] P. Artzner, F. Delbaen, J.-M. Eber, and D. Heath. Coherent measures of risk. *Math. Finance*, 9(3):203–228, 1999.

[6] M. Avellaneda, A. Levy, and A. Paras. Pricing and hedging derivative securities in markets with uncertain volatilities. *Appl. Math. Finance*, 2(1):73–88, 1995.

[7] Basel Committee on Banking Supervision. *Range of Practices and Issues in Economic Capital Modelling*, 2008.

[8] S. Basu and A. Dassios. *Modelling Interest Rates and Calculating Bond Prices*. London School of Economics, 1997.

[9] M.W. Baxter and A.J.O. Rennie. *Financial Calculus: An Introduction to Derivative Pricing*. Cambridge University Press, Cambridge, 1996.

[10] Y.Z. Bergman. Option pricing with differential interest rates. *Rev. Financ. Stud.*, 8(2):475–500, 1995.

[11] F. Black. How to use the holes in (b)lack-(s)choles. *J. Applied Corporate Finance*, 1(4):67–73, 1989.

[12] F. Black and M. Scholes. The pricing of options and corporate liabilities. *J. Polit. Econ.*, 81:637–654, 1973.

[13] Z. Bodie and R.C. Merton. *Finance*. Prentice Hall, Upper Saddle River, NJ, 2000.

[14] E.V. Boguslavskaya. Exact solution of a problem of the optimal control of investments in a diffusion model. *Uspekhi Mat. Nauk*, 52(2(314)):157–158, 1997.

[15] N.L. Bowers, H.U. Gerber, J.C. Hickman, D.A. Jones, and C.J. Nesbitt. *Actuarial Mathematics*. Society of Actuaries, 1986.

[16] P.P. Boyle and T. Vorst. Option replicating in discrete time with transaction costs. *J. Finance*, 47(1):271–293, 1992.

[17] M.J. Brennan and E.S. Schwartz. The pricing of equity-linked life insurance policies with an asset value guarantee. *J. Financial Economics*, 3(3):195–213, 1976.

[18] A. Černý and J. Kallsen. Hedging by sequential regressions revisited. *Math. Finance*, 19(4):591–617, 2009.

[19] L.B. Chang and K. Palmer. Smooth convergence in the binomial model. *Finance Stoch.*, 11(1):91–105, 2007.

[20] J.C. Cox, R.A. Ross, and M. Rubinstein. Option pricing: a simplified approach. *J. Finan. Econ.*, 7(3):145–166, 1979.

[21] H. Cramér. *Collective Risk Theory: A Survey of the Theory from the Point of View of the Theory of Stochastic Processes*. Skandia Insurance Company, Stockholm, 1955. Reprinted from the Jubilee Volume of Försäkringsaktiebolaget Skandia.

[22] M. Crouhy, D. Galai, and R. Mark. *Risk Management*. McGraw-Hill, New York, 2001.

[23] J. Cvitanić and I. Karatzas. Generalized Neyman-Pearson lemma via convex duality. *Bernoulli*, 7(1):79–97, 2001.

[24] R.C. Dalang, A. Morton, and W. Willinger. Equivalent martingale measures and no-arbitrage in stochastic securities market models. *Stochastics Stochastics Rep.*, 29(2):185–201, 1990.

[25] F. Delbaen and J. Haezendonck. A martingale approach to premium calculation principles in an arbitrage free market. *Insurance Math. Econom.*, 8(4):269–277, 1989.

[26] F. Delbaen and W. Schachermayer. *The Mathematics of Arbitrage*. Springer Finance. Springer-Verlag, Berlin, 2006.

[27] F. Diener and M. Diener. Asymptotics of the price oscillations of a European call option in a tree model. *Math. Finance*, 14(2):271–293, 2004.

[28] A.K. Dixit and R.S. Pindyck. *Investment under Uncertainty*. Princeton University Press, Princeton, 1993.

[29] K. Dowd. Adjusting for risk: an improved (s)harpe ratio. *International Review of Economics and Finance*, 9(3):209–222, 2000.

[30] R.J. Elliott and P.E. Kopp. *Mathematics of Financial Markets*. Springer Finance. Springer-Verlag, New York, 1999.

[31] S. Emmer, C. Klüppelberg, and R. Korn. Optimal portfolios with bounded capital at risk. *Math. Finance*, 11(4):365–384, 2001.

[32] H. Föllmer and P. Leukert. Quantile hedging. *Finance Stoch.*, 3(3):251–273, 1999.

[33] H. Föllmer and P. Leukert. Efficient hedging: cost versus shortfall risk. *Finance Stoch.*, 4(2):117–146, 2000.

[34] H. Föllmer and A. Schied. *Stochastic Finance*, volume 27 of *de Gruyter Studies in Mathematics*. Walter de Gruyter & Co., Berlin, 2002. An introduction in discrete time.

[35] H. Föllmer and D. Sondermann. Hedging of nonredundant contingent claims. In *Contributions to mathematical economics*, pages 205–223. North-Holland, Amsterdam, 1986.

[36] A. Frolova, Y. Kabanov, and S. Pergamenshchikov. In the insurance business risky investments are dangerous. *Finance Stoch.*, 6(2):227–235, 2002.

[37] P.N. Ghauri and P.J. Buckley. International mergers and acquisitions: past, present and future. In *Advances in Mergers & Acquisitions*, pages 207–229. Emerald Group Publishing Limited, Bingley, 2003.

[38] M. Hardy. *Investment Guarantees: Modeling and Risk Management for Equity-Linked Life Insurance*. John Wiley & Sons Inc., Hoboken, NJ, 2003.

[39] J.M. Harrison and D.M. Kreps. Martingales and arbitrage in multiperiod securities markets. *J. Econom. Theory*, 20(3):381–408, 1979.

[40] J.M. Harrison and S.R. Pliska. Martingales and stochastic integrals in the theory of continuous trading. *Stochastic Process. Appl.*, 11(3):215–260, 1981.

[41] S. Heston. A closed-form solution for options with stochastic volatility with applications to bond and currency options. *Rev. Financ. Stud.*, 6(2):327–343, 1993.

[42] S. Heston and G. Zhou. On the rate of convergence of discrete-time contingent claims. *Math. Finance*, 10(1):53–75, 2000.

[43] T.S.Y. Ho and S.-B. Lee. Term structure movements and pricing interest rate contingent claims. *J. Finance*, 41(5):1011–1029, 1986.

[44] J. Hull. *Options, Futures, and Other Derivative Securities*. Prentice Hall, Upper Saddle River, NJ, 2002.

[45] M.-W. Hung and Y.-H. Liu. Pricing vulnerable options in incomplete markets. *J. Futures Markets*, 25(2):135–170, 2005.

[46] F. Jamshidian. An exact bond option formula. *J. Finance*, 44(1):205–209, 1989.

[47] È. Kamke. *Handbook of Ordinary Differention Equations*. Nauka, Moscow, 1976.

[48] I. Karatzas and S.E. Shreve. *Methods of Mathematical Finance*, volume 39 of *Applications of Mathematics*. Springer-Verlag, New York, 1998.

[49] S.G. Kellison. *The Theory of Interest*. McGraw-Hill, New York, 1991.

[50] A.N. Kolmogorov. *Foundations of the Theory of Probability*. Chelsea Publishing Co., New York, 1956. Translation edited by Nathan Morrison, with an added bibliography by A. T. Bharucha-Reid.

[51] R. Korn. Contingent claim valuation in a market with different interest rates. *ZOR—Math. Methods Oper. Res.*, 42(3):255–274, 1995.

[52] N.V. Krylov. *Controlled Diffusion Processes*, volume 14 of *Applications of Mathematics*. Springer-Verlag, New York, 1980.

[53] D. Lamberton and B. Lapeyre. *Introduction to Stochastic Calculus Applied to Finance*. Chapman & Hall, London, 1996.

[54] E. Lehmann and J.P. Romano. *Testing Statistical Hypotheses*. Springer Texts in Statistics. Springer-Verlag, Berlin, third edition, 2005.

[55] H.E. Leland. Option pricing and replication with transaction costs. *J. Finance*, 40(5):1283–1301, 1985.

[56] D. Lien. A note on the relationships between some risk-adjusted performance measures. *J. Futures Markets*, 22(5):483–495, 2002.

[57] A.W. Lo. The statistics of (s)harpe ratios. *Financial Analysts Journal*, 58(4):36–52, 2002.

[58] W. Margrabe. The value of an option to exchange one asset for another. *J. Finance*, 33(1):177–186, 1978.

[59] A. Melnikov and Yu. Romanyuk. Evaluating the performance of Gompertz, Makeham and Lee-Carter mortality models for risk management with unit-linked contracts. *Insurance Math. Econom.*, 39(3):310–329, 2006.

[60] A. Melnikov and Yu. Romanyuk. Efficient hedging and pricing of equity-linked life insurance contracts on several risky assets. *Int. J. Theor. Appl. Finance*, 11(3):295–323, 2008.

[61] A. Melnikov and V. Skornyakova. Quantile hedging and its application to life insurance. *Statist. Decisions*, 23(4):301–316, 2005.

[62] A.V. Melnikov. A binomial financial market in the context of the algebra of stochastic exponentials and martingales. *Teor. Ĭmovīr. Mat. Stat.*, (62):96–104, 2000.

[63] A.V. Melnikov. On the unity of quantitative methods of pricing in finance and insurance. *Proc. Steklov Inst. Math.*, 237:50–72, 2002.

[64] A.V. Melnikov, M.M. Moliboga, and V.S. Skornyakova. Valuation of flexible insurance contracts. *Teor. Ĭmovīr. Mat. Stat.*, (73):97–103, 2005.

[65] A.V. Melnikov, N.V. Popova, and V.S. Skornyakova. *Mathematical Methods of Financial Analysis*. Ankil, Moscow, 2006.

[66] A.V. Melnikov, S.N. Volkov, and M.L. Nechaev. *Mathematics of Financial Obligations*, volume 212 of *Translations of Mathematical Monographs*. American Mathematical Society, Providence, RI, 2002.

[67] R.C. Merton. Theory of rational option pricing. *Bell J. Econom. Management Sci.*, 4:141–183, 1973.

[68] T. Moeller. Risk-minimizing hedging strategies for unit-linked life-insurance contracts. *Austin Bull.*, 28:17–47, 1998.

[69] J.J. Murphy. *Technical Analysis of Financial Markets: A Comprehensive Guide to Trading Methods and Applications*. Prentice Hall, Upper Saddle River NJ, 1999.

[70] M. Musiela and M. Rutkowski. *Martingale Methods in Financial Modelling*, volume 36 of *Applications of Mathematics*. Springer-Verlag, Berlin, 1997.

[71] P. A. Mykland. The interpolation of options. *Finance Stoch.*, 7(4):417–432, 2003.

[72] S.K. Nawalkha, N.A. Beliaeva, and G.M. Soto. *Dynamic Term Structure Modeling: The Fixed Income Valuation Course*. John Wiley & Sons Inc., Hoboken, NJ, second edition, 2007.

[73] B. Øksendal. *Stochastic Differential Equations*. Universitext. Springer-Verlag, Berlin, sixth edition, 2003. An introduction with applications.

[74] R.S. Pindyck. Investment of uncertain cost. *J. Financial Economics*, 34:53–76, 1993.

[75] E. Platen and D. Heath. *A Benchmark Approach to Quantitative Finance.* Springer Finance. Springer-Verlag, Berlin, 2006.

[76] S.R. Pliska. *Introduction to Mathematical Finance.* Blackwell Publishers, Oxford, 1997.

[77] S.T. Rachev, F.J. Fabozzi, and S.V. Stoyanov. *Advanced Stochastic Models, Risk Assessment, and Portfolio Optimization: The Ideal Risk, Uncertainty, and Performance Measures.* John Wiley & Sons Inc., Hoboken, NJ, 2008.

[78] R. Rebonato. *Volatility and Correlation: The Perfect Hedger and the Fox.* John Wiley & Sons Inc., Hoboken, NJ, second edition, 2004.

[79] R.T. Rockafellar, S. Uryasev, and M. Zabarankin. Generalized deviations in risk analysis. *Finance Stoch.*, 10(1):51–74, 2006.

[80] L.C.G. Rogers and Z. Shi. The value of an Asian option. *J. Appl. Probab.*, 32(4):1077–1088, 1995.

[81] T. Rolski, H. Schmidli, V. Schmidt, and J. Teugels. *Stochastic Processes for Insurance and Finance.* Wiley Series in Probability and Statistics. John Wiley & Sons Ltd., Chichester, 1999.

[82] S. Ross. *A First Course in Probability.* Macmillan Co., New York, second edition, 1984.

[83] V.I. Rotar. *Actuarial Models: the Mathematics of Insurance.* CRC Press, Boca Raton, FL, 2006.

[84] M. Schweizer. Variance-optimal hedging in discrete time. *Math. Oper. Res.*, 20(1):1–32, 1995.

[85] M. Schweizer. From actuarial to financial valuation principles. *Insurance Math. Econom.*, 28(1):31–47, 2001.

[86] B.A. Sevastyanov. *Branching Processes.* Nauka, Moscow, 1971.

[87] A.N. Shiryaev. *Probability*, volume 95 of *Graduate Texts in Mathematics.* Springer-Verlag, New York, second edition, 1996.

[88] A.N. Shiryaev. *Essentials of Stochastic Finance*, volume 3 of *Advanced Series on Statistical Science & Applied Probability.* World Scientific Publishing Co. Inc., River Edge, NJ, 1999. Facts, models, theory.

[89] A.N. Shiryaev. Quickest detection problems in the technical analysis of the financial data. In H. Geman et al., editor, *Mathematical Finance Bachelier Congress 2000*, pages 487–521, New York, 2002. Springer-Verlag.

[90] E.S.W. Shiu. The probability of eventual ruin in the compound binomial model. *Austin Bulletin*, 19(2):179–190, 1989.

[91] M. Steffensen. A no arbitrage approach to Thiele's differential equation. *Insurance Math. Econom.*, 27(2):201–214, 2000.

[92] M.S. Taqqu and W. Willinger. The analysis of finite security markets using martingales. *Adv. in Appl. Probab.*, 19(1):1–25, 1987.

[93] J.M. Vanden. Exact superreplication strategies for a class of derivative assets. *Appl. Math. Finance*, 13(1):61–87, 2006.

[94] A. White and J. Hull. The pricing of options on assets with stochastic volatilities. *J. Finance*, 42(2):281–300, 1987.

[95] G.E. Willmot. Ruin probabilities in the compound binomial model. *Insur. Math. Econom.*, 12(2):133–142, 1993.

[96] M. Xu. Risk measure pricing and hedging in incomplete markets. *Annals of Finance*, 2(1):51–71, 2006.

Glossary of Notation

$:=$	equality by definition
a.s.	almost surely
$\emptyset$	the empty set
$\square$	the end of proof

$\{x \in A \mid Z\}$	the subset of A whose elements possess property Z
$A \times B$	the cartesian product of sets A and B
I_A	the indicator function of set A
$f\|_A$	the restriction of function $f : X \to Y$ to the subset A of X
$(a_k), (a_k)_{k=1}^{\infty}$	the sequence $a_1, \dots, a_k, \dots$
$\mathbb{N}, \mathbb{Z}, \mathbb{R}$	the sets of natural numbers, integers and real numbers
$\mathbb{R}^N$	the set of all real N-tupels $(r_1, \dots, r_n)$
2^A	the set of all subsets of A

$f(x) =_{x \to a} \mathcal{O}(g(x))$	$\|f(x)\| \leq \text{const} \|g(x)\|$ in a neighborhood of a
$o(x)$	a function satisfying $\|o(x)/x\| \to 0$ as $x \to 0$
$[\![x]\!]$	the integer part of $x \in \mathbb{R}$
$x \wedge y$	$:= \min\{x, y\}$
$C^n[0, \infty)$	the space of n-times continuously differentiable functions on $[0, \infty)$

$P(A)$	the probability of event A
$P(A\|B)$	the conditional probability of event A assuming event B
$P(A\|\mathcal{F})$	the conditional probability of A with respect to a σ-algebra $\mathcal{F}$
$\widetilde{P}$	a martingale probability
$\mathcal{M}(S_n/B_n)$	the collection of all martingale probabilities
$E(X)$	the expectation of a random variable X
$Var(X)$	the variance of a random variable X
$\mathcal{N}(m, \sigma^2)$	a Gaussian (normal) random variable

	with mean value m and variance σ^2
$E(X\|Y)$	the conditional expectation of a random variable X with respect to a random variable Y
$E(X\|\mathcal{F})$	the conditional expectation of a random variable X with respect to a σ-algebra $\mathcal{F}$
$Cov(X,Y)$	the covariance of X and Y
$(X)^+$	$:= \max\{X,0\}$
$\mathbb{F}$	a filtration (information flow)
$\langle M,M \rangle$	the quadratic variation of a martingale M
$H * m_n$	a discrete stochastic integral
$(\varphi * W)_t$	stochastic integral
$\varepsilon_n(U)$	stochastic exponential
$\mathcal{E}_t(Y)$	stochastic exponential
SF	the collection of all self-financing portfolios
$\mathcal{M}_0^N$	the collection of all stopping times
$Q \ll P$	measure Q is absolutely continuous with respect to measure P

Index

For Product Safety Concerns and Information please contact our EU
representative GPSR@taylorandfrancis.com
Taylor & Francis Verlag GmbH, Kaufingerstraße 24, 80331 München, Germany

www.ingramcontent.com/pod-product-compliance
Ingram Content Group UK Ltd.
Pitfield, Milton Keynes, MK11 3LW, UK
UKHW021626240425
457818UK00019B/738